CLUSTERING

A Data Recovery Approach

Second Edition

Chapman & Hall/CRC
Computer Science and Data Analysis Series

The interface between the computer and statistical sciences is increasing, as each discipline seeks to harness the power and resources of the other. This series aims to foster the integration between the computer sciences and statistical, numerical, and probabilistic methods by publishing a broad range of reference works, textbooks, and handbooks.

SERIES EDITORS
David Blei, Princeton University
David Madigan, Rutgers University
Marina Meila, University of Washington
Fionn Murtagh, Royal Holloway, University of London

Proposals for the series should be sent directly to one of the series editors above, or submitted to:

Chapman & Hall/CRC
4th Floor, Albert House
1-4 Singer Street
London EC2A 4BQ
UK

Published Titles

Bayesian Artificial Intelligence, Second Edition
Kevin B. Korb and Ann E. Nicholson

Clustering for Data Mining: A Data Recovery Approach, Second Edition
Boris Mirkin

Computational Statistics Handbook with MATLAB®, Second Edition
Wendy L. Martinez and Angel R. Martinez

Correspondence Analysis and Data Coding with Java and R
Fionn Murtagh

Design and Modeling for Computer Experiments
Kai-Tai Fang, Runze Li, and Agus Sudjianto

Exploratory Data Analysis with MATLAB®, Second Edition
Wendy L. Martinez, Angel R. Martinez, and Jeffrey L. Solka

Exploratory Multivariate Analysis by Example Using R
François Husson, Sébastien Lê, and Jérôme Pagès

Introduction to Data Technologies
Paul Murrell

Introduction to Machine Learning and Bioinformatics
Sushmita Mitra, Sujay Datta, Theodore Perkins, and George Michailidis

Microarray Image Analysis: An Algorithmic Approach
Karl Fraser, Zidong Wang, and Xiaohui Liu

Pattern Recognition Algorithms for Data Mining
Sankar K. Pal and Pabitra Mitra

R Graphics
Paul Murrell

R Programming for Bioinformatics
Robert Gentleman

Semisupervised Learning for Computational Linguistics
Steven Abney

Statistical Computing with R
Maria L. Rizzo

Statistical Learning and Data Science
Mireille Gettler Summa, Léon Bottou, Bernard Goldfarb, Fionn Murtagh, Catherine Pardoux, and Myriam Touati

Computer Science and Data Analysis Series

CLUSTERING
A Data Recovery Approach
Second Edition

Boris Mirkin

CRC Press
Taylor & Francis Group
Boca Raton   London   New York

CRC Press is an imprint of the
Taylor & Francis Group, an **informa** business
A CHAPMAN & HALL BOOK

CRC Press
Taylor & Francis Group
6000 Broken Sound Parkway NW, Suite 300
Boca Raton, FL 33487-2742

© 2013 by Taylor & Francis Group, LLC
CRC Press is an imprint of Taylor & Francis Group, an Informa business

No claim to original U.S. Government works

Printed in the United States of America on acid-free paper
Version Date: 2012912

International Standard Book Number: 978-1-4398-3841-9 (Hardback)

Library of Congress Cataloging-in-Publication Data

Mirkin, B. G. (Boris Grigor'evich)
　　Clustering : a data recovery approach / Boris Mirkin. -- 2nd ed.
　　　　p. cm. -- (Chapman & Hall/CRC computational science series)
　　Earlier ed. published under title: Clustering for data mining.
　　Includes bibliographical references and index.
　　ISBN 978-1-4398-3841-9 (hardcover : alk. paper)
　　1. Data mining. 2. Cluster analysis. 3. Data recovery (Computer science) I. Mirkin, B.
G. (Boris Grigor'evich) Clustering for data mining. II. Title.

QA76.9.D343M57 2012
006.3'12--dc23
                                                                    2012028180

Visit the Taylor & Francis Web site at
http://www.taylorandfrancis.com

and the CRC Press Web site at
http://www.crcpress.com

Contents

Preface to the Second Edition

One of the goals of the first edition of this book back in 2005 was to present a coherent theory for K-Means partitioning and Ward hierarchical clustering. This theory leads to effective data pre-processing options, clustering algorithms and interpretation aids as well as to firm relations to other areas of data analysis.

The goal of this second edition is to consolidate, strengthen and extend this island of understanding in the light of recent developments. Here are examples of newly added material for each of the objectives:

- Consolidating:
  - Five equivalent formulations for K-Means criterion in Section 7.2.5
  - Usage of split base vectors in hierarchical clustering
  - Similarities between the clustering data recovery models and singular/eigenvalue decompositions
- Strengthening:
  - Experimental evidence to support the PCA-like anomalous pattern clustering as a tool to initialize K-Means
  - Weighting variables with Minkowski metric three-step K-Means
  - Effective versions of least-squares divisive clustering
- Extending:
  - Similarity and network clustering including additive and spectral clustering approaches
  - Consensus clustering

Moreover, the material on validation and interpretation of clusters is updated with a system better reflecting the current state of the art and with our recent "lifting in taxonomies" approach.

The structure of the book has been streamlined by adding two chapters: "Similarity Clustering" and "Validation and Interpretation," while removing two chapters: "Different Clustering Approaches" and "General Issues." The chapter on mathematics of the data recovery approach, in a much extended version, almost doubled in size, now concludes the book. Parts of the removed chapters are integrated within the new structure. The changes have added a hundred pages and a couple of dozen examples to the text and, in fact, transformed it into a different species of a book. In the first edition, the book had a Russian doll structure, with a core and a couple of nested shells around. Now it is a linear structure presentation of the data recovery clustering.

This book offers advice to students, educated users and application developers regarding clustering goals and ways to achieve them. This advice involves methods that are compatible with the data recovery framework and experimentally tested. Fortunately, this embraces the most popular approaches including the most recent ones. The emphasis on the data recovery framework sets this book apart from the other books on clustering that try to inform the reader of as many approaches as possible with no regard for their properties.

The second edition includes a number of recent results of the joint work with T. Fenner, S. Nascimento, M. MingTso Chiang, and R. Amorim. Unpublished results of joint work with students of the NRU HSE Moscow, E. Kovaleva and A. Shestakov are included, too. This work has been supported by the Laboratory of Decision Choice and Analysis NRU HSE Moscow and, also, the Laboratory of Algorithms and Technologies for Networks Analysis NRU HSE Nizhny Novgorod (by means of RF government grant, ag. 11.G34.31.0057), as well as by grants from the Academic Fund of the NRU HSE in 2010–2012.

MATLAB® is the trademark of The MathWorks, Inc. For product information, please contact:

The MathWorks, Inc.
3 Apple Hill Drive
Natick, MA 01760-2098 USA
Tel: 508-647-7000
Fax: 508-647-7001
E-mail: info@mathworks.com
Web: www.mathworks.com

Preface to the First Edition

Clustering is a discipline devoted to finding and describing cohesive or homogeneous chunks in data, the clusters.

Some examples of clustering problems are

- Finding common surf patterns in a set of web users

- Automatically finding meaningful parts in a digitalized image

- Partition of a set of documents in groups by similarity of their contents

- Visual display of the environmental similarity between regions on a country's map

- Monitoring socio-economic development of a system of settlements via a small number of representative settlements

- Finding protein sequences in a database that are homologous to a query protein sequence

- Finding anomalous patterns of gene expression data for diagnostic purposes

- Producing a decision rule for separating potentially bad-debt credit applicants

- Given a set of preferred vacation places, finding out what features of the places and vacationers attract each other

- Classifying households according to their furniture purchasing patterns and finding the groups' key characteristics to optimize furniture marketing and production

Clustering is a key area in data mining and knowledge discovery, which are activities oriented toward finding non-trivial patterns in data collected in databases. It is an important part of machine learning in which clustering is considered as unsupervised learning. Clustering is part of information retrieval where it helps to properly shape responses to queries. Currently, clustering is becoming a subject of its own, in its capacity to organize, shape, relate and maintain knowledge of phenomena and processes.

Earlier developments of clustering techniques have been associated, primarily, with three areas of research: factor analysis in psychology [79], numerical taxonomy in biology [191], and unsupervised learning in pattern recognition [38].

Technically speaking, the idea behind clustering is rather simple: introduce a measure of similarity between entities under consideration and combine similar entities into the same clusters while keeping dissimilar entities in different clusters. However, implementing this idea is less than straightforward.

First, too many similarity measures and clustering techniques have been invented with virtually no support to a non-specialist user for choosing among them. The trouble with this is that different similarity measures and/or clustering techniques may, and frequently do, lead to different results. Moreover, the same technique may also lead to different cluster solutions depending on the choice of parameters such as the initial setting or the number of clusters specified. On the other hand, some common data types, such as questionnaires with both quantitative and categorical features, have been left virtually without any sound similarity measure.

Second, use and interpretation of cluster structures may become an issue, especially when available data features are not straightforwardly related to the phenomenon under consideration. For example, certain data of customers available at a bank, such as age and gender, typically are not very helpful in deciding whether to grant a customer a loan or not.

Researchers acknowledge specifics of the subject of clustering. They understand that the clusters to be found in data may very well depend not only on the data, but also on the user's goals and degree of granulation. They frequently consider clustering as art rather than science. Indeed, clustering has been dominated by learning from examples rather than theory-based instructions. This is especially visible in texts written for inexperienced readers, such as [7,45,175].

The general opinion among researchers is that clustering is a tool to be applied at the very beginning of investigation into the nature of a phenomenon under consideration, to view the data structure and then decide upon applying better-suited methodologies. Another opinion of researchers is that methods for finding clusters as such should constitute the core of the discipline; related questions of data pre-processing, such as feature quantization and standardization, definition and computation of similarity, and post-processing, such as interpretation and association with other aspects of the phenomenon, should be left beyond the scope of the discipline because they are motivated by external considerations related to the substance of the phenomenon under investigation. The author would like to share the former opinion and disagree with the latter because it is at odds with the former: on the very first steps of knowledge discovery, substantive considerations are quite shaky, and it is unrealistic to expect that they alone could lead to properly defining the issues of pre- and post-processing.

Such a dissimilar opinion has led me to think that the discovered clusters must be treated as an "ideal" representation of the data that could be used for recovering the original data back from the ideal format. This is the idea of the data recovery approach: not only use data for finding clusters, but also use clusters for recovering the data. In a general situation, the data recovered from the summarized clusters cannot fit the original data exactly, which can be due to various factors such as presence of noise in data or inadequate features. The difference can be used for evaluation of the quality of clusters: the better the fit, the better the clusters. This perspective would also lead to addressing of issues in pre- and post-processing, which is going to become possible because parts of the data that are well explained by clusters can be separated from those that are not.

The data recovery approach is common in more traditional data mining and statistics areas such as regression, analysis of variance and factor analysis, where it works, to a great extent, due to the Pythagorean decomposition of the data scatter into "explained" and "unexplained" parts. Why not apply the same approach in clustering?

In this book, most popular clustering techniques, K-Means for partitioning, Ward's method for hierarchical clustering, similarity clustering and consensus partitions are presented in the framework of the data recovery approach. This approach leads to the Pythagorean decomposition of the data scatter into parts explained and unexplained by the found cluster structure. The decomposition has led to a number observations that amount to a theoretical framework in clustering. The framework appears to be well suited for extensions of the methods to different data types such as mixed scale data including continuous, nominal and binary features. In addition, a bunch of both conventional and original interpretation aids have been derived for both partitioning and hierarchical clustering based on contributions of features and categories to clusters and splits. One more strain of clustering techniques, one-by-one clustering which is becoming increasingly popular, naturally emerges within the framework to give rise to intelligent versions of K-Means, mitigating the need for user-defined setting of the number of clusters and their hypothetical prototypes. Moreover, the framework leads to a set of mathematically proven properties relating classical clustering with other clustering techniques such as conceptual clustering, spectral clustering, consensus clustering, and graph theoretic clustering as well as with other data analysis concepts such as decision trees and association in contingency data tables.

These are all presented in this book, which is oriented toward a reader interested in the technical aspects of clustering, be they a theoretician or a practitioner. The book is especially well suited for those who want to learn WHAT clustering is by learning not only HOW the techniques are applied, but also WHY. In this way the reader receives knowledge which should allow them not only to apply the methods but also to adapt, extend and modify them according to the reader's own goals.

This material is organized into seven chapters presenting a unified theory along with computational, interpretational and practical issues of real-world data analysis with clustering:

- What Is Clustering? (Chapter 1)
- What Is Data? (Chapter 2)

- What Is K-Means? (Chapter 3)
- What Is Least-Squares Hierarchical Clustering? (Chapter 4)
- What Is Similarity and Consensus Clustering? (Chapter 5)
- What Computation Tools are Available to Validate and Interpret Clusters? (Chapter 6)
- What Is the Data Recovery Approach? (Chapter 7)

This material is intermixed with popular related approaches such as SOM (self-organizing maps), EM (expectation-maximization), Laplacian transformation and spectral clustering, Single link clustering, and so on.

This structure is intended, first, to introduce popular clustering methods and their extensions to problems of current interest, according to the data recovery approach, without learning the theory (Chapters 1 through 5), then to review the trends in validation and interpretation of clustering results (Chapter 6), and then describe the theory leading to these and related methods (Chapter 7).

At present, there are two types of literature on clustering, one leaning toward providing general knowledge and the other giving more instruction. Books of the former type are Gordon [62] targeting readers with a degree of mathematical background and Everitt et al. [45] that does not require mathematical background. These include a great deal of methods and specific examples but leave rigorous data analysis instruction beyond the prime contents. Publications of the latter type are chapters in data-mining books such as Dunham [40] and monographs such as in Ref. [218]. They contain selections of some techniques described in an *ad hoc* manner, without much concern for relations between them, and provide detailed instruction on algorithms and their parameters.

This book combines features of both approaches. However, it does so in a rather distinct way. The book does contain a number of algorithms with detailed instructions, as well as many application examples. But the selection of methods is based on their compatibility with the data recovery approach rather than just popularity. The approach allows to cover some issues in pre- and post-processing matters that are usually left with no instruction at all.

In the book, I had to clearly distinguish between four different perspectives: (a) classical statistics, (b) machine learning, (c) data mining, and (d) knowledge discovery, as those leading to different answers to the same questions.

The book assumes that the reader may have no mathematical background beyond high school: all necessary concepts are defined within the text. However, it does contain some technical material needed for shaping and explaining a technical theory. Thus, it might be of help if the reader is acquainted with basic notions of calculus, statistics, matrix algebra, graph theory, and set theory.

To help the reader, the book conventionally includes references and index. Each individual chapter is preceded by a set of goals and a dictionary of key concepts. Summarizing overviews are supplied to Chapters 3 through 7. Described methods are accompanied with numbered computational examples showing the working of the methods on relevant data sets from those presented in Chapter 1. The datasets are freely available for download in the author's webpage http://www.dcs.bbk.ac.uk/mirkin/clustering/Data. Computations have been carried out with in-house programs for MATLAB®, the technical computing tool developed by The MathWorks (see its Internet website: www.mathworks.com).

The material has been used to teach data clustering courses to MSc and BSc CS students in several colleges across Europe. Based on these experiences, different teaching options can be suggested depending on the course objectives, time resources, and students' background.

If the main objective is teaching clustering methods and there is little time available, then it would be advisable to pick up first the material on generic K-Means in Sections 3.1.1 and 3.1.2, and then review a couple of related methods such as PAM in Section 3.5, iK-Means in Section 3.3, Ward agglomeration in Section 4.2 and division in Section 4.3, and single linkage in Section 5.5. Given some more time, similarity, spectral and consensus clustering should be taken care of. A review of cluster validation techniques should follow the methods. In a more relaxed regime, issues of interpretation should be brought forward as described in Chapter 6.

The materials in Chapters 4 through 7 can be used as a complement to a recent textbook by the author [140].

Acknowledgments

Too many people contributed to the approach and to this book to list them all. However, I would like to mention those researchers whose support was important for channeling my research efforts: Dr. E. Braverman, Dr. V. Vapnik, Professor Y. Gavrilets, and Professor S. Aivazian in Russia; Professor F. Roberts, Professor F. McMorris, Professor P. Arabie, Professor T. Krauze, and Professor D. Fisher in the United States; Professor E. Diday, Professor L. Lebart, and Professor B. Burtschy in France; Professor H.-H. Bock, Dr. M. Vingron, and Dr S. Suhai in Germany. The structure and contents of this book have been influenced by the comments of Dr. I. Muchnik (Rutgers University, New Jersey, United States), Professor M. Levin (Higher School of Economics, Moscow, Russia), Dr. S. Nascimento (University Nova, Lisbon, Portugal), and Professor F. Murtagh (Royal Holloway, University of London, United Kingdom).

Acknowledgments

Examples

1

What Is Clustering?

After reading this chapter, the reader will have a general understanding of

1. What is clustering?
2. Elements of a clustering problem
3. Clustering goals
4. Quantitative and categorical features
5. Popular cluster structures: partition, hierarchy, and single cluster
6. Different perspectives on clustering: statistics, machine learning, data mining, and knowledge discovery

A set of small but real-world clustering case study problems will be presented.

Key Concepts

Association Finding interrelations between different aspects of a phenomenon by matching descriptions of the same clusters made using feature spaces corresponding to the aspects.

Classification An actual or ideal arrangement of entities under consideration in classes to shape and keep knowledge; specifically, to capture the structure of phenomena, and relate different aspects of the phenomenon in question to each other. This term is also used in a narrow sense referring to a classifier action in assigning entities to prespecified classes.

Cluster A set of similar entities.

Cluster representative An element of a cluster to represent its "typical" properties. This is used for interpretation in those domains knowledge of which is not well formulated.

Cluster structure A representation of an entity set I as a set of clusters that form either a partition of I or hierarchy on I or a not so well-structured clustering of I.

Cluster tendency A description of a cluster in terms of the normal or average values of relevant features.

Clustering Finding and describing cluster structures in a data set.

Clustering goal A problem of associating or structuring or describing or generalizing or visualizing, that is being addressed with clustering.

Clustering criterion A formal definition or scoring function that can be used in clustering algorithms as a measure of success.

Conceptual description A statement characterizing a cluster or cluster structure in terms of relevant features.

Data In this text, this can be either an entity-to-feature data table or a square matrix of pairwise similarity or dissimilarity index values.

Data-mining perspective In data mining, clustering is a tool for finding interesting patterns and regularities in data sets.

Generalization Making general statements about data and, potentially, about the phenomenon the data relate to.

Knowledge discovery perspective In knowledge discovery, clustering is a tool for updating, correcting, and extending the existing knowledge. In this regard, clustering is but empirical classification.

Machine-learning perspective In machine learning, clustering is a tool for prediction.

Statistics perspective In statistics, clustering is a method to fit a prespecified probabilistic model of the data-generating mechanism.

Structuring Representing data using a cluster structure.

Visualization Mapping data onto a well-known "ground" image in such a way that properties of the data are reflected in the structure of the ground image. The potential users of the visualization must be well familiar with the ground image as, for example, with the coordinate plane or a genealogy tree.

1.1 Case Study Problems

Clustering is a discipline devoted to finding and describing homogeneous groups of entities, that is, clusters, in data sets. Why would one need this? Here is a list of potentially overlapping objectives for clustering.

1. *Structuring*, that is, representing data as a set of clusters.

2. *Description* of clusters in terms of features, not necessarily involved in finding them.

3. *Association*, that is, finding interrelations between different aspects of a phenomenon by matching descriptions of the same clusters in terms of features related to the aspects.

4. *Generalization*, that is, making general statements about the data structure and, potentially, the phenomena the data relate to.

5. *Visualization*, that is, representing cluster structures visually over a well-known ground image.

These goals are not necessarily mutually exclusive, nor do they cover the entire range of clustering goals; they rather reflect the author's opinion on the main applications of clustering. A number of real-world examples of data and related clustering problems illustrating these goals will be given further on in this chapter. The data sets are small so that the reader could see their structures with the naked eye.

1.1.1 Structuring

Structuring is finding principal groups of entities in their specifics. The cluster structure of an entity set can be looked at through different glasses. One user may wish to look at the set as a system of nonoverlapping classes; another user may prefer to develop a taxonomy as a hierarchy of more and more abstract concepts; yet another user may wish to focus on a cluster of "core" entities considering the rest as merely a nuisance. These are conceptualized in different types of cluster structures, such as partition, hierarchy, and single subset.

1.1.1.1 Market Towns

Table 1.1 represents a small portion of a list of 1300 English market towns characterized by the population and services listed in the following box.

MARKET TOWN FEATURES

P	Population resident in 1991 Census
PS	Primary schools
Do	Doctor surgeries
Ho	Hospitals
Ba	Banks and building societies
SM	National chain supermarkets
Pe	Petrol stations
DIY	Do-it-yourself shops
SP	Public swimming pools
PO	Post offices
CA	Citizen's advice bureaux (free legal advice)
FM	Farmers' markets

For the purposes of social monitoring, the set of all market towns should be partitioned into similarity clusters in such a way that a representative town from each of the clusters may be used as a unit of observation. Those characteristics of the clusters that distinguish them from the rest should be used to properly select representative towns.

TABLE 1.1

Market Towns

Town	P	PS	Do	Ho	Ba	SM	Pe	DIY	SP	PO	CA	FM
Ashburton	3660	1	0	1	2	1	2	0	1	1	1	0
Bere Alston	2362	1	0	0	1	1	0	0	0	1	0	0
Bodmin	12,553	5	2	1	6	3	5	1	1	2	1	0
Brixham	15,865	7	3	1	5	5	3	0	2	5	1	0
Buckfastleigh	2786	2	1	0	1	2	2	0	1	1	1	1
Bugle/Stenalees	2695	2	0	0	0	0	1	0	0	2	0	0
Callington	3511	1	1	0	3	1	1	0	1	1	0	0
Dartmouth	5676	2	0	0	4	4	1	0	0	2	1	1
Falmouth	20,297	6	4	1	11	3	2	0	1	9	1	0
Gunnislake	2236	2	1	0	1	0	1	0	0	3	0	0
Hayle	7034	4	0	1	2	2	2	0	0	2	1	0
Helston	8505	3	1	1	7	2	3	0	1	1	1	1
Horrabridge/Yel	3609	1	1	0	2	1	1	0	0	2	0	0
Ipplepen	2275	1	1	0	0	0	1	0	0	1	0	0
Ivybridge	9179	5	1	0	3	1	4	0	0	1	1	0
Kingsbridge	5258	2	1	1	7	1	2	0	0	1	1	1
Kingskerswell	3672	1	0	0	0	1	2	0	0	1	0	0
Launceston	6466	4	1	0	8	4	4	0	1	3	1	0
Liskeard	7044	2	2	2	6	2	3	0	1	2	2	0
Looe	5022	1	1	0	2	1	1	0	1	3	1	0
Lostwithiel	2452	2	1	0	2	0	1	0	0	1	0	1
Mevagissey	2272	1	1	0	1	0	0	0	0	1	0	0
Mullion	2040	1	0	0	2	0	1	0	0	1	0	0
Nanpean/Foxhole	2230	2	1	0	0	0	0	0	0	2	0	0
Newquay	17,390	4	4	1	12	5	4	0	1	5	1	0
Newton Abbot	23,801	13	4	1	13	4	7	1	1	7	2	0
Padstow	2460	1	0	0	3	0	0	0	0	1	1	0
Penryn	7027	3	1	0	2	4	1	0	0	3	1	0
Penzance	19,709	10	4	1	12	7	5	1	1	7	2	0
Perranporth	2611	1	1	0	1	1	2	0	0	2	0	0
Porthleven	3123	1	0	0	1	1	0	0	0	1	0	0
Saltash	14,139	4	2	1	4	2	3	1	1	3	1	0
South Brent	2087	1	1	0	1	1	0	0	0	1	0	0
St Agnes	2899	1	1	0	2	1	1	0	0	2	0	0
St Austell	21,622	7	4	2	14	6	4	3	1	8	1	1
St Blazey/Par	8837	5	2	0	1	1	4	0	0	4	0	0
St Columb Major	2119	1	0	0	2	1	1	0	0	1	1	0
St Columb Road	2458	1	0	0	0	1	3	0	0	2	0	0
St Ives	10,092	4	3	0	7	2	2	0	0	4	1	0
St Just	2092	1	0	0	2	1	1	0	0	1	0	0
Tavistock	10,222	5	3	1	7	3	3	1	2	3	1	1
Torpoint	8238	2	3	0	3	2	1	0	0	2	1	0
Totnes	6929	2	1	1	7	2	1	0	1	4	0	1
Truro	18,966	9	3	1	19	4	5	2	2	7	1	1
Wadebridge	5291	1	1	0	5	3	1	0	1	1	1	0

Note: Market towns in the West Country, England.

As one would imagine, the number of services available in a town in general follows the town size. Then the clusters can be mainly described in terms of the population size. According to the clustering results, this set of towns, as well as the complete set of almost 1300 English market towns, can be structured in seven clusters. These clusters can be described as falling in the four tiers of population: large towns of about 17–20,000 inhabitants, two clusters of medium-sized towns (8–10,000 inhabitants), three clusters of small towns (about 5000 inhabitants), and a cluster of very small settlements with about 2500 inhabitants. The clusters in the same population tier are defined by the presence or absence of this or that service. For example, each of the three small town clusters is characterized by the presence of a facility, which is absent in the two others: a Farmer's Market, a Hospital or a Swimming Pool, respectively. The number of clusters can be determined in the process of computations (see Sections 3.3 and 6.3.2).

This data set is analyzed on pp. 22, 46, 53, 71, 103, 112, 210, 212, 214, and 216.

1.1.1.2 Primates and Human Origin

Table 1.2 presents genetic distances between human and three genera of great apes. The Rhesus monkey is added as a distant relative to signify a starting divergence event. Allegedly, the humans and chimpanzees diverged approximately 5 million years ago, after they had diverged from the other great apes. Are clustering results compatible with this knowledge?

The data are a square matrix of dissimilarity values between the species from Table 1.2 taken from [135, p. 30]. (Only sub-diagonal distances are shown since the table is symmetric.) An example of the analysis of this matrix is given on p. 196.

The query: what species are in the humans cluster? This obviously can be treated as a single cluster problem: one needs only one cluster to address the issue. The structure of the data is so simple that the cluster consisting of the chimpanzee, gorilla, and the human can be seen without any big theory: distances within this subset are much similar, all about the average 1.51, and by far smaller than the other distances in the matrix.

In biology, this problem is traditionally addressed by using the evolutionary trees, which are analogous to the genealogy trees except that species stand

TABLE 1.2

Primates

Genus	Human	Chimpanzee	Gorilla	Orangutan
Chimpanzee	1.45			
Gorilla	1.51	1.57		
Orangutan	2.98	2.94	3.04	
Rhesus monkey	7.51	7.55	7.39	7.10

Note: Distances between four Primate species and Rhesus monkey.

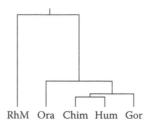

RhM   Ora   Chim   Hum   Gor

FIGURE 1.1
A tree representing pair-wise distances between the Primate species from Table 1.2.

there instead of relatives. An evolutionary tree built from the data in Table 1.2 is presented in Figure 1.1. The closest relationship between the human and chimpanzee is quite obvious. The subject of human evolution is treated in depth with data-mining methods in [23].

1.1.1.3 Gene Presence–Absence Profiles

Evolutionary analysis is an important tool not only to understand evolution, but also to analyze gene functions in humans and other relevant organisms. The major assumption underlying the analysis is that all the species are descendants of the same ancestral species, so that the subsequent evolution can be depicted in terms of the events of divergence only, as in the evolutionary tree in Figure 1.1. Although frequently debated, this assumption allows to make sense of many data sets otherwise incomprehensible.

The terminal nodes, referred to as the leaves, correspond to the species under consideration, and the root denotes the hypothetical common ancestor. The interior nodes represent other hypothetical ancestral species, each being the last common ancestor to the set of organisms in the leaves of the sub-tree rooted in it. An evolutionary tree is frequently supplemented by data on the gene content of the extant species corresponding to leaves as exemplified in Table 1.3. Here, the columns correspond to 18 simple, unicellular organisms, bacteria and archaea (collectively called prokaryotes), and a simple eukaryote, yeast *Saccharomyces cerevisiae*. The list of species is given in Table 1.4.

The rows in Table 1.3 correspond to individual genes represented by the so-called clusters of orthologous groups (COGs) that supposedly include genes originating from the same ancestral gene in the common ancestor of the respective species. COG names reflecting the functions of the respective genes in the cell are given in Table 1.5. These tables present but a small part of the publicly available COG database currently including 66 species and 4857 COGs (see website www.ncbi.nlm.nih.gov/COG).

The pattern of presence–absence of a COG in the analyzed species is shown in Table 1.3, with zeros and ones standing for the absence and presence, respectively. Therefore, a COG can be considered an attribute which is either present or absent in every species. Two of the COGs, in the top two rows,

TABLE 1.3

Gene Profiles

No.	COG						Species											
		y	a	o	m	p	k	z	q	v	d	r	b	c	e	f	g	s	j
1	COG0090	1	1	1	1	1	1	1	1	1	1	1	1	1	1	1	1	1	1
2	COG0091	1	1	1	1	1	1	1	1	1	1	1	1	1	1	1	1	1	1
3	COG2511	0	1	1	1	1	1	1	0	0	0	0	0	0	0	0	0	0	0
4	COG0290	0	0	0	0	0	0	0	1	1	1	1	1	1	1	1	1	1	1
5	COG0215	1	1	1	0	1	1	1	1	1	1	1	1	1	1	1	1	1	1
6	COG2147	1	1	1	1	1	1	1	0	0	0	0	0	0	0	0	0	0	0
7	COG1746	0	1	1	1	1	1	1	0	0	0	0	0	0	0	0	0	0	0
8	COG1093	1	1	1	1	1	1	1	0	0	0	0	0	0	0	0	0	0	0
9	COG2263	0	1	1	1	1	1	1	0	0	0	0	0	0	0	0	0	0	0
10	COG0847	1	1	0	0	1	0	0	1	1	1	1	1	1	1	1	1	1	1
11	COG1599	1	1	1	1	1	1	1	0	0	0	0	0	0	0	0	0	0	0
12	COG3066	0	0	0	0	0	0	0	0	0	0	0	0	0	1	0	1	0	0
13	COG3293	0	0	0	0	0	0	0	0	0	1	1	0	1	0	0	0	0	1
14	COG3432	0	1	0	1	1	0	1	0	0	0	0	0	0	0	0	0	0	0
15	COG3620	0	1	0	0	1	1	0	0	0	0	0	0	0	0	0	0	0	0
16	COG1709	0	1	1	1	1	1	1	0	0	0	0	0	0	1	0	0	0	0
17	COG1405	1	1	1	1	1	1	1	0	0	0	0	0	0	0	0	0	0	0
18	COG3064	0	0	0	0	0	0	0	0	0	0	0	0	0	1	0	1	0	0
19	COG2853	0	0	0	0	0	0	0	0	0	0	0	0	0	1	1	1	1	1
20	COG2951	0	0	0	0	0	0	0	0	0	0	0	0	0	1	1	1	1	1
21	COG3114	0	0	0	0	0	0	0	0	0	0	0	0	0	1	1	1	0	0
22	COG3073	0	0	0	0	0	0	0	0	0	0	0	0	0	1	1	1	1	0
23	COG3026	0	0	0	0	0	0	0	0	0	0	0	0	0	1	1	1	0	0
24	COG3006	0	0	0	0	0	0	0	0	0	0	0	0	0	1	0	1	0	0
25	COG3115	0	0	0	0	0	0	0	0	0	0	0	0	0	1	1	1	1	0
26	COG2414	0	1	0	1	1	1	1	0	0	0	0	0	0	1	0	0	0	0
27	COG3029	0	0	0	0	0	0	0	0	0	0	1	0	0	1	0	1	0	0
28	COG3107	0	0	0	0	0	0	0	0	0	0	0	0	0	1	1	1	1	0
29	COG3429	0	0	0	0	0	0	0	0	0	1	1	0	1	0	0	0	0	0
30	COG1950	0	0	0	0	0	0	0	0	0	1	1	1	1	0	0	0	0	0

Note: Presence–absence profiles of 30 COGs in a set of 18 genomes.

are present at each of the 18 genomes, whereas the others are only present in some of the species.

An evolutionary tree must be consistent with the presence–absence patterns. Specifically, if a COG is present in two species, then it should be present in their last common ancestor and, thus, in all the other descendants of the last common ancestor. However, in most cases, the presence–absence pattern of a COG in extant species is far from the "natural" one: many of

TABLE 1.4

Species

Species	Code	Species	Code
Saccharomyces cerevisiae	y	*Deinococcus radiodurans*	d
Archaeoglobus fulgidus	a	*Mycobacterium tuberculosis*	r
Halobacterium sp. NRC-1	o	*Bacillus subtilis*	b
Methanococcus jannaschii	m	*Synechocystis*	c
Pyrococcus horikoshii	k	*Escherichia coli*	e
Thermoplasma acidophilum	p	*Pseudomonas aeruginosa*	f
Aeropyrum pernix	z	*Vibrio cholera*	g
Aquifex aeolicus	q	*Xylella fastidiosa*	s
Thermotoga maritima	v	*Caulobacter crescentus*	j

Note: List of 18 species (one eukaryota, six archaea and 11 bacteria) represented in
         Table 1.3.

the genes are dispersed over several subtrees. According to comparative genomics, this may happen because of a multiple loss and horizontal transfer of genes. The hierarchy should be constructed in such a way that the number of inconsistencies is minimized.

The so-called principle of maximum parsimony (MP) is a straightforward formalization of this idea. Unfortunately, MP does not always lead to appropriate solutions because of both intrinsic Inconsistencies and computational issues. A number of other approaches have been proposed including the hierarchical cluster analysis (see [163]).

Especially appealing in this regard is the divisive cluster analysis. This approach begins by splitting the entire data set in two parts, thus imitating the evolutionary divergence of the last universal common ancestor (LUCA) producing its two descendants. The same process then applies to each of the split parts until a stop-criterion is invoked to halt the division steps. The divisive clustering imitates the process of evolutionary divergence to an extent. To further approximate the real evolutionary process, the characters involved in the divergence should be discarded immediately after the division [142]. Gene profile data are analyzed on pp. 145, 155, and 249.

1.1.1.4 *Knowledge Structure: Algebraic Functions*

An education research team in Russia has proposed a different methodology for knowledge control and testing. Instead of conventional exam questions directly testing the knowledge, they propose to test the structure of the knowledge. According to this approach, a set of notions that are fundamental for the subject, the knowledge of which is being tested, is extracted first, and then, each student is asked to score the similarity between the notions. The idea is that there exists a structure of semantic relations among the concepts. This structure is to be acquired in the learning process. Then, by extracting

TABLE 1.5

COG Names and Functions

Code	Name
COG0090	Ribosomal protein L2
COG0091	Ribosomal protein L22
COG2511	Archaeal Glu-tRNAGln
COG0290	Translation initiation factor IF3
COG0215	Cysteinyl-tRNA synthetase
COG2147	Ribosomal protein L19E
COG1746	tRNA nucleotidyltransferase (CCA-adding enzyme)
COG1093	Translation initiation factor eIF2alpha
COG2263	Predicted RNA methylase
COG0847	DNA polymerase III epsilon
COG1599	Replication factor A large subunit
COG3066	DNA mismatch repair protein
COG3293	Predicted transposase
COG3432	Predicted transcriptional regulator
COG3620	Predicted transcriptional regulator with C-terminal CBS domains
COG1709	Predicted transcriptional regulators
COG1405	Transcription initiation factor IIB
COG3064	Membrane protein involved
COG2853	Surface lipoprotein
COG2951	Membrane-bound lytic murein transglycosylase B
COG3114	Heme exporter protein D
COG3073	Negative regulator of sigma E
COG3026	Negative regulator of sigma E
COG3006	Uncharacterized protein involved in chromosome partitioning
COG3115	Cell division protein
COG2414	Aldehyde:ferredoxin oxidoreductase
COG3029	Fumarate reductase subunit C
COG3107	Putative lipoprotein
COG3429	Uncharacterized BCR, stimulates glucose-6-P dehydrogenase activity
COG1950	Predicted membrane protein

the structure underlying the student's scoring, one can evaluate the student's knowledge by comparing it with an expert-produced structure: the greater the discrepancy, the worse the knowledge [179]. A similarity score matrix between elementary algebraic functions is presented in Table 1.6.

This matrix has been filled in by a high-school teacher of mathematics who used a eight-point system for scoring: the 0 score means no relation at all and 7 score means a full relation, that is, coincidence or equivalence (see also [135], p. 42). What is the cognitive structure manifested in the scores? How that can be extracted?

TABLE 1.6

Algebraic Functions

Function	e^x	lnx	$1/x$	$1/x^2$	x^2	x^3	$\sqrt{x}$	$\sqrt[3]{x}$
lnx	6							
$1/x$	1	1						
$1/x^2$	1	1	6					
x^2	2	2	1	1				
x^3	3	2	1	1	6			
$\sqrt{x}$	2	4	1	1	5	4		
$\sqrt[3]{x}$	2	4	1	1	5	3	5	
$|x|$	2	3	1	1	5	2	3	2

Note: Similarities between nine elementary functions scored by a high-school mathematics teacher in the 7-rank scale. (Analyzed: p. 180.)

Conventionally, multidimensional-scaling techniques are applied to extract a spatial arrangement of the functions in such a way that the axes correspond to meaningful clusters.

Another, perhaps more adequate structure would be represented by elementary attributes shared, with potentially different intensities, by various parts of the set. Say, functions $e^x, lnx, x^2, x^3, x^{1/2}, x^{1/3}$, and x are all monotone increasing (intensity 1), functions x^3 and x^2 are fast growing (intensity 4), all the functions are algebraic (intensity 1), and so on. Assuming that the similarity between two entities is the sum of the intensities of the attributes shared by them, one arrives at the model of additive clusters [133,140,185]. According to this approach, each of the attributes is represented by a cluster of the entities that share the attribute, along with its intensity. Therefore, the problem is reciprocal: given a similarity matrix, like that in Table 1.6, find a structure of additive clusters that approximates the matrix as closely as possible.

1.1.2 Description

The problem of description is of automatically supplying clusters found by a clustering algorithm with a conceptual description. A good conceptual description can be used for better understanding and better predicting.

1.1.2.1 Describing Iris Genera

Table 1.7 presents probably the most popular data set in the communities related to data analysis and machine learning: 150 Iris specimens, each measured on four morphological variables: sepal length (w1), sepal width (w2), petal length (w3), and petal width (w4), as collected by botanist E. Anderson and published in a founding paper of celebrated British statistician R. Fisher in 1936 [15]. It is said that there are three species in the table, (I) *Iris setosa* (diploid), (II) *Iris versicolor* (tetraploid), and (III) *Iris virginica* (hexaploid), each represented by 50 consecutive rows in the data table.

TABLE 1.7

Iris

Entity in a Class	Class I Iris setosa				Class II Iris versicolor				Class III Iris virginica			
	w1	w2	w3	w4	w1	w2	w3	w4	w1	w2	w3	w4
1	5.1	3.5	1.4	0.3	6.4	3.2	4.5	1.5	6.3	3.3	6.0	2.5
2	4.4	3.2	1.3	0.2	5.5	2.4	3.8	1.1	6.7	3.3	5.7	2.1
3	4.4	3.0	1.3	0.2	5.7	2.9	4.2	1.3	7.2	3.6	6.1	2.5
4	5.0	3.5	1.6	0.6	5.7	3.0	4.2	1.2	7.7	3.8	6.7	2.2
5	5.1	3.8	1.6	0.2	5.6	2.9	3.6	1.3	7.2	3.0	5.8	1.6
6	4.9	3.1	1.5	0.2	7.0	3.2	4.7	1.4	7.4	2.8	6.1	1.9
7	5.0	3.2	1.2	0.2	6.8	2.8	4.8	1.4	7.6	3.0	6.6	2.1
8	4.6	3.2	1.4	0.2	6.1	2.8	4.7	1.2	7.7	2.8	6.7	2.0
9	5.0	3.3	1.4	0.2	4.9	2.4	3.3	1.0	6.2	3.4	5.4	2.3
10	4.8	3.4	1.9	0.2	5.8	2.7	3.9	1.2	7.7	3.0	6.1	2.3
11	4.8	3.0	1.4	0.1	5.8	2.6	4.0	1.2	6.8	3.0	5.5	2.1
12	5.0	3.5	1.3	0.3	5.5	2.4	3.7	1.0	6.4	2.7	5.3	1.9
13	5.1	3.3	1.7	0.5	6.7	3.0	5.0	1.7	5.7	2.5	5.0	2.0
14	5.0	3.4	1.5	0.2	5.7	2.8	4.1	1.3	6.9	3.1	5.1	2.3
15	5.1	3.8	1.9	0.4	6.7	3.1	4.4	1.4	5.9	3.0	5.1	1.8
16	4.9	3.0	1.4	0.2	5.5	2.3	4.0	1.3	6.3	3.4	5.6	2.4
17	5.3	3.7	1.5	0.2	5.1	2.5	3.0	1.1	5.8	2.7	5.1	1.9
18	4.3	3.0	1.1	0.1	6.6	2.9	4.6	1.3	6.3	2.7	4.9	1.8
19	5.5	3.5	1.3	0.2	5.0	2.3	3.3	1.0	6.0	3.0	4.8	1.8
20	4.8	3.4	1.6	0.2	6.9	3.1	4.9	1.5	7.2	3.2	6.0	1.8
21	5.2	3.4	1.4	0.2	5.0	2.0	3.5	1.0	6.2	2.8	4.8	1.8
22	4.8	3.1	1.6	0.2	5.6	3.0	4.5	1.5	6.9	3.1	5.4	2.1
23	4.9	3.6	1.4	0.1	5.6	3.0	4.1	1.3	6.7	3.1	5.6	2.4
24	4.6	3.1	1.5	0.2	5.8	2.7	4.1	1.0	6.4	3.1	5.5	1.8
25	5.7	4.4	1.5	0.4	6.3	2.3	4.4	1.3	5.8	2.7	5.1	1.9
26	5.7	3.8	1.7	0.3	6.1	3.0	4.6	1.4	6.1	3.0	4.9	1.8
27	4.8	3.0	1.4	0.3	5.9	3.0	4.2	1.5	6.0	2.2	5.0	1.5
28	5.2	4.1	1.5	0.1	6.0	2.7	5.1	1.6	6.4	3.2	5.3	2.3
29	4.7	3.2	1.6	0.2	5.6	2.5	3.9	1.1	5.8	2.8	5.1	2.4
30	4.5	2.3	1.3	0.3	6.7	3.1	4.7	1.5	6.9	3.2	5.7	2.3
31	5.4	3.4	1.7	0.2	6.2	2.2	4.5	1.5	6.7	3.0	5.2	2.3
32	5.0	3.0	1.6	0.2	5.9	3.2	4.8	1.8	7.7	2.6	6.9	2.3
33	4.6	3.4	1.4	0.3	6.3	2.5	4.9	1.5	6.3	2.8	5.1	1.5
34	5.4	3.9	1.3	0.4	6.0	2.9	4.5	1.5	6.5	3.0	5.2	2.0
35	5.0	3.6	1.4	0.2	5.6	2.7	4.2	1.3	7.9	3.8	6.4	2.0
36	5.4	3.9	1.7	0.4	6.2	2.9	4.3	1.3	6.1	2.6	5.6	1.4
37	4.6	3.6	1.0	0.2	6.0	3.4	4.5	1.6	6.4	2.8	5.6	2.1
38	5.1	3.8	1.5	0.3	6.5	2.8	4.6	1.5	6.3	2.5	5.0	1.9
39	5.8	4.0	1.2	0.2	5.7	2.8	4.5	1.3	4.9	2.5	4.5	1.7
40	5.4	3.7	1.5	0.2	6.1	2.9	4.7	1.4	6.8	3.2	5.9	2.3
41	5.0	3.4	1.6	0.4	5.5	2.5	4.0	1.3	7.1	3.0	5.9	2.1
42	5.4	3.4	1.5	0.4	5.5	2.6	4.4	1.2	6.7	3.3	5.7	2.5
43	5.1	3.7	1.5	0.4	5.4	3.0	4.5	1.5	6.3	2.9	5.6	1.8
44	4.4	2.9	1.4	0.2	6.3	3.3	4.7	1.6	6.5	3.0	5.5	1.8
45	5.5	4.2	1.4	0.2	5.2	2.7	3.9	1.4	6.5	3.0	5.8	2.2
46	5.1	3.4	1.5	0.2	6.4	2.9	4.3	1.3	7.3	2.9	6.3	1.8
47	4.7	3.2	1.3	0.2	6.6	3.0	4.4	1.4	6.7	2.5	5.8	1.8
48	4.9	3.1	1.5	0.1	5.7	2.6	3.5	1.0	5.6	2.8	4.9	2.0
49	5.2	3.5	1.5	0.2	6.1	2.8	4.0	1.3	6.4	2.8	5.6	2.2
50	5.1	3.5	1.4	0.2	6.0	2.2	4.0	1.0	6.5	3.2	5.1	2.0

Note: Anderson–Fisher data on 150 Iris specimens.

The classes are defined by the genotype; the features are of the phenotype. Is it possible to describe the classes using the features? It is well known from previous analyzes that taxa II and III are not well separated in the feature space (e.g. specimens 28, 33, and 44 from class II are more similar to specimens 18, 26, and 33 from class III than to specimens of the same species, see Figure 1.10 on p. 25). This leads to the idea of deriving new features from those available with the view of finding better descriptions of the classes.

Some non-linear machine-learning techniques such as Neural Nets [77] and Support Vector Machines [59] can tackle the problem and produce a decent decision rule involving non-linear transformation of the features. Unfortunately, rules derived with these methods are not comprehensible to the human mind and, thus, difficult to use for interpretation and description. The human mind needs somewhat more tangible logics which can reproduce and extend botanists' observations that, for example, the petal area provides for much better resolution than the original linear sizes. A method for building cluster descriptions of this type will be described in Section 6.4.3.

The Iris data set is analyzed on pp. 99, 118, 190, and 240.

1.1.2.2 Body Mass

Table 1.8 presents data of the height and weight of 22 male individuals p1–p22, of which the individuals p13–p22 are considered overweight and p1–p12 normal. As Figure 1.2 clearly shows, a line separating these two sets should run along the elongated cloud formed by the entity points. The groups have been defined according to the so-called body mass index, BMI. The BMI is defined as the ratio of the weight, in kilograms, to the squared height, in meters. The individuals whose BMI is 25 or over are considered overweight. The problem is to make a computer to automatically transform the current height–weight feature space into a representation that would allow one to clearly distinguish between the overweight and normally built individuals.

Can the BMI-based decision rule be derived computationally? One would obviously have to consider whether a linear description could suffice. A linear rule of thumb does exist indeed: a man is considered overwheight if the difference between his height in centimeters and weight in kilograms is greater than 100. A man 175 cm in height should normally weigh 75 kg or less, according to this rule.

Once again it should be pointed out that non-linear transformations supplied by machine-learning tools for better prediction may be not necessarily usable for the purposes of description.

The body mass data set is analyzed on pp. 115 and 233.

1.1.3 Association

Finding associations between different aspects of phenomena is one of the most important goals of classification. Clustering, as a classification of the empirical data, should also do the job. A relation between different aspects of

TABLE 1.8

Body Mass

Individual	Height (cm)	Weight (kg)
p1	160	63
p2	160	61
p3	165	64
p4	165	67
p5	164	65
p6	164	62
p7	157	60
p8	158	60
p9	175	75
p10	173	74
p11	180	79
p12	185	84
p13	160	67
p14	160	71
p15	170	73
p16	170	76
p17	180	82
p18	180	85
p19	175	78
p20	174	77
p21	171	75
p22	170	75

Note: Height and weight of 22 individuals.

a phenomenon in question can be established if the aspects are represented by different sets of variables so that the same clusters are well described in terms of each of the sets. Different descriptions of the same cluster are linked as those referring to the same contents, though possibly with different errors.

1.1.3.1 Digits and Patterns of Confusion between Them

The rectangle in the upper part of Figure 1.3 is used to draw numeral digits in a styled manner of the kind used in digital electronic devices. Seven binary presence/absence variables e1, e2,..., e7 in Table 1.9 correspond to the numbered segments on the rectangle in Figure 1.3.

Although the digit character images might be arbitrary, finding patterns of similarity in them can be of interest in training operators of digital devices.

Results of a psychological experiment on confusion between the segmented numerals are in Table 1.10. A respondent looks at a screen at which a numeral digit appears for a very short time (stimulus), then the respondent's report

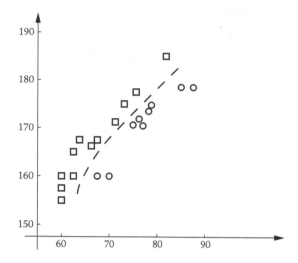

FIGURE 1.2
Twenty-two individuals at the height–weight plane.

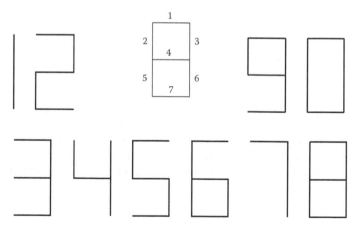

FIGURE 1.3
Styled digits formed by segments of the rectangle.

of what is that is recorded (response). The frequencies of responses versus shown stimuli stand in the rows of Table 1.10 [135].

The problem is to find general patterns of confusion if any. Interpreting them in terms of the segment presence–absence variables in Digits data Table 1.9 is part of the problem. Interpretation of the clusters in terms of the drawings, if successful, would allow one to look for a relationship between the patterns of drawing and confusion.

TABLE 1.9

Digits

Digit	e1	e2	e3	e4	e5	e6	e7
1	0	0	1	0	0	1	0
2	1	0	1	1	1	0	1
3	1	0	1	1	0	1	1
4	0	1	1	1	0	1	0
5	1	1	0	1	0	1	1
6	1	1	0	1	1	1	1
7	1	0	1	0	0	1	0
8	1	1	1	1	1	1	1
9	1	1	1	1	0	1	1
0	1	1	1	0	1	1	1

Note: Segmented numerals presented with seven binary variables corresponding to presence/absence of the corresponding edge in Figure 1.3.

TABLE 1.10

Confusion

Stimulus	Response									
	1	2	3	4	5	6	7	8	9	0
1	877	7	7	22	4	15	60	0	4	4
2	14	782	47	4	36	47	14	29	7	18
3	29	29	681	7	18	0	40	29	152	15
4	149	22	4	732	4	11	30	7	41	0
5	14	26	43	14	669	79	7	7	126	14
6	25	14	7	11	97	633	4	155	11	43
7	269	4	21	21	7	0	667	0	4	7
8	11	28	28	18	18	70	11	577	67	172
9	25	29	111	46	82	11	21	82	550	43
0	18	4	7	11	7	18	25	71	21	818

Note: Confusion between the segmented numeral digits.

Indeed, there can be distinguished four major confusion clusters in the Digits data, as will be found in Section 4.5.2 and described in Section 6.4 (see also pp. 75, 153, 157, 158, 165, and 191). In Figure 1.4, these four clusters are presented along with distinctive segments on the drawings. We can see that all relevant features are concentrated on the left and down the rectangle. It remains to be seen if there is any physio-psychological mechanism behind this and how it can be utilized.

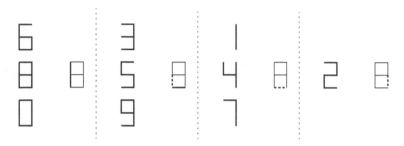

FIGURE 1.4
Visual representation of four Digits confusion clusters: solid and dotted lines over the rectangle show distinctive features that must be present in or absent from all entities in the cluster.

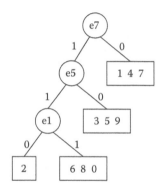

FIGURE 1.5
The conceptual tree of digits.

Moreover, it appears the attributes in Table 1.9 are quite relevant on their own, pinpointing the same patterns that have been identified as those of confusion. This can be clearly seen in Figure 1.5, which illustrates a classification tree for digits found using an algorithm for conceptual clustering presented in Section 4.4. On this tree, clusters are the terminal boxes and interior nodes are labeled by the features involved in the classification. The fact that the edge-based clusters coincide with the confusion clusters indicates a strong link between the drawings and the confusion, which is hardly surprising, after all. The features involved are the same in both Figures 1.4 and 1.5.

1.1.3.2 Colleges

The illustrative data set in Table 1.11 relates to eight fictitious colleges in the United Kingdom. Two quantitative features of college sizes are

1. Stud—The number of students in the scale of full-time students.
2. Acad—The number of academics in the scale of full-time teaching staff.

TABLE 1.11

Colleges

College	Stud	Acad	NS	EL	Course Type
Soliver	3800	437	2	No	MSc
Sembey	5880	360	3	No	MSc
Sixpent	4780	380	3	No	BSc
Etom	3680	279	2	Yes	MSc
Efinen	5140	223	3	Yes	BSc
Enkee	2420	169	2	Yes	BSc
Aywar	4780	302	4	Yes	Certif.
Annaty	5440	580	5	Yes	Certif.

Note: Eight colleges: the first three mostly in Science, the next three in Engineering and the last two in Arts. These categories are not supposed to be part of the data: they are present in a subtle way, via first letters of the names S, E, and A, respectively, and, also, being put in different boxes in the table.

Three features of teaching environment:

3. NS—The number of different schools in the college.
4. EL—Yes or No depending on whether the college provides distant e-learning courses or not.
5. Course type—The course type is a categorical feature with three categories: (a) Certificate, (b) Master, (c) Bachelor, depending on the mainstream degree provided by the college.

As we have seen already with the digits data, features are not necessarily quantitative. They also can be categorical, such as EL, a binary variable or Course type, a nominal variable.

The data in Table 1.11 can be utilized to advance two of the clustering goals:

1. *Structurization:* To cluster the set of colleges and intensionally describe clusters in terms of the features. We expect the clusters to be concordant with the three main areas: Science, Engineering and Arts.
2. *Association:* To analyze interrelations between two aspects of the colleges: (a) the size (presented by the numbers of STud and Acad) and (b) the structure (the other three variables). For instance, we may find clusters in the size features space and conceptually describe them by using features of the structure. The number of entities that do not satisfy the description will score the extent of dissimilarity between the

spaces: the greater the error, the lower the correlation. We expect, in this particular case, to have a high correlation between these aspects, since both must be related to the same cause (the college specialty area) which is absent from the feature list (see p. 220).

This data set is used for illustration of many concepts and methods described further on; see pp. 52, 62, 64, 68, 89, 91, 92, 93, 97, 101, 105, 119, 123, 140, 144, 148, 172, 174, 175, 188, 198, and so on.

1.1.4 Generalization

Generalization, or overview, of data is a (set of) statement(s) about properties of the phenomenon reflected in the data under consideration. To make a generalization with clustering, one may need to do a multistage analysis: at first, structure the entity set; second, describe clusters; third, find associations between different aspects.

An exciting application of this type can be found in the area of text mining. With the abundance of text information flooding every Internet user, the discipline of text mining is flourishing. A traditional paradigm in text mining is underpinned by the concept of key word. The key word is a string of symbols (typically corresponding to a language word or phrase) which is considered important for the analysis of a pre-specified collection of texts. Thus, first comes a collection of texts defined by a meaningful query such as "recent mergers among insurance companies" or "medieval Britain." (Keywords can be produced by human experts in the domain or from the statistical analysis of the collection.) Then a virtual or real text-to-keyword table can be created with keywords treated as features. Each of the texts (entities) can be represented by the number of occurrences of each of the keywords. Clustering of such a table may lead to finding subsets of texts covering similar aspects of the subject.

This approach is being pursued by a number of research and industrial groups, some of which have built clustering engines on top of Internet search engines: given a query, such a clustering engine singles out several dozen of the most relevant web pages, resulting from a search by a search engine such as Google, finds keywords or phrases in the corresponding texts, clusters web pages according to the keywords used as features, and then describes clusters in terms of the most relevant keywords or phrases. Two such websites are iBoogie at http://iboogie.tv and Toxseek at http://toxseek.nlm.nih.gov. The former maintains several dozen languages and presents a hierarchical classification of selected web pages. The latter is a meta-search and clustering engine for Environmental Health and Toxicology at the National Library of Medicine of the United States covering a number of related websites. On May 4, 2011, in response to the query "clustering" iBoogie analyzed 100 web pages dividing them in a number of categories and subcategories. Of these, the most numerous are not necessarily non-overlapping categories "Cluster server," "Data clustering," and "Hierarchical clustering" of 19, 15, and 10 websites, respectively. The latter analyzed 121 websites, dividing them in categories "clustering" (42 websites), "cluster analysis" (13 websites), "gene expression data" (12 websites), and so on.

The activity of generalization so far mainly relies on human experts who supply understanding of a substantive area behind the text corpus. Human experts develop a text-to-feature data table that can be further utilized for generalization. Such is a collection of 55 articles on Bribery cases from central Russian newspapers 1999–2000 presented in Table 1.13 according to [145]. The features reflect the following fivefold structure of bribery situations: two interacting sides—the office and the client, their interaction, the corrupt service rendered and the environment in which it all occurs.

These structural aspects can be characterized by 11 features that have been manually recovered from the newspaper articles; they are presented in Table 1.12.

TABLE 1.12

List of 11 Features I–XI and Their Categories with Respect to Five Aspects of a Bribery Situation

Actor	Service	Interaction	Environment
I. Office	**III. Type of service**	**V. Initiator**	**IX. Condition**
1. Enterprise	1. Obstr. of justice	1. Client	1. Regular routine
2. City	2. Favors	2. Official	2. Monitoring
3. Region	3. Extortion		3. Sloppy regulations
4. Federal	4. Category change		4. Irregular
	5. Cover-up		
II. Client	**IV. Occurrence**	**VI. Bribe level**	**X. Branch**
1. Individual	1. Once	1. $10K or less	1. Government
2. Business	2. Multiple	2. Up to $100K	2. Law enforcement
		3. ≥$100K	3. Other
		VII. Type	**XI. Punishment**
		1. Infringement	1. None
		2. Extortion	2. Administrative
			3. Arrest
		VIII. Network	4. Arrest followed by release
		1. None	5. Arrest with imprisonment
		2. Within office	
		3. Between offices	
		4. Clients	

To show how these features can be applied to a newspaper article, let us quote an article that appeared in the "Kommersant" newspaper on March 20, 1999 (translated from Russian):

MAYOR OF A COAL TOWN UNDER ARREST

Thursday this week, Mr Evgeny Parshukov, Mayor of town Belovo near Kemerovo, was arrested under a warrant issued by the region attorney, Mr Valentin Simuchenkov. The mayor is accused of receiving a bribe and abusing his powers for wrongdoing. Before having been elected to the mayoral post in June 1997, he received a credit of 62,000 roubles from Belovo Division of KUZBASS Transports Bank to support his election campaign. The Bank then cleared up both the money and interest on it, allegedly because after his election Mr Parshukov ordered the Finance Department of the town administration, as well as all municipal organizations in Belovo, to move their accounts into the Transports Bank. Also, the attorney office claims that in 1998 Mr Parshukov misspent 700,000 roubles from the town budget. The money had come from the Ministry of Energy to create new jobs for mine workers made redundant because of mine closures. However, Mr Parshukov ordered to lend the money at a high interest rate to the Municipal Transports agency. Mr Parshukov does not deny the facts. He claims however that his actions involve no crime.

A possible coding of the 11 features in this case constitutes the contents of row 29 in Table 1.13. The table presents 55 cases that have been more or less unambiguously coded (from the original 66 cases [145]).

The prime problem here is similar to those in the market towns and digits data: to see if there are any patterns at all. To generalize, one has to make sense of patterns in terms of the features. In other words, we are interested in getting a synoptic description of the data in terms of clusters which are to be found and described.

On the first glance, no structure can be seen in the data, nor that the scientists specializing in the research of corruption could see any. However, after applying an intelligent version of the algorithm K-Means as described later in Example 3.13, Section 3.3, a rather simple core structure was detected. The structure is defined by just two features. "It is the branch of government that determines which of the five types of corrupt services are rendered: Local government → Favors or Extortion; Law enforcement → Obstruction of Justice or Cover-Up; Other → Category Change," see examples on pp. 113, 223, and 273 for specifics.

TABLE 1.13

Bribery

Case	Of	Cl	Serv	Occ	Ini	Br	Typ	Net	Con	Branch	Pun
1	2	2	2	2	1	2	1	3	3	1	5
2	2	1	5	2	1	1	1	3	2	1	5
3	2	2	2	1	1	3	1	1	3	1	4
4	1	2	3	1	2	1	2	1	3	1	3
5	1	1	4	2	2	1	1	4	3	3	3
6	1	2	3	1	2	2	2	1	3	1	5
7	3	2	1	1	2	3	2	3	2	2	5
8	2	2	4	2	1	1	1	2	1	2	5
9	1	2	3	1	2	1	2	1	1	1	5
10	1	2	3	1	2	1	2	1	1	1	5
11	2	2	5	1	2	3	2	2	2	1	5
12	2	2	1	1	2	2	1	1	4	2	5
13	3	2	1	1	1	3	1	1	4	2	2
14	2	1	4	1	2	1	1	2	1	2	5
15	3	2	2	1	2	1	1	2	3	1	5
16	2	1	4	2	2	1	1	1	1	3	3
17	4	2	2	1	1	2	1	1	4	1	5
18	2	2	5	1	1	2	2	2	3	2	5
19	2	2	5	1	2	1	2	1	3	2	5
20	2	2	1	1	1	2	1	4	3	2	5
21	1	2	3	1	2	2	2	1	3	1	5
22	1	2	2	2	2	2	1	1	4	1	3
23	1	1	4	1	2	1	1	1	1	3	5
24	3	2	5	1	2	1	1	2	2	2	2
25	2	1	2	2	2	1	2	1	3	3	5
26	1	2	5	1	2	2	1	1	3	1	3
27	1	1	4	2	1	2	1	2	4	3	5
28	1	1	5	1	2	1	1	1	2	3	5
29	2	2	2	1	2	2	1	1	3	1	5
30	2	2	3	1	2	2	2	3	3	1	5
31	2	2	3	1	2	1	2	3	3	1	5
32	4	2	2	1	2	1	1	1	3	1	5
33	3	1	1	1	2	1	1	1	4	2	3
34	3	2	1	2	1	1	1	2	4	2	5
35	3	1	3	2	2	1	2	3	4	2	3
36	2	1	3	1	2	2	1	3	1	2	5
37	2	2	5	1	2	3	1	1	2	2	5
38	2	1	1	1	2	2	1	3	4	2	4
39	2	2	1	1	1	1	1	4	4	2	5
40	1	2	3	2	2	1	2	1	3	2	5
41	1	1	4	2	2	1	2	1	2	3	5
42	1	1	4	2	2	1	1	1	3	3	5
43	2	1	1	1	1	1	1	4	4	2	5
44	3	2	5	2	2	1	1	1	2	2	5
45	2	1	1	1	1	1	1	2	4	2	5
46	3	2	5	2	2	1	1	2	2	2	5
47	2	1	1	1	2	1	2	2	4	2	1
48	2	1	1	1	2	1	1	1	4	2	5
49	3	1	2	1	1	2	1	3	4	1	5
50	3	1	1	1	2	1	1	1	4	2	5
51	2	1	2	1	2	1	1	1	4	2	5
52	3	2	1	1	2	1	2	1	2	2	5
53	2	2	5	1	2	2	2	2	2	2	5
54	2	2	5	2	2	2	1	2	2	2	5
55	2	2	5	1	2	3	2	1	2	2	2

Note: Data with features from Table 1.12.

1.1.5 Visualization of Data Structure

Visualization is a rather vague area involving psychology, cognitive sciences, and other disciplines, which is rapidly developing. The subject of data visualization is conventionally defined as creation of mental images to gain insight and understanding [120,197]. This, however, seems too wide and includes too many non-operational images such as realistic, surrealistic, and abstract paintings. For computational purposes, we consider that data visualization is mapping data onto a known ground image such as a coordinate plane, geography map or a genealogy tree in such a way that properties of the data are reflected in the structure of the ground image.

Among ground images, the following are the most popular: geographical maps, networks, 2D displays of one-dimensional objects such as graphs or pie-charts or histograms, 2D displays of two-dimensional objects, and block structures. Sometimes, the very nature of the data suggests what ground image should be used. All of these can be used with clustering, and we are going to review most of them except for geographical maps.

1.1.5.1 One-Dimensional Data

1D data over pre-specified groups or found clusters can be of two types: (a) the distribution of entities over groups and (b) values of a feature within clusters. Accordingly, there can be two types of visual support for these.

Consider, for instance, groups of the market town data defined by the population. According to Table 1.1, the population ranges approximately between 2000 and 24,000 habitants. Let us divide the range in five equal intervals, usually referred to as bins, that are defined thus to have size $(24{,}000 - 2000)/5 = 4400$ and bound points 6400, 10,800, 15,200, and 19,600.

In Table 1.14, the data of the groups are displayed: their absolute and relative sizes and also the average numbers of banks and the standard deviations within them. For the definitions of the average and standard deviation, see Section 2.1.2.

TABLE 1.14

Population Groups

Group	Size	Frequency (%)	Banks	Std Banks
I	25	55.6	1.80	1.62
II	11	24.4	4.82	2.48
III	2	4.4	5.00	1.00
IV	3	6.7	12.00	5.72
V	4	8.9	12.50	1.12
Total	45	100	4.31	4.35

Note: Data of the distribution of population groups and numbers of banks within them.

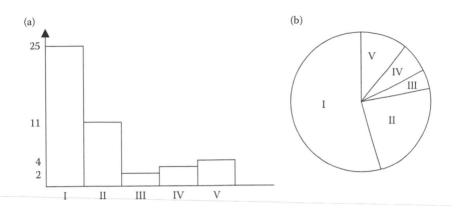

FIGURE 1.6
(a) Histogram and (b) pie-chart presenting the distribution of population over five equally sized bins in Market data.

Figure 1.6 shows two conventional displays for the distribution: a *histogram* (part (a)) in which bars are proportional to the group sizes and a *pie-chart* in which a pie is partitioned into slices proportional to the group sizes (part (b)). These relate to different visual effects. The histogram positions the categories along the horizontal axis, thus providing for a possibility to see the distribution's shape, which can be useful when the categories have been created as interval bins of a quantitative feature, as is this case. The pie-chart sums the group sizes to the total so that one can see what portions of the pie account for different categories.

1.1.5.2 One-Dimensional Data within Groups

To visualize a quantitative feature within pre-specified groups, *box-plots* and *stick-plots* are utilized. They show within-cluster central values and their dispersion, which can be done in different ways. Figure 1.7 presents a box-plot (a) and stick-plot (b) of feature Bank within the five groups defined above as Population bins. The box-plot in Figure 1.7a represents each group as a box bounded by its 10% percentile values separating extreme 10% cases both on the top and on the bottom of the feature Bank range. The within group ranges are shown by "whiskers" that can be seen above and below the boxes of groups I and II; the other groups have no whiskers because of too few entities in each of them. A line within each box shows the within-group average. The stick-plot in Figure 1.7b represents the within-group averages by "sticks," with their "whiskers" proportional to the standard deviations.

Since the displays are in fact two-dimensional, both features and distributions can be shown on a box-plot simultaneously. Figure 1.8 presents a box-plot of the feature Bank over the five bins with the box widths made

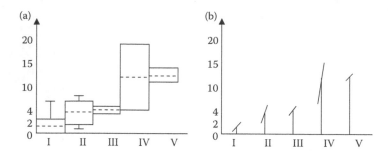

FIGURE 1.7
Box plot (a) and stick plot (b) presenting the feature Bank over the five bins in Market data.

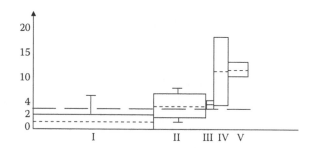

FIGURE 1.8
Box plot presenting the feature Bank over five bins in Market data along with bin sizes.

proportional to the group sizes. This time, the grand mean is also shown by the horizontal dashed line.

A similar box plot for the three genera in the Iris data is presented in Figure 1.9. This time the percentiles are taken at 20%. (In MATLAB®, the percentiles are 25% by default, and the central line is of the median rather than the mean.)

1.1.5.3 Two-Dimensional Display

A conventional two-dimensional display of two quantitative features is the so-called *scatter-plot*, representing all the entities by the corresponding points on a plane of the features. A scatter-plot at the plane of two variables can be seen in Figure 1.2 for the body mass data on p. 14. A scatter plot in the space of two first principal components is presented in Figure 1.10: the Iris specimens are labeled by the class number (1, 2, or 3); centroids are gray circles; the most deviate entities (30 in class 1, 32 in class 2 and 39 in class 3) are shown in boxes. For an explanation of the principal components, see Section 7.1.4. The scatter plot illustrates that two of the classes are intermixed.

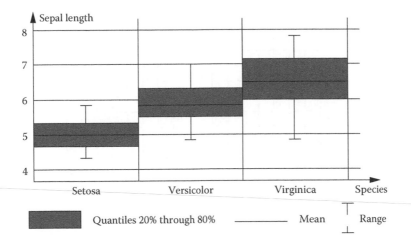

FIGURE 1.9
Box plot of three classes of Iris specimens from Table 1.7 over the sepal length w1; the classes are presented by both the percentile boxes and within cluster range whiskers; the choice of percentiles can be adjusted by the user.

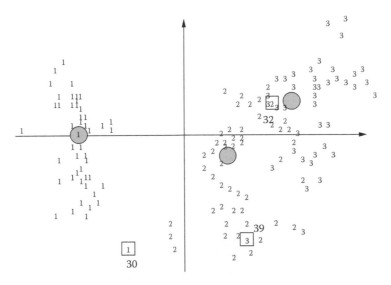

FIGURE 1.10
Scatter plot of Iris specimens in the plane of the first two principal components.

1.1.5.4 Block Structure

A block structure is a representation of the data table as organized in larger blocks of a specific pattern with transpositions of rows and/or columns. In principle one can imagine various block patterns [71], of which the most common is a pattern formed by the largest entry values.

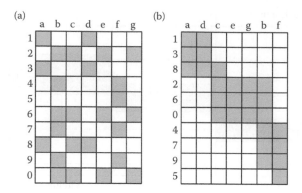

FIGURE 1.11

The matrix block structure: hidden on (a) and visible on (b).

Figure 1.11 presents an illustrative example similar to that in [197]. In part (a), results of seven treatments (denoted by letters from a to g) applied to each of 10 crops denoted by numerals are presented: gray represents a success and blank space, a failure. The pattern of gray seems rather chaotic in Figure 1.11a. However, it becomes very much orderly when rows and columns are appropriately rearranged. Figure 1.11b clearly demonstrates a visible block structure in the matrix. The structure maps subsets of treatments to different sets of crops, which can be used, for instance, in specifying adjacent locations for crops.

Figure 1.12 presents a somewhat more realistic example. Its part (a) displays a matrix of value transfers between nine industries during a year: the (i, j)th entry is gray if the transfer from industry i to industry j is greater than a specified threshold, and blank otherwise. Figure 1.12b shows a block structure pattern that becomes visible when the order of industries from 1 to 9 changes for the order 1–6–9–4–2–5–8–3–7. The reordering is made simultaneously on

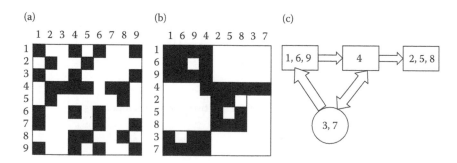

FIGURE 1.12

Value transfer matrix represented using those entries that are greater than a threshold (a); the same matrix, with rows and columns simultaneously reordered, is shown in (b); and the structure is represented by a graph in (c).

both rows and columns because both represent the same industries, as simultaneously the sources (rows) and targets (columns) of the value transfer. There can be discerned four blocks of different patterns (1–6–9, 4, 2–5–8, 3–7) in Figure 1.12b. The structure of the transfers between the blocks can be represented by a graph in Figure 1.12c.

1.1.5.5 Structure

A simple structure such as a chain or a tree or just a small graph, whose vertices (nodes) correspond to clusters and edges to associations between them, is a frequent visual image.

Two examples are presented in Figure 1.13: a tree structure over clusters reflecting a common origin is shown in part (a) and a graph corresponding to the block structure of Figure 1.12c is shown in part (b) to reflect links between clusters of industries in the production process.

A similar tree structure is presented in Figure 1.5 on p. 16 to illustrate a classification tree for Digits. Tree leaves, the terminal boxes, show clusters as entity sets; the features are shown along corresponding branches; the entire structure illustrates the relation between clusters in such a way that any combination of the segments can be immediately identified and placed into a corresponding cluster.

1.1.5.6 Visualization Using an Inherent Topology

In many cases, the entities come from an image themselves—such as in the cases of analysis of satellite images or topographic objects. For example, consider the Digit data set: all the integer symbols are associated with segments of the generating rectangle in Figure 1.3, p. 14 that can be used thus as the ground image.

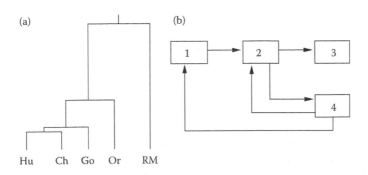

FIGURE 1.13
Visual representation of relations between clusters: (a) the evolutionary structure of the Primate genera according to distances in Table 1.2 and (b) interrelation between the clusters of industries according to Figure 1.12c.

Figure 1.4 visualizes clusters of digits along with their defining features resulting from analyzes conducted later in Example 4.12 (p. 158) as parts of the generating rectangle. There are four major confusion clusters in the digits data of Figure 1.3 that are presented with distinctive features shown using segments defining the drawing of digits.

1.2 Bird's-Eye View

This section contains general remarks on clustering and can be skipped on the first reading.

1.2.1 Definition: Data and Cluster Structure

After looking through the series of case study examples in the previous section, we can give a more formal definition of clustering than that in the Preface: Clustering is a discipline devoted to revealing and describing cluster structures in data sets.

To animate this definition, one needs to specify the four concepts involved:

   a. Data

   b. Cluster structure

   c. Finding a cluster structure

   d. Describing a cluster structure

1.2.1.1 Data

The concept of data refers to any recorded and stored information such as satellite images or time series of prices of certain stocks or survey questionnaires filled in by respondents. Two types of information are associated with data: the data entries themselves, for example, recorded prices or answers to questions, and metadata, that is, legends to rows and columns giving meaning to entries. The aspect of developing and maintaining databases of records, taking into account the relations stored in metadata, is part of data mining [40,47,76].

In this text, we begin with the so-called entity-to-variable table whose entries represent values of pre-specified variables at pre-specified entities.

The variables are synonymously called attributes, features, characteristics, characters and parameters. Such terms as case, object, observation, instance and record are in use as synonyms to the term "entity" accepted here.

The data table often arises directly from experiments or observations, from surveys, and from industrial or governmental statistics. This also is a conventional form for representing database records. Other data types such as signals or images can be modeled in this format, too, via digitalized representation.

However, a digital representation, typically, involves much more information than can be kept in a data table format. Such is the spatial arrangement of pixels, which is not maintained in the concept of data table. The contents of a data table are assumed to be invariant under permutations of rows and columns and corresponding metadata.

Another data type traditionally considered in clustering is the similarity or dissimilarity between entities (or features). The concept of similarity is most important in clustering: similar objects are to be put into the same cluster and those dissimilar into different clusters. There have been invented dozens of (dis)similarity indices. Some of them nicely fit into theoretical frameworks and will be considered further in the text.

One more data type considered in this text is the co-occurrence or redistribution tables that represent the same substance distributed between different categories. Such is the Confusion data in Table 1.10 in which the substance is the counts of individuals. An important property of this type of data is that any part of the data table, related to a subset of rows and/or a subset of columns, can be meaningfully summed to form the part of the total flow within the subset of rows (and/or the subset of columns). The sums represent the total flow to, from, and within the subset(s). Thus, problems of clustering and aggregating are naturally linked here. Until recently, these types of data appeared as a result of data analysis rather than input to it. Currently it has become one of the major data formats. Examples are: distributions of households purchasing various commodities or services across postal districts or other regional units, counts of telephone calls across areas and counts of visits in various categories of websites.

1.2.1.2 Cluster Structure

The concept of *cluster* typically refers to a set of entities that is cohesive in such a way that entities within it are more similar to each other than to the outer entities.

Three major types of cluster structures are: (a) a single cluster considered against the rest or whole of the data, (b) a partition of the entity set in a set of clusters, and (c) a (nested) hierarchy of clusters.

Of these three, partition is the most conventional format, probably because it is relevant to both science and management, the major forces behind scientific developments. The scientist, as well as the manager, would want an unequivocal control over the entire universe under consideration. This is why they may wish to partition the entity set into a set of nonoverlapping clusters.

In some situations, there is no need for a total partition. The user may be satisfied with just a single (or few) cluster(s) and leaving the rest unclustered. Examples:

1. A bank manager wants to learn how to distinguish potential fraudsters from the clients or

2. A marketing researcher separates a segment of customers prone to purchase a particular product or

3. A bioinformatician seeks a set of proteins homologous to a query protein sequence.

Incomplete clustering is a recently recognized addition to the body of clustering approaches, very suitable not only at the situations above but also as a tool for conventional partitioning via cluster-by-cluster procedures such as those described in Sections 7.2.6 and 7.4.2.

The hierarchy is the oldest and probably least understood of the cluster structures. To see how important it is, it should suffice to recall that the Aristotelian approach to classification encapsulated in library classifications and biological taxonomies is always based on hierarchies. Moreover, hierarchy underlies most advanced data-processing tools such as wavelets and quadtrees. Currently, this concept underlies major developments in computational ontologies [173]. It is ironic then that as a cluster structure in its own right, the concept of hierarchy rarely features in clustering, especially when clustering is confined to partitioning of geometric points in cohesive groupings.

1.2.2 Criteria for Obtaining a Good Cluster Structure

To *find a cluster structure* in a data table means to find such clusters that allow the characteristics of individual entities to be reasonably well substituted by characteristics of clusters. A cluster structure is found by a *method* according to a *criterion* of how well the data are represented by clusters. Criteria and methods are, to an extent, independent from each other so that the same method such as agglomeration or splitting can be used with different criteria.

Criteria usually are formulated in terms of (dis)similarity between entities. This helps in formalizing the major idea that entities within clusters should be similar to each other and those between clusters dissimilar. Dozens of similarity-based criteria developed so far can be categorized in three broad classes:

1. Definition-based

2. Index-based

3. Computation-based

The first category comprises methods for finding clusters according to an explicit definition of a cluster. An example: A cluster is a subset S of entities such that for all i, j in S the similarity between i and j is greater than the similarities between these and any k outside S. Such a property must hold for all entities with no exceptions. This may be hard to satisfy because well-isolated clusters are rather rare in real world data. However, when the definition of cluster is relaxed to include less isolated clusters, too many may then appear. This is why definition-based methods are not popular in practical clustering.

A criterion in the next category involves an index, that is, a numerical function that scores different cluster structures and, in this way, may guide the process of choosing the best. However, not all indices are suitable for obtaining reasonable clusters. Those derived from certain model-based considerations tend to be computationally hard to optimize. Optimizing methods are thus bound to be local and, therefore, heavily reliant on the initial settings, which involve, as in the case of K-Means clustering, pre-specifying the number of clusters and the location of their central points. Accordingly, the found cluster structure may be rather far from the global optimum and, therefore, must be *validated*. Cluster validation may be done according to internal criteria such as that involved in the optimization process or external criteria comparing the clusters found with those known from external considerations or according to its stability with respect to randomly resampling entities/features. These will be outlined in Section 6.2 and exemplified in Section 6.2.3.

The third category comprises computation methods involving various heuristics for individual entities to be added to or removed from clusters, for merging or splitting clusters, and so on. Since operations of this type are necessarily local, they resemble local search optimization algorithms, though they typically have no guiding scoring index to follow and thus can include various tricks making them flexible. However, such flexibility is associated with an increase in the number of ad hoc parameters such as various similarity thresholds and, in this way, turning clustering from a reproducible activity into a kind of magic. Validation of a cluster structure found with a heuristic-based algorithm becomes a necessity.

In this book, we adhere to an index-based principle, which scores a cluster structure against the data from which it has been built. The cluster structure here is used as a device for reconstructing the original data table; the closer the reconstructed data are to the original ones, the better the structure. It is this principle that is referred to as the *data recovery approach* in this book. Many index-based and computation-based clustering methods can be reinterpreted according to this principle, so that interrelations between different methods and concepts can be found. Novel methods can be derived from the principle too (see Chapter 5, especially Sections 5.4 through 5.6). It should be noted, though, that we are going to use only the most straightforward rules for decoding the data from cluster structures.

1.2.3 Three Types of Cluster Description

Cluster descriptions help in understanding, explaining and predicting clusters. These may come in different formats of which the most popular are the following three: (a) representative, (b) tendency, (c) conceptual description.

A *representative*, or a *prototype*, is an object such as a literary character or a kind of wine or mineral, representing the most typical features of a cluster. A representative is useful as a "naive" representation of a set of entities that are easily available empirically but difficult to describe conceptually. It is said that some aggregate language constructs, such as "fruit," are mentally maintained via prototypes, such as "apple" [107]. In clustering, the representative is usually a most central entity in its cluster.

A *tendency* expresses a cluster's most likely features such as its way of behavior or pattern. It is usually related to the center of gravity of the cluster and its differences from the average. In this respect, the tendency models the concept of type in classification studies.

A *conceptual description* may come in the form of a classification tree built for predicting a class or partition. Another form of conceptual description is an *association*, or *production, rule*, stating that if an object belongs to a cluster then it must have such and such properties. Or, if an object satisfies the premise, then it belongs in the cluster. Probably the simplest conceptual description of a cluster is a predicate like this: "the cluster is characterized by the feature A being between values a1 and a2." The existence of a feature A, which suffices to distinctively describe a cluster in this way is rather rare in data analysis. Typically, features in data are rather superficial and do not express their essential properties; therefore, they cannot supply straightforward descriptions.

The task of conceptual cluster description has something to do with supervised machine learning and pattern recognition. Indeed, given a cluster, its description may be used to predict, for new objects, whether they belong to the cluster or not, depending on how much they satisfy the description. On the other hand, a decision rule obtained with a machine-learning procedure, such as a classsification tree, can be considered a cluster description usable for the interpretation purposes. Still the goals—interpretation in clustering and prediction in machine learning—do differ.

1.2.4 Stages of a Clustering Application

Typically, clustering as a data analysis activity involves the following five stages:

a. Developing a data set

b. Data preprocessing and standardizing

c. Finding clusters in data

d. Interpreting clusters

e. Drawing conclusions

To develop a data set, one needs to define a substantive problem or issue, however vague it may be, and then determine what data set related to the issue can be extracted from an existing database or set of experiments or survey, and so on.

Data preprocessing is the stage of preparing data to processing it by a clustering algorithm; typically, it includes developing a uniform data set, frequently called a "flat" file, from a database, checking for missing and unreliable entries, rescaling and standardizing variables, deriving a unified similarity measure, etc.

The cluster-finding stage involves application of a clustering algorithm and results in a (series of) cluster structure(s) to be presented, along with interpretation aids, to substantive specialists for an expert judgment and interpretation in terms of features, both those used for clustering (internal features) and those not used (external features). At this stage, the expert may see difficulties in the interpretation of the results and suggest a modification of the data by adding/removing features and/or entities. The modified data are subject to the same processing procedure. The final stage is the drawing of conclusions, with respect to the issue in question, from the interpretation of the results. The more focused are the regularities implied by the findings, the better the quality of conclusions.

There is a commonly held opinion among specialists in data analysis that the discipline of clustering concerns only the proper clustering stage C while the other four are of concern to the specialists in the application domain. Indeed, usually, clustering results cannot and are not supposed to solve the entire substantive problem, but rather relate to an aspect of it.

On the other hand, clustering algorithms are supposedly most applicable to situations and issues in which the user's knowledge of the domain is more superficial than profound. What are the choices regarding the issues of data preprocessing, initial setting and interpretation of the results? These are especially difficult for a laymen user who has an embryonic knowledge of the domain. The advice for conducting more studies and experiments is not practical in most cases. Sometimes a more viable strategy would be to use such clustering methods that help the user by translating the choice from the language of cluster parameters to a more intuitive terminology.

Currently, no model-based recommendations can be made about the initial and final stages, A and E. However, the data recovery approach does allow us to use the same formalisms for tackling not stage C only, but also B and D; see Sections 2.4, 4.4 and 6.4 for related prescriptions and discussions.

1.2.5 Clustering and Other Disciplines

Clustering is a multidisciplinary activity, regardless of its many applications. In particular,

1. *Data* relates to database, data structure, measurement, similarity and dissimilarity, statistics, matrix theory, metric and linear spaces, graphs, data analysis, data mining, and so on.

2. *Cluster structure* relates to discrete mathematics, abstract algebra, cognitive science, graph theory, and so on.

3. *Extracting* cluster structures relates to algorithms, matrix analysis, optimization, computational geometry, and so on.

4. *Describing* clusters relates to machine learning, pattern recognition, mathematical logic, knowledge discovery, and so on.

1.2.6 Different Perspectives of Clustering

Clustering is a discipline that can be viewed from different angles, which may be sometimes confusing because different perspectives may contradict each other. A question such as "How many clusters are out there?," which is legitimate in one perspective, can be meaningless in the other. Similarly, the issue of validation of clusters may have different solutions in different frameworks. The author finds it useful to distinguish between the perspectives supplied by the classical statistics, machine learning, data analysis (or, data mining) and conceptual classification.

1.2.6.1 Classical Statistics Perspective

Classical statistics tends to view any data table as a sample from a probability distribution whose properties or parameters are to be estimated using the data. In the case of clustering, clusters are supposed to be associated with different probabilistic distributions which are intermixed in the data and should be recovered from it.

Within this approach, such questions as "How many clusters are out there?" and "How to preprocess the data?" are well substantiated and can be dealt with according to the assumptions of the underlying model.

In many cases, the statistical paradigm suits the phenomenon in question quite well and should be applied as the one corresponding to what is called the scientific method: make a hypothesis of the phenomenon in question, then look for relevant data and check how well they fit the hypothesis.

A trouble with this approach is that in most cases clustering is applied to phenomena of which almost nothing is known, not only of their underlying mechanisms but of the very features measured or to be measured. Then any modeling assumptions of the data generation would be necessarily rather arbitrary and so would the conclusions based on them.

Moreover in many cases the set of entities is rather unique and less than natural to be considered a sample from a larger population, e.g., the set of European countries or single malt whisky brands.

Sometimes the very concept of a cluster as a probabilistic distribution is at odds with intuition. Look, for example, at a bell-shaped Gaussian distribution which is a good approximation for such features as the height or weight of young male individuals of the same ethnicity so that they form a single cluster indeed. However, when confronted with a task of dividing this group, for example, according to their fighting capabilities (such as in the sport of boxing), the group cannot be considered homogeneous anymore and must be further partitioned into more homogeneous chunks. Some say that there must be a border between "natural" clusters that exist by themselves and "artificial" clusters that are drawn on purpose; that a bell-shape distribution corresponds to a natural cluster and a boxing weight category to an artificial one. However, it is not always easy to tell which is which. There will be always situations when a cluster of potentially weak fighters (or bad customers, or homologous proteins) must be cut out from the rest.

1.2.6.2 Machine-Learning Perspective

Machine learning tends to view the data as a device for learning how to predict prespecified or newly created categories. The entities are considered as coming one at a time so that the machine can learn adaptively in a supervised manner. To conceptualize this, the stream of data must be assumed to come from a probabilistic population, an assumption which has much in common with the classical statistics approach. Yet it is the prediction rather than the model fitting which is the central issue in machine learning.

Such a shift in the perspective has led to the development of strategies for predicting categories such as decision trees and support vector machines as well as resampling methods such as the bootstrap and cross-validation.

1.2.6.3 Data-Mining Perspective

Data mining is not much interested in reflection on where the data entries have come from nor how they have been collected. It is assumed that a data set or database has been collected already and, however bad or good it is, the major concern is in finding patterns and regularities within the data as they are. Machine learning and classical statistics methods are welcome here—not because of their theoretical underpinnings, but for their capacity to do the job.

This view, expressed as early as in the 1960s and 1970s in many countries including France, Russia, and Japan in such subjects as analysis of questionnaires and of inter-industrial transfers, was becoming more and more visible, but it did not make it into prominence until the nineties. By that time, big warehouse databases became available, which led to the discovery of patterns of transactions with the so-called association search methods [40]. The patterns proved themselves correct when superstores increased profits by changing their policies accordingly.

It should be added that the change of the paradigm from modeling of mechanisms of data generation to data analysis and data mining has drastically changed requirements to methods and programmes. According to the classical statistics approach, the user must know the models and methods they use; if a method is applied wrongly, the results can be wrong too. Thus, applications of statistical methods are to be confined within a small circle of experts. In data analysis and data mining, it is the patterns not methods that matter. This shifts the focus of computations from the statistical methods to the user's substantive area and requires a user-friendly software.

Similarly, the validation objectives seem to diverge here too: in statistics and machine learning the stress goes on the consistency of the algorithms, which is not quite so important in data analysis, in which the consistency of patterns, rather than algorithms, matters most.

1.2.6.4 Classification and Knowledge-Discovery Perspective

The classification perspective is rarely mentioned in data analysis and statistics. The term "classification" itself is usually referred to in a very limited sense: as an activity of assigning prespecified categories (class labels) to entities, in contrast to clustering which assigns entities with newly created categories (clusters)—the supervised learning versus unsupervised learning.

Yet there is a long standing tradition in sciences that applies the term "classification" to an actual or ideal arrangement of entities under consideration in classes aimed at:

1. Shaping and keeping knowledge
2. Capturing the structure of phenomena
3. Relating different aspects of a phenomenon in question to each other

These make the concept of classification a specific mechanism for knowledge discovery and handling. Consider, for instance, the Periodic Chart of chemical elements. Its rows correspond to numbers of electron shells in the atoms, and its columns to the numbers of electrons in the external shell, thus capturing the structure of the phenomenon. These also relate to most important physical properties and chemical activities of the elements, thus relating different aspects of the phenomenon. And the chart is a compact form of representing the knowledge; moreover, historically it is this form itself, developed rather ad hoc, that made possible rather fast progress of the sciences to the current theories of the matter.

In spite of the fact that the notion of classification as part of scientific enquiry was introduced very early (Aristotle and the like), the very term "classification" seems a missing item in the vocabulary of the current scientific discourse. This may have happened because in conventional sciences, classifications are defined within well-developed substantive theories according to variables defined within the theories. Thus, there has been no need in specific theories for classification.

Clustering should be considered as classification in this sense, based on empirical data in a situation when clear theoretical concepts and definitions are absent and the regularities are unknown. The clustering goals should relate to the classification goals above. This brings one more aspect to clustering. Consider, for example, how one can judge whether a clustering is good or bad? According to the classification/knowledge-discovery view, this is easy and has nothing to do with statistics: just look at how well clusters fit within the existing knowledge of the domain, how well they allow updating, correcting, and extending the knowledge.

Somewhat simplistically, one might say that two of the points stressed in this book, that of the data recovery approach and the need to not only find, but also to describe clusters, fit well into the two perspectives, the former into data analysis and data mining, and the latter into classification as knowledge discovery.

2

What Is Data?

After reading through this chapter, the reader will know of

1. Three types of data tables: (a) feature-to-entity, (b) similarity/dissimilarity, and (c) contingency/flow tables, and how to standardize them
2. Quantitative, categorical, and mixed data, and ways to pre-process and standardize them
3. Characteristics of feature spread and centrality
4. Bi-variate distributions over mixed data, correlation and association, and their characteristics
5. Visualization of association in contingency tables with Quetelet coefficients
6. Multidimensional concepts of distance and inner product
7. Similarity data and types of similarity data
8. The concept of data scatter

Key Concepts

Affinity data A similarity matrix derived from an entity-to-feature table by using, usually, the so-called Gaussian kernel $a(x,y) = \exp(-d(x,y)/D)$ where $d(x,y)$, squared Euclidean distance between multivariate points x and y, and D a user-defined value.

Average The average value of a feature over a subset of entities. If the feature is binary and corresponds to a category, the average is the category frequency in the subset. The average over the entire entity set is referred to as the grand mean.

Contingency coefficient A summary index of statistical association between two sets of categories in a contingency table. The greater it is, the closer the association to a conceptual one.

Contingency table Given two sets of categories corresponding to rows and columns, respectively, this table presents counts of entities co-occurring at the intersection of each pair of categories from the two sets. When categories within each of the sets are mutually disjoint, the contingency table can be aggregated by summing up relevant entries.

Correlation The shape of a scatter plot showing the extent to which two features can be considered mutually related. The (product-moment) correlation coefficient captures the extent at which one of the features can be expressed as a linear function of the other.

Data scatter The sum of squared entries of the data matrix; it is equal to the sum of feature contributions or the summary distance from entities to zero. At redistribution tables, this is modified to take into account the weights of rows and columns.

Data standardization This is a data transformation usually related to shifting feature origins and changing their scales. Currently each feature is standardized independently of the others. The new origin serves as a backdrop with respect to which the data are analyzed. The new scales should provide a unified scale for different features.

Data table Also referred to as *flat file* (in databases) or *vector space data* (in information retrieval), this is a two-dimensional array whose rows correspond to entities, columns to features, and entries to feature values at the entities.

Distance Given two vectors of the same size, the (Euclidean squared) distance is the sum of squared differences of corresponding components, $d(x,y) = \sum_i (x_i - y_i)^2$. It is closely related to the inner product: $d(x,y) = <x - y, x - y>$.

Entity Also referred to as *observation* (in statistics) or *case* (in social sciences) or *instance* (in artificial intelligence) or *object*, this is the main item of clustering corresponding to a data table row.

Feature Also referred to as *variable* (in statistics) or *character* (in biology) or *attribute* (in logic), this is another major data item corresponding to a data table column. It is assumed that feature values can be compared to each other, at least, whether they coincide or not (categorical features), or even averaged over any subset of entities (quantitative feature case).

Inner product Given two vectors of the same size, the inner product is the sum of products of corresponding components, $<x,y> = \sum_i x_i y_i$. The inner product is closely related to the distance: $d(x,y) = <x,x> + <y,y> - 2<x,y>$.

Kernel A symmetric function of two multivariate variables used for deriving a similarity matrix between entities from an entity-to-feature table. In a situation when the kernel is semipositive definite, it imitates the operation of inner product over a set of features found with a non-linear transformation of the original features according to the so-called eigen-functions of the kernel.

Quetelet index In contingency tables: A value showing the change in frequency of a row category when a column category becomes known.

The greater the value, the greater the association between the column and row categories.

Range The interval in which a feature takes its values; the difference between the feature maximum and minimum over a data set.

Redistribution data A data table whose entries can be meaningfully summed across the table because they express parts of their total. One can think that the sums of rows in such a table represent the quantities that are being redistributed from rows to columns, so that the sums of columns represent the resulting quantities.

Scatter plot A graph representing entities as points on the plane formed by two quantitative features.

Variance The average of squared deviations of feature values from the average.

2.1 Feature Characteristics

2.1.1 Feature Scale Types

The Colleges data in Table 1.11 will be used to illustrate data-handling concepts in this section. For the reader's convenience, the table is reprinted here as Table 2.1.

A data table of this type represents a unity of the set of rows, always denoted as I further on, the set of columns denoted by V and the table X itself, the set of values x_{iv} in rows $i \in I$ and columns $v \in V$. The number of rows, or cardinality of I, $|I|$, will be denoted by N, and the number of columns, the

TABLE 2.1

Colleges

College	Stud	Acad	NS	EL	Course Type
Soliver	3800	437	2	No	MSc
Sembey	5880	360	3	No	MSc
Sixpent	4780	380	3	No	BSc
Etom	3680	279	2	Yes	MSc
Efinen	5140	223	3	Yes	BSc
Enkee	2420	169	2	Yes	BSc
Aywar	4780	302	4	Yes	Certif.
Annaty	5440	580	5	Yes	Certif.

Note: Eight colleges: the first three mostly in Science, the next three in Engineering, and the last two in Arts. These subjects are not part of the data: they are present in a subtle way, through first letters of the names, S, E, and A, respectively.

cardinality of V, $|V|$, by M. Rows will always correspond to entities, columns to features.

It is assumed that each feature $v \in V$ has a measurement scale assigned to it, and of course a name. All within-column entries are supposed to have been measured in the same scale and thus comparable within the scale; this is not so over rows in X. Three different types of scales that are present in Table 2.1 and will be dealt with in the remainder are quantitative (Stud, Acad, and NS), nominal (Course type), and binary (EL).

Let us comment on these scale types:

1. *Quantitative*: A feature is quantitative if the operation of taking its average is meaningful.

   It is quite meaningful to compute and compare the average values of feature Stud or Acad for different subjects in Table 2.1. Somewhat less convincing is the case of NS which must be an integer; some authors even consider such "counting" features a different scale type. Still, we can safely say that on average Certificate colleges have more schools than those of MSc or BSc degrees. This is why counting features are considered also quantitative in this text.

2. *Nominal*: A categorical feature is said to be nominal if its categories are (i) disjoint, that is, no entity can fall in more than one of them and (ii) not ordered, that is, they only can be compared with respect to whether they coincide or not. Course type, in Table 2.1, is such a feature.

   Categorical features maintaining (ii) but not (i) are referred to as multi-choice variables. For instance, Colleges data might include a feature that presents a list of classes taught. Since a class can be taught in several colleges simultaneously, that would produce a one-to-many mapping of the entities to the categories. There is no problem in treating this type of data within the framework described here. For instance, the Digit data table may be treated as that representing the only, multi-choice, variable "Segment" which has the set of seven segments as its categories.

   Categorical features that maintain (i) but have their categories ordered are called ranked, or order scale, variables. Variable Bribe level in the Bribery data of Tables 1.12 and 1.13 is rank: its three categories are obviously ordered according to the bribe size. Traditionally, it is assumed for the rank variables that only the order of categories does matter, wheras the intervals between the categories are irrelevant. According to this view, the rank categories may accept any quantitative coding which is compatible with the order. Say, categories a, b, and c can be assigned values 1, 2, and 3; or, equivalently, 1, 1000000, and 1000001; or, equivalently, 1, 2, 1000000. Such an arbitrariness in the scaling would make the rank features difficult to deal with in the context of mixed data tables. We maintain a different view, going a hundred years back to C. Spearman: the ranks are treated as numerical values so that the rank variables are considered quantitative and processed accordingly. In particular, seven of the 11 variables in Bribery data (II. Client, IV. Occurrence, V. Initiator, VI. Bribe, VII. Type, VIII. Network, and XI. Punishment) will be considered ranked with ranks assigned in Table 1.12 and treated as quantitative values.

There are two approaches to the issue of handling categorical features in cluster analysis. According to one, more conventional, approach, categorical variables are considered non-treatable quantitatively. The only quantitative operation admitted for categories is counting the number or frequency of its occurrences at various subsets. To conduct cluster analysis, categorical data, according to this view, can only be utilized for deriving an entity-to-entity (dis)similarity measure. Then this measure can be used for finding clusters.

This text attends to a different approach: a category defines a quantitative zero–one variable on entities, a dummy, marking the presence of the category by unity and absence, by zero. This is further treated as a quantitative coding. It will be seen, later in the text, that this view, in fact, is not at odds with the former perspective but nicely fits into it with specifications which are sound both geometrically and statistically.

3. *Binary*: A categorical feature is said to be binary if it has two categories which can be thought of as Yes or No answer to a question such as feature EL in Table 2.1. A two-category feature can be considered either a nominal or binary one, depending on the context. For instance, feature "Gender" of a human should be considered a nominal feature, whereas the question "Are you female?" a binary feature, because the latter assumes that it is the "female," not "male," category which is of interest. Operationally, the difference between these two types will amount to how many binary features should be introduced to represent the feature under consideration in full. Feature "Gender" cannot be represented by one column with Yes or No categories: two are needed, one for "Female" and one for "Male."

2.1.2 Quantitative Case

As mentioned, we maintain that a defining property of a quantitative variable is that taking its average is a meaningful operation. Given a feature $v \in V$ whose values y_{iv}, $i \in I$, constitute a column in the data table, its *average* over entity subset $S \subseteq I$ is defined by the formula

$$c_v(S) = (1/N_S) \sum_{i \in S} y_{iv} \tag{2.1}$$

where N_S is the number of entities in S.

The average $c_v = c_v(I)$ of $v \in V$ over the entire set I is sometimes referred to as *grand mean*. After the grand mean c_v of $v \in V$ has been subtracted from all elements of the column feature $v \in V$, the grand mean of v becomes zero. Such a variable is referred to as *centered*.

It should be mentioned that there exists a different approach to defining the quantitative scale type. The abstract measurement theory defines the scale type by using the so-called admissible transformations $y = \phi(x)$. The scale type is claimed to depend on the set of transformations ϕ considered admissible, that is, not changing the scale contents. To be a quantitative feature scale, the feature's admissible transformations are to be affine transformations defined by equations $y = ax + b$ for some real a and b. Such a transformation converts the x values into the y values by changing the scale factor a times and shifting the scale origin to $-b$ after this. For example, the temperature Celsius scale x is transformed into the Fahrenheit scale with $\phi(x) = 1.8x + 32$. Standardizations of data with affine transformations are at the heart of our approach to clustering.

Our definition is compatible with the abstract measurement theory approach. Indeed, if a feature x admits affine transformations, it is meaningful to compare its average values over various entity sets. Let x_J and x_K be the averages of sets $\{x_j : j \in J\}$ and $\{x_k : k \in K\}$ respectively, and, say, $x_J \leq x_K$. Does the same automatically hold for the averages of $y = ax + b$ over J and K? To address this question, we consider values $y_j = ax_j + b, j \in J$ and $y_k = ax_k + b$, $k \in K$, and calculate their averages, y_J and y_K. It is easy to prove that $y_K = ax_K + b$ and $y_J = ax_J + b$ so that any relation between x_J and x_K remains the same for y_J and y_K, up to the obvious reversal when a is negative (which means that rescaling involves change of the direction of the scale). A reverse statement holds as well: if a differentiable mapping $y = \phi(x)$ does not change the order of the averages, then it is an affine transformation.

Other concepts of "centrality" have been considered too; the most popular of them are

i. Midrange, point in the middle of the range; it is equi-distant from the minimum and maximum values of the feature.

ii. Median, the middle item in the series of elements of column v sorted in the ascending (or descending) order.

iii. Mode, "the most likely" value, which is operationally defined by partitioning the feature range in a number of bins (intervals of the same size) and determining at which of the bins the number of observations is maximum: the center of this bin is the mode, up to the error related to the bin size.

Each of these has its advantages and drawbacks as a centrality measure. The median is the most stable with regard to changes in the sample and, especially, to the presence of outliers. An outlier can drastically change the average, but it would not affect the median. However, the calculation of the median requires sorting the entity set, which may be costly sometimes. Midrange is insensitive to the shape of the distribution and is highly sensitive to outliers. The mode is of interest when distribution of the feature is far from uniform.

These may give one a hint with respect to what measure should be used in a specific situation. For example, if the data to be analyzed have no specific properties at all, the average should be utilized. When outliers or data errors are expected, the median would be a better bet.

The average, median and midrange all fit within the following approximation model which is at the heart of the data recovery approach. Given a number of reals, $x_1, x_2, \ldots, x_N$, find a unique real a that can be used as their "consensus" value so that for each i, a approximates x_i up to a small residual ϵ_i: $x_i = a + \epsilon_i, i \in I$. The smaller the residuals, the better the consensus. To minimize the residuals $\epsilon_i = x_i - a$, they should be combined into a scalar criterion such as $L_1 = \sum_i |x_i - a|$, $L_\infty = \max_i |x_i - a|$, or $L_2 = \sum_i |x_i - a|^2$. It appears, L_1 is minimized by the median, L_∞ by midrange and L_2 by the average. The average fits best because it solves the least-squares approximation problem, and the least-square criterion is the base of all the mathematics underlying the data recovery approach in this volume.

A number of characteristics have been defined to measure the features' dispersion or spread. Probably the simplest of them is the feature's *range*, the difference between its maximum and minimum values. This measure should be used cautiously as it may be overly sensitive to changes in the entity set. For instance, removal of Truro from the set of entities in the Market town data immediately reduces the range of variable Banks to 14 from 19. Further removal of St Blazey/Par further reduces the range to 13. Moreover, the range of variable DIY shrinks to 1 from 3 when these two towns are removed. Obviously, no such drastic changes would emerge when all 1300 of the English Market towns are present.

A somewhat more elaborate characteristic of dispersion is the so-called (empirical) variance of $v \in V$ which is defined as

$$s_v^2 = \sum_{i \in I} (y_{iv} - c_v)^2 / N \qquad (2.2)$$

where c_v is the grand mean. That is, s_v^2 is the average squared deviation of y_{iv} from c_v.

The standard deviation of $v \in V$ is defined as just $s_v = \sqrt{s_v^2}$ which has also a statistical meaning as the square-average deviation of the variable's values from its grand mean. The standard deviation is zero, $s_v = 0$, if and only if the variable is constant, that is, all the v entries are equal to each other.

In some packages, especially statistical ones, denominator $N - 1$ is used instead of N in the definition (2.2) because of a probabilistic model assumed considerations (see any text on mathematical statistics). This should not much affect results because N is assumed constant here and, moreover, $1/N$ and $1/(N - 1)$ do not much differ when N is large.

TABLE 2.2

Summary Characteristics of the Market Town Data

	P	PS	Do	Ho	Ba	Su	Pe	DIY	SP	PO	CAB	FM
Mean	7351.4	3.0	1.4	0.4	4.3	1.9	2.0	0.2	0.5	2.6	0.6	0.2
Std	6193.2	2.7	1.3	0.6	4.4	1.7	1.6	0.6	0.6	2.1	0.6	0.4
Range	21,761.0	12.0	4.0	2.0	19.0	7.0	7.0	3.0	2.0	8.0	2.0	1.0

For the Market town data, with $N = 45$ and $n = 12$, the summary characteristics are in Table 2.2.

The standard deviations in Table 2.2 are at least as twice as small as the ranges, which is true for all data tables (see Statement 2.2).

The values of the variance s_v^2 and standard deviation s_v obviously depend on the variable's spread measured by its range. Multiplying the column $v \in V$ by $\alpha > 0$ obviously multiplies its range and standard deviation by α, and the variance by α^2.

The quadratic index of spread, s_v^2, depends not only on the scale, but also on the character of the feature's distribution within its range. Can we see how?

Let as consider all quantitative variables defined on N entities $i \in I$ and ranged between 0 and 1 inclusive, and analyze at what distributions the variance attains its maximum and minimum values.

It is not difficult to see that any such feature v that minimizes the variance s_v^2 is equal to 0 at one of the entities, 1 at another entity and $y_{iv} = c_v = 1/2$ at all other entities $i \in I$. The minimum variance s_v^2 is $1/2N$ then.

Among the distributions under consideration, the maximum value of s_v^2 is reached at a feature v which is binary, that is, has only boundary points, 0 and 1, as its values. Indeed, if v has any other value at an entity i, then the variance will only increase if we redefine v in such a way that it becomes 0 or 1 at i depending on whether y_{iv} is smaller or greater than v's average c_v. For a binary v, let us specify proportion p of values y_{iv} at which the feature is larger than its grand mean, $y_{iv} > c_v$. Then, obviously, the average $c_v = 0 * (1 - p) + 1 * p = p$ and, thus, $s_v^2 = (0 - p)^2 * (1 - p) + (1 - p)^2 * p = p(1 - p)$.

The choice of the left and right bounds of the range, 0 and 1, does have an effect on the values attained by the extremal variable but not on the conclusion of its binariness. That means that the following is proven.

Statement 2.1. With the range and proportion p of values that are smaller than the average pre-specified, the distribution at which the variance reaches its maximum is the distribution of a binary feature having p values at the left bound and $1 - p$ values at the right bound of the range.

Among the binary features, the maximum variance is reached at $p = 1/2$, the maximum uncertainty. This implies one more property.

Statement 2.2. For any feature, its standard deviation is at least as twice as small as its range.

Proof: Indeed, with the range being unity between 0 and 1, the maximum variance is $p(1-p) = 1/4$ at $p = 1/2$ leading to the maximum standard deviation of just half of the unity range, q.e.d.

From the intuitive point of view, the range being the same, the greater the variance the better the variable suits the task of clustering.

2.1.3 Categorical Case

Let us first consider binary features and then nominal ones.

To quantitatively recode a binary feature, its Yes category is converted into 1 and No into 0. The grand mean of the obtained zero/one variable will be p_v, the proportion of entities falling in the category Yes. Its variance will be $s_v^2 = p(1-p)$.

In statistics, two types of probabilistic mechanisms for generating zero/one binary variables are popular, Bernoulli/binomial and Poisson. Each relies on having the proportion of ones, p, pre-specified. However, the binomial distribution assumes that every single entry has the same probability p of being unity, whereas Poisson distribution does not care about individual entries: just that the proportion p of entries randomly thrown into a column must be unity. This subtle difference makes the variance of the Poisson distribution greater: the variance of the binomial distribution is equal to $s_v^2 = p(1-p)$ and the variance of the Poisson distribution is equal to $\pi_v = p$. Thus, the variance of a one–zero feature considered as a quantitative feature corresponds to the statistical model of binomial distribution.

Turning to the case of nominal variables, let us denote the set of categories of a nominal variable l by V_l. Any category $v \in V_l$ is conventionally characterized by its frequency, the number of entities, N_v, falling in it. The sum of frequencies is equal to the total number of entities in I, $\sum_{v \in V_l} N_v = N$. The *relative frequencies*, $p_v = N_v/N$, sum to unity. The vector $p = (p_v)$, $v \in V_l$, is referred to as the *distribution* of l (over I). A category with the largest frequency is referred to as the distribution's mode. The dispersion of a nominal variable l is frequently measured by the so-called *Gini coefficient*, or qualitative variance:

$$G = \sum_{v \in V_l} p_v(1 - p_v) = 1 - \sum_{v \in V_l} p_v^2 \tag{2.3}$$

This is zero if all the entities fall in one of the categories only. G is maximum when the distribution is *uniform*, that is, when all category frequencies are the same, $p_v = 1/|V_l|$ for all $v \in V_l$.

A similar measure referred to as *entropy* and defined as

$$H = - \sum_{v \in V_l} p_v \log p_v \tag{2.4}$$

with the logarithm's base 2 is also quite popular [29]. Entropy reaches its minimum and maximum values at the same distributions as the Gini coefficient. Moreover, the Gini coefficient can be thought of as a linearized version of entropy since $1 - p_v$ linearly approximates $\log p_v$ at p_v close to 1. In fact, both can be considered averaged information measures, just that one uses $-\log p_v$ and the other $1 - p_v$ to express the information contents.

There exists a general formula to express the diversity of a nominal variable as $S_q = (1 - \sum_{v \in V_l} p_v^q)/(q - 1)$, $q > 0$. The entropy and Gini index are special cases of S_q since $S_2 = G$ and $S_1 = H$ assuming S_1 to be the limit of S_q when q tends to 1.

A nominal variable l can be converted into a quantitative format by assigning a zero/one feature to each of its categories $v \in V_l$ coded by 1 or 0 depending on whether an entity falls into the category or not. These binary features are referred to sometimes as dummy variables.

Unlike a binary feature, a two-category nominal feature such as "Gender" is to be enveloped into two columns, each corresponding to one of the categories, "Male" and "Female" of "Gender." This way of quantization is quite convenient within the data recovery approach as will be seen further in Section 4.4 and others. However, it is also compatible with the conventional view of quantitative measurement scales as expressed in terms of admissible transformations. Indeed, for a nominal scale x, any one-to-one mapping $y = \phi(x)$ is admissible. When there are only two categories, x_1 and x_2, they can be recoded into any y_1 and y_2 with an appropriate rescaling factor a and shift b so that the transformation of x to y can be considered an affine one, $y = ax + b$. Obviously, $a = (y_1 - y_2)/(x_1 - x_2)$ and $b = (x_1 y_2 - x_2 y_1)/(x_1 - x_2)$ will do the recoding. In other words, nominal features with two categories can be considered quantitative. The binary features, in this context, are those with category Yes coded by 1 and No by 0 for which transformation $y = ax + b$ is meaningful only when $a > 0$ and $b = 0$.

The vector of averages of the dummy category features, p_v, $v \in V_l$, is but the distribution of l. Moreover, the Gini coefficient appears to be but the summary Bernoullian variance of the dummy category features, $G = \sum_{v \in V_l} p_v (1 - p_v)$. In the case when l has only two categories, this becomes just the variance of any of them doubled. Thus, the transformation of a nominal variable into a bunch of zero-one dummies conveniently converts it into a quantitative format which is compatible with the traditional treatment of nominal features.

2.2 Bivariate Analysis

Statistical science in the pre-computer era developed a number of tools for the analysis of interrelations between variables, which will be useful in the sequel. In the remainder of this section, a review is given of the three cases emerging from the pair-wise considerations, with emphasis on the measurement scales: (a) quantitative-to-quantitative, (b) categorical-to-quantitative, and (c) categorical-to-categorical variables. The discussion of the latter case follows that in [138].

2.2.1 Two Quantitative Variables

Mutual interrelations between two quantitative features can be caught with a scatter plot such as in Figure 1.10, p. 24. Two indices for measuring association between quantitative variables have attracted considerable attention in statistics and data mining: those of covariance and correlation.

The covariance coefficient between the variables x and y considered as columns in a data table, $x = (x_i)$ and $y = (y_i)$, $i \in I$, can be defined as

$$\text{cov}(x, y) = (1/N) \sum_{i \in I} (x_i - \bar{x})(y_i - \bar{y}) \tag{2.5}$$

where $\bar{x}$ and $\bar{y}$ are the average values of x and y, respectively.

Obviously, $\text{cov}(x, x) = s_x^2$, the variance of x defined in Section 2.1.3.

The covariance coefficient changes proportionally when the variable scales are changed. A scale-invariant version of the coefficient is the correlation coefficient (sometimes referred to as the Pearson product–moment correlation coefficient) which is the covariance coefficient normalized by the standard deviations:

$$r(x, y) = \text{cov}(x, y) / (s(x) s(y)) \tag{2.6}$$

A somewhat simpler formula for the correlation coefficient can be obtained if the data are first standardized by subtracting their average and dividing the results by the standard deviation: $r(x, y) = \text{cov}(x', y') = \langle x', y' \rangle / N$ where $x_i' = (x_i - \bar{x})/s_x$, $y_i' = (y_i - \bar{y})/s_y$, $i \in I$; this standardization is referred to sometimes as z-scoring. Thus, the correlation coefficient is but the mean of the component-to-component, that is, inner, product of feature vectors when both of the scales are standardized as above.

The coefficient of correlation can be interpreted in different theoretic frameworks. These require some preliminary knowledge of mathematics and can be omitted at the first reading, which is reflected in using a smaller font for explaining them.

1. *Cosine.* A geometric approach, relying on linear algebra concepts introduced later in Section 2.3.2, offers the view that the covariance coefficient as the inner product of feature column-vectors is related to the angle between the vectors so that $\langle x, y \rangle = ||x|| \, ||y|| \cos(x, y)$. This can be illustrated with Figure 2.1; norms $||x||$ and $||y||$ are Euclidean lengths of intervals from 0 to x and y, respectively. The correlation coefficient is the inner product of the corresponding normalized variables, that is, the cosine of the angle between the vectors.

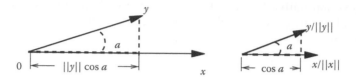

FIGURE 2.1
Geometrical meaning of the inner product and correlation coefficient.

2. *Linear slope and variance.* The data recovery approach suggests that one of the features is modeled as a linear (affine) function of the other, say, y as $ax + b$ where a and b are chosen to minimize the norm of the difference, $||y - ax - b||$. It appears, the optimal slope a is proportional to $r(x, y)$ and, moreover, the square $r(x, y)^2$ expresses that part of the variance of y that is taken into account by $ax + b$ (see details in Section 7.1.3).

3. *Parameter in Gaussian distribution.* The correlation coefficient has a very clear meaning in the framework of probabilistic bivariate distributions. Consider, for the sake of simplicity, features x and y normalized so that the variance of each is unity. Denote the matrix formed by the two features as its columns by $z = (x, y)$ and assume a unimodal distribution over z, controlled by the so-called Gaussian, or normal, density function (see Section 3.5.5) which is proportional to the exponent of $-z^T \Sigma^{-1} z / 2$ where Σ is a 2×2 matrix equal to $\Sigma = \begin{pmatrix} 1 & r \\ r & 1 \end{pmatrix}$. The parameter r determines the distance between the foci of the ellipse $z^T \Sigma^{-1} z = 1$: the greater r, the greater the distance. At $r = 0$, the distance is zero so that the ellipsis is a circle and at r tending to 1 or -1 the distance tends to the infinity so that the ellipse degenerates into a straight line. It appears $r(x, y)$ is a sample-based estimate of this parameter.

These three frameworks capture different pieces of the correlation "elephant." That of cosine is the most universal framework: one may always take that measure to see to what extent two features go in concert, that is, to what extent their highs and lows co-occur. As any cosine, the correlation coefficient is between -1 and 1, the boundary values corresponding to the coincidence of the normalized variables or to a linear relation between the original features. The correlation coefficient being zero corresponds to the right angle between the vectors: the features are not correlated! Does that mean they must be independent in this case? Not necessarily. The linear slope approach allows one to see how this may happen: just the line best fitting the scatter plot must be horizontal. According to this approach, the square of the correlation coefficient shows to what extent the relation between variables, as observed, is owed to linearity. The Gaussian distribution view is the most

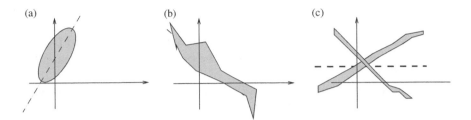

FIGURE 2.2
Three cases of correlation: (a) positive correlation, (b) negative correlation, and (c) no correlation.
The shaded area is randomly covered by entity points.

demanding: it requires a properly defined distribution function, a unimodal
one, if not normal.

Three examples of scatter plots in Figure 2.2 illustrate some potential cases
of correlation: (a) strong positive, (b) strong negative, and (c) zero correlation.
Yet be reminded: the correlation coefficient cannot cover all the relationships
or the lack of those between variables; it does capture only those of linear
relation and those close enough to that.

2.2.2 Nominal and Quantitative Variables

How should one measure association between a nominal feature and a quantitative feature? By measuring an extent at which specifying a category can
lead to a better prediction of the quantitative feature.

Let $S = \{S_1, \ldots, S_K\}$ be the partition of the entity set I corresponding to categories of the nominal variable so that subset S_k consists of N_k entities falling
in k-th category of the variable. The quantitative variable will be denoted by
y with its values y_i for $i \in I$. The box-plot such as in Figure 1.7 on page 24 is
a visual representation of the relationship between the nominal variable represented by partition S and the quantitative variable y represented by boxes
and whiskers.

Let us introduce a framework at which one can measure the extent of prediction of y values. Let the predicted y value for any entity be the grand
mean $\bar{y}$ if no other information is supplied, or $\bar{y}_k = \sum_{i \in S_k} y_i / N_k$, the within-
class average, if the entity is known to belong to S_k. The average error of these
predictions can be estimated in terms of the variances. To do this, one should
relate within-class variances of y to its total variance s^2: the greater the change,
the lesser the error and the closer the association between S and y.

An index, referred to as the correlation ratio, measures the proportion of the
total feature variance that is explained by the within-class average predictions.
Consider

$$s_k^2 = \sum_{i \in S_k} (y_i - \bar{y}_k)^2 / N_k \tag{2.7}$$

where N_k is the number of entities in S_k and $\bar{y}_k$ the feature's within cluster average. This is within class k variance of variable y. Let the proportion of S_k in I be p_k, so that $p_k = N_k/N$. Then the average variance within partition $S = \{S_1, \ldots, S_K\}$ will be $\sum_{k=1}^{K} p_k s_k^2$. This can be proven to never be greater than the total variance s^2.

However, the average within class variance can be as small as zero—when all values y_i coincide with $\bar{y}_k$ within each category S_k, that is, when y is piecewise constant across S. All the cluster boxes of a box-plot of y over S degenerate into straight lines in this case. In such a situation, partition S is said to perfectly match y. The smaller the difference between the average within-class variance and s^2, the worse the match between S and y. The relative value of the difference,

$$\eta^2 = \frac{s^2 - \sum_{k=1}^{K} p_k s_k^2}{s^2} \tag{2.8}$$

is referred to as the correlation ratio.

The correlation ratio is between 0 and 1, the latter corresponding to the perfect match case. The minimum value, zero, is reached when the average within class variance coincide with the total variance.

2.2.3 Two Nominal Variables Cross Classified

Association between two nominal variables is represented with the so-called contingency table. A contingency, or cross-classification, data table corresponds to two sets of disjoint categories, such as the Course type and Subject in the College data, which, respectively, form rows and columns of Table 2.3.

Entries of the contingency table are co-occurrences of row and column categories, that is, counts of numbers of entities that fall simultaneously in the corresponding row and column categories such as in Table 2.3.

In a general case, with the row categories denoted by $t \in T$ and column categories by $u \in U$, the co-occurrence counts are denoted by N_{tu}. The frequencies of row and column categories usually are called marginals (since

TABLE 2.3

Cross Classification of the Eight Colleges According to the Subject and Course Type in the Count/Proportion Format

Subject	Course Type			
	MSc	BSc	Certificate	Total
Science	2/0.250	1/0.125	0/0	3/0.375
Engineering	1/0.125	2/0.250	0/0	3/0.375
Arts	0/0	0/0	2/0.250	2/0.250
Total	3/0.375	3/0.375	2/0.375	8/1.000

TABLE 2.4

A Contingency Table or Cross Classification of Two Sets of Categories, $t \in T$ and $u \in U$ on the Entity Set I

Category	1	2	$\cdots$	$|U|$	Total
1	N_{11}	N_{12}	$\cdots$	$N_{1|U|}$	N_{1+}
2	N_{21}	N_{22}	$\cdots$	$N_{2|U|}$	N_{2+}
$\cdots$	$\cdots$	$\cdots$	$\cdots$	$\cdots$	$\cdots$
$|T|$	$N_{|T|1}$	$N_{|T|2}$	$\cdots$	$N_{|T||U|}$	$N_{|T|+}$
Total	N_{+1}	N_{+2}	$\cdots$	$N_{+|U|}$	N

Note: Marginal frequencies are in row and column labeled "Total."

they are presented on margins of contingency tables as in Table 2.3). They will be respectively denoted by N_{t+} and N_{+u} since, when the categories within each of the two sets do not overlap, they are sums of co-occurrence entries, N_{tu}, in rows, t, and columns, u, respectively. The proportions $p_{tu} = N_{tu}/N$, $p_{t+} = N_{t+}/N$, and $p_{+u} = N_{+u}/N$ are also frequently used as contingency table entries. Using this notation, a contingency table can be presented as Table 2.4.

Contingency tables can be also considered for pre-categorized quantitative features, as demonstrated in the following example.

Example 2.1: A Cross-Classification of Market Towns

Let us partition the Market town set in four classes according to the number of Banks and Building Societies (feature Ba): class T_1 to include towns with Ba equal to 10 or more; T_2 with Ba equal to 4 to 9; T_3 with Ba equal to 2 or 3; and T_4 to consist of towns with one bank or no bank at all. Let us cross classify partition $T = \{T_v\}$ with feature FM, presence or absence of a Farmers' market in the town. In a corresponding contingency table with columns corresponding to classes T_v, rows to presence or absence of Farmers' markets, the entries show the co-occurrence (see Table 2.5).

The matrix of frequencies, or proportions, $p_{tu} = N_{tu}/N$ for Table 2.5 can be obtained by dividing all its entries by $N = 45$ (see Table 2.6). ∎

TABLE 2.5

Cross Classification of the Bank Partition with FM Feature at Market Towns

FMarket	Number of Banks				Total
	10+	4+	2+	1−	
Yes	2	5	1	1	9
No	4	7	13	12	36
Total	6	12	14	13	45

TABLE 2.6

Frequencies, Per Cent, in the Bivariate Distribution of the
Bank Partition and FM at Market Towns

FMarket	Number of Banks				Total
	10+	4+	2+	1−	
Yes	4.44	11.11	2.22	2.22	20.00
No	8.89	15.56	28.89	26.67	80.00
Total	13.33	26.67	31.11	28.89	100.00

A contingency table gives a portrayal of association between two categorical features, or partitions corresponding to them. This portrayal may be rather blurred sometimes. Let us make the portrayal in Table 2.5 sharper by removing from the sample those thirteen towns that fall in the less populated cells of Table 2.5 (see Table 2.7).

Table 2.7 shows a clear-cut association between two features on the trimmed subsample: the Farmers' markets are present only in towns in which the number of banks is 4 or greater. A somewhat subtler relation: the medium numbers of banks are more closly associated with the presence of a Farmers' market than the higher ones. Yet such a data trimming is not an option because it borders with cheating.

One can think that the extent of association between two features can be captured by counting mismatches. There are 13 mismatches in Table 2.5—those removed from Table 2.7. Table 2.3 shows that Course type is quite close to Subject on the colleges sample; yet there are two mismatching entities, one from Science and the other from Engineering. The sheer numbers of mismatching entities measure the differences between category sets rather well when the distribution of entities within each category is more or less uniform, as it is in Table 2.3. When the frequencies of different categories drastically differ, as in Table 2.5, the numbers of mismatching entities should be weighted according to the category frequencies. Can we discover the relation in Table 2.5 without removal of entities?

TABLE 2.7

Cross Classification of the Bank Partition with FM Feature
at a Trimmed Subsample of Market Towns

FMarket	Number of Banks				Total
	10+	4+	2+	1−	
Yes	2	5	0	0	7
No	0	0	13	12	25
Total	2	5	13	12	32

To measure the association between categories according to a contingency table, a founding father of the science of statistics, A. Quetelet, proposed the relative or absolute change of the conditional probability of a category (1832). The conditional probability $p(u/t) = N_{tu}/N_{t+} = p_{tu}/p_{t+}$ measures the proportion of category u in category t. Quetelet coefficients measure the difference between $p(u/t)$ and the average rate p_{+u} of $u \in U$. The Quetelet absolute probability change is defined as

$$g_{tu} = p(u/t) - p_{+u} = (p_{tu} - p_{t+}p_{+u})/p_{t+} \qquad (2.9)$$

and the Quetelet relative change

$$q_{tu} = g_{tu}/p_{+u} = p_{tu}/(p_{t+}p_{+u}) - 1 = (p_{tu} - p_{t+}p_{+u})/(p_{t+}p_{+u}) \qquad (2.10)$$

If, for example, t is a risk factor for an illness condition, such as "exposure to certain allergens" and u is an allergic reaction such as asthma, and $p_{tu} = 0.001, p_{t+} = 0.01, p_{+u} = 0.02$, that would mean that 10% of those people who have been exposed to the allergens, $p(u/t) = p_{tu}/p_{t+} = 0.001/0.1 = 0.1$, would contract the disease, while on average only 2% have the condition. Thus, the exposure to the allergens multiplies risk of contracting the condition fivefold, that is, increases the probability of contracting it by 400%. This is exactly the value of $q_{tu} = 0.001/0.0002 - 1 = 4$. The value of g_{tu} expresses the absolute difference between $p(u/t) = 0.1$ and $p_{+u} = 0.02$, which is not that dramatic, just 0.08.

The summary Quetelet coefficients (weighted by the co-occurrence values) can be considered as summary measures of association between two category sets especially when their distributions are far from uniform:

$$G^2 = \sum_{t \in T} \sum_{u \in U} p_{tu} g_{tu} = \sum_{t \in T} \sum_{u \in U} \frac{p_{tu}^2}{p_{t+}} - \sum_{u \in U} p_{+w}^2 \qquad (2.11)$$

and

$$Q^2 = \sum_{t \in T} \sum_{u \in U} p_{tu} q_{tu} = \sum_{t \in T} \sum_{u \in U} \frac{p_{tu}^2}{p_{t+}p_{+u}} - 1 \qquad (2.12)$$

The right-hand 1 in Equation 2.12 comes from $\sum_{t \in T} \sum_{u \in U} p_{tu} = 1$ when the categories t, as well as categories u, are mutually exclusive and cover the entire set I. In this case, Q^2 is equal to the well-known Pearson chi-squared coefficient χ^2 defined by a different formula:

$$\chi^2 = \sum_{t \in T} \sum_{u \in U} \frac{(p_{tu} - p_{t+}p_{+u})^2}{p_{t+}p_{+u}} \qquad (2.13)$$

The fact that $Q^2 = \chi^2$ can be proven with simple algebraic manipulations. Indeed, take the numerator in χ^2: $(p_{tu} - p_{t+}p_{+u})^2 = p_{tu}^2 - 2p_{tu}p_{t+}p_{+u} + p_{t+}^2 p_{+u}^2$.

Divided by the denominator $p_{t+}p_{+u}$, this becomes $p_{tu}^2/p_{t+}p_{+u} - 2p_{tu} + p_{t+}p_{+u}$. Summing the first item over all u and t leads to $Q^2 + 1$. The second item sums to -2 and the third item to 1, which proves the statement.

The chi-squared coefficient is by far the most popular association coefficient. There is a probability-based theory describing what values of $N\chi^2$ can be explained by fluctuations of the random sampling from the population.

The difference between equivalent Expressions (2.12) and (2.13) for the relative Quetelet coefficient Q^2 reflects deep differences in perspectives. Pearson introduced the coefficient in the format of $N\chi^2$ with χ^2 in Equation 2.13 to measure the deviation of the bivariate distribution in an observed contingency table from the model of statistical independence. Two partitions (categorical variables) are referred to as statistically independent if any entry in their relative contingency table is equal to the product of corresponding marginal proportions; that is, in the current notation,

$$p_{tu} = p_{t+}p_{+u} \tag{2.14}$$

for all $t \in T$ and $u \in U$.

Expression (2.13) is a quadratic measure of deviation of the contingency table entries from the model of statistical independence. Under the hypothesis that the features are indeed statistically independent in the population so that the difference of the observed data is just due to a random sampling bias, the distribution of $N\chi^2$ has been proven to converge to the chi-squared probabilistic distribution with $(|T| - 1)(|U| - 1)$ degrees of freedom, at N tending to infinity. This is used for statistical hypotheses testing. Texts in mathematical statistics and manuals claim that, without relating the observed contingency counts to the model of statistical independence, there is no point in considering χ^2. This perspective sets a very restrictive framework for using χ^2 as an association measure. In particular, the presence of zero entries (such as in Tables 2.3 and 2.7) in a contingency table is at odds with the independence hypothesis and is subject to a separate treatment.

Expression (2.12) sets a very different framework which has nothing to do with the statistical independence. In this framework, χ^2 is Q^2, the average relative change of the probability of categories u when categories t become known. This is an association measure, with no restriction on using $\chi^2 = Q^2$ in this capacity, with zero co-occurrences or not.

Let us take a deeper look in the nature of the measure by looking at its extreme values. It is not difficult to prove that the summary coefficient Q^2 reaches its maximum value

$$\max \chi^2 = \min(|U|, |T|) - 1 \tag{2.15}$$

in tables with the structure of Table 2.7, at which only one element is not zero in every column (or row, if the number of rows is greater than the number of columns) [138]. Such a structure suggests a conceptual one-to-many or even one-to-one relation between categories of the two features, which means that the coefficient is good in measuring association indeed. For example, according to Table 2.7, Farmers' markets are present if and only if the number of banks or building societies is 4 or greater, and $Q^2 = 1$ in this table.

The minimum value of Q^2, zero, is reached in the case of statistical independence between the features, which obviously follows from the "all squared" form of the coefficient in Equation 2.13.

Formula (2.12) suggests a way for visualization of dependencies in a "blurred" contingency table by putting the additive items $p_{tu}q_{tu}$ as (t, u) entries of a show-case table. The proportional but greater values $Np_{tu}q_{tu} = N_{tu}q_{tu}$ can be used as well, since they sum to the $N\chi^2$ considered in the probabilistic framework.

Example 2.2: Highlighting Positive Contributions to the Total Association

The table of the relative Quetelet coefficients q_{tu} for Table 2.5 is presented in Table 2.8 and that of items $N_{tu}q_{tu}$ in Table 2.9.

It is easy to see that the highlighted positive entries in both of the tables express the same pattern as in Table 2.7 but on the entire data set, without trimming it by removing entities from the table.

Table 2.9 demonstrates one more property of the items $p_{tu}q_{tu}$ summed in the chi-square coefficient: their within-row or within-column sums are always positive. ∎

TABLE 2.8

Relative Quetelet Coefficients, Per Cent, for Table 2.5

FMarket	Number of Banks			
	10+	4+	2+	1−
Yes	66.67	108.33	−64.29	−61.54
No	−16.67	−27.08	16.07	15.38

TABLE 2.9

Items Summed in the Chi-Squared Contingency Coefficient (Times N) in the Quetelet Format (2.12) for Table 2.5

FMarket	Number of Banks				Total
	10+	4+	2+	1−	
Yes	1.33	5.41	−0.64	−0.62	5.48
No	−0.67	−1.90	2.09	1.85	1.37
Total	0.67	3.51	1.45	1.23	6.86

Highlighting the positive entries $p_{tu}q_{tu} > 0$ (or $q_{tu} > 0$) can be used for visualization of the pattern of association between any categorical features [139].

Similar to Equation 2.13, though asymmetric, expression can be derived for G^2:

$$G^2 = \sum_{t \in T} \sum_{u \in U} \frac{(p_{tu} - p_{t+}p_{+u})^2}{p_{t+}} \tag{2.16}$$

Although it also can be considered a measure of deviation of the contingency table from the model of statistical independence, G^2 has been always considered in the literature as a measure of association. A corresponding definition involves the Gini coefficient considered in Section 2.1.3, $G(U) = 1 - \sum_{u \in U} p_{+u}^2$. Within category t, the variation is equal to $G(U/t) = 1 - \sum_{u \in U} (p_{tu}/p_{t+})^2$, which makes, on average, the qualitative variation that cannot be explained by T: $G(U/T) = \sum_{t \in T} p_{t+} G(U/t) = 1 - \sum_{t \in T} \sum_{u \in U} p_{tu}^2 / p_{t+}$.

The difference $G(U) - G(U/T)$ represents that part of $G(U)$ that is explained by T, and this is exactly G^2 in Equation 2.11.

2.2.4 Relation between the Correlation and Contingency Measures

Let us elaborate on the interrelation between the correlation and contingency measures. K. Pearson addressed the issue by proving that, given two quantitative features whose ranges have been divided into a number of bins, under some standard mathematical assumptions, the value of $\sqrt{\chi^2/(1 + \chi^2)}$ converges to that of the correlation coefficient when the number of bins tends to infinity [92].

To define a framework for experimentally exploring the issue in the context of a pair of mixed scale features, let us consider a quantitative feature A and a nominal variable A_t obtained by partitioning the range of A into t categories, with respect to a pre-specified partition $S = \{S_k\}$. The relation between S and A can be captured by comparing the correlation ratio $\eta^2(S, A)$ with corresponding values of contingency coefficients $G^2(S, A_t)$ and $Q^2(S, A_t)$. The choice of the indexes is not random. As proven in Section 7.2.3, $\eta^2(S, A)$ is equal to the contribution of A and clustering S to the data scatter. In the case of A_t, analogous roles are played by indexes G^2 and $\chi^2 = Q^2$.

Relations between η^2, G^2, and χ^2 can be quite complex depending on the bivariate distribution of A and S. However, when the distribution is organized in such a way that all the within-class variances of A are smaller than its overall variance, the pattern of association expressed in G^2 and χ^2 generally follows that expressed in η^2.

To illustrate this, let us set an experiment according to the data in Table 2.10: within each of four classes, S_1, S_2, S_3, and S_4, a pre-specified number of observations is randomly generated with a pre-specified mean and

TABLE 2.10

Setting of the Experiment

Class	S_1	S_2	S_3	S_4
Number of observations	200	100	1000	1000
Variance	1.0	1.0	4.0	0.1
Initial mean	0.5	1.0	1.5	2.0
Final mean	10	20	30	40

variance. The totality of 2300 generated observations constitutes the quantitative feature A for which the correlation ratio $\eta^2(S, A)$ is calculated. Then, the range of A is divided in $t = 5$ equally spaced intervals, bins, constituting categories of the corresponding attribute A_t, which is cross classified with S to calculate G^2 and χ^2. This setting follows that described in [139].

The initial within-class means are less than the corresponding variances. Multiplying each of the initial means by the same factor value, $f = 1, 2, \ldots, 20$, the means are step by step diverged in such a way that the within-class samples become more and more separated from each other, thus increasing the association between S and A. The final means in Table 2.10 correspond to $f = 20$.

This is reflected in Figure 2.3 where the horizontal axis corresponds to the divergence factor, f, and the vertical axis represents values of the three coefficients for the case when the within class distribution of A is uniform (on the left) or Gaussian, or normal (on the right). We can see that the patterns follow each other rather closely in the case of a uniform distribution. There are small diversions from that in the case of a normal distribution. The product–moment correlation between G^2 and χ^2 is always about 0.98–0.99, whereas they both correlate with η^2 on the level of 0.90. The difference in values of G^2, χ^2, and η^2 is caused by two factors: first, by the coarse categorical character of A_t versus the fine-grained quantitative character of A, and, second, by the difference in their contributions to the data scatter. The second factor scales G^2 down and χ^2 up, to the maximum value 3 according to Equation 2.15.

2.2.5 Meaning of the Correlation

Correlation is a phenomenon which may be observed between two features co-occurring over the same observations: the features are co-related in such a way that change in one of them accords with a corresponding change in the other.

These are frequently asked questions: Given a high correlation or association, is there any causal relation involved? Given a low correlation, are the features independent from each other? If there is a causal relation, should it translate into a higher correlation value?

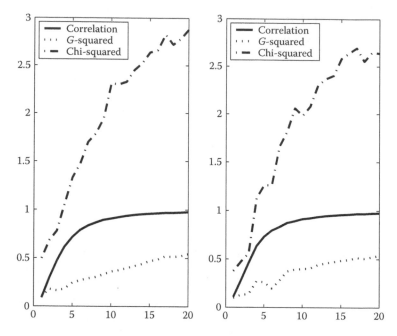

FIGURE 2.3
Typical change of the correlation ratio (solid line), G-squared (dotted line) and chi-squared (dash-dotted line) with the increase of the class divergence factor in the case of uniform (left) and normal (right) within-class distributions of the quantitative variable A.

The answer to each is: no, not necessarily.

To make our point less formal, let us refer to a typical statistics news nugget brought to life by newspapers and BBC Ceefax June 25, 2004: "Children whose mothers eat fish regularly during the pregnancy develop better language and communication skills. The finding is based on the analysis of eating habits of 7400 mothers ... by the University of North Carolina, published in the journal Epidemiology."

At the face value, the case is simple: eat more fish while pregnant and your baby will be better off in the contest of language and communication skills. The real value is a cross classification of mother's eating habits and her baby's skills over the set of 7400 mother–baby couples at which the cell combining "regular fish eating" with "better language skills" has accounted for a considerably greater number of observations than it would be expected under the statistical independence, that is, the corresponding Quetelet coefficient q is positive. So what? Could it be just because of the fish? Very possibly: some say that the phosphorus which is abundant in fish is a building material for the brain. Yet some say that the phosphorus diet has nothing to do with the brain development. They think that the correlation is just a manifestation of a deeper relation between family income, not accounted for in the data, and the two features: in richer families it is both the case that mothers eat more expensive food, fish included, and babies have better language skills. The conclusion: more research is needed to see which of these two explanations better fits the data. And more research may bring further unaccounted for and unforeseen factors and observations.

Example 2.3: False Correlation and Independence

To illustrate the emergence of "false" correlations and non-correlations, let us dwell on the mother–baby example above involving the following binary features: A—"more fish eating mother," B—"baby's better language skills," and C—"healthier baby." Table 2.11 presents artificial data on 2000 mother–baby couples relating A with B and C.

According to Table 2.11, the baby's language skills (B) are indeed positively related to mother's fish eating (A): 520 observations at cell AB rather than 400 expected if A and B were independent, which is supported by a positive Quetelet coefficient $q(B/A) = 30\%$. In contrast, no relation is observed between fish eating (A) and baby's health (C): all A/C cross classifying entries on the right of Table 2.11 are proportional to the products of marginal frequencies. For instance, with $p(A) = 0.4$ and $p(C) = 0.7$ their product $p(A)p(C) = 0.28$ accounts for 28% of 2000 observations, that is, 560, which is exactly the entry at cell AC.

However, if we take into account one more binary feature, D, which assigns Yes to better off families, and break down the sample according to D, the data may show a different pattern (see Table 2.12). All has turned upside down in Table 2.12: what has been independent in Table 2.11, A and C, has become associated within both D and not-D categories, and what was correlated in Table 2.11, A and B, became independent within both D and not-D categories!

Specifically, with these illustrative data, one can see that A accounts for 600 cases within D category and 200 within not-D category. Similarly, B accounts for 800 cases within D and only 200 within not-D. Independence between A and B within each stratum brings the numbers of AB to 480 in D and only 40 in not-D. In this way, the mutually independent A and B within each stratum become correlated in the combined sample, because both A and B are concentrated mostly within D.

TABLE 2.11

Association between Mother's Fish Eating (A) and Her Baby's Language Skills (B) and Health (C)

Feature	B	$\bar{B}$	C	$\bar{C}$	Total
A	520	280	560	240	800
$\bar{A}$	480	720	840	360	1200
Total	1000	1000	1400	600	2000

TABLE 2.12

Association between Mother's Fish Eating (A) and Baby's Language
Skills (B) and Health (C) with Income (D) Taken into Account

Feature D	Feature A	B	B̄	C	C̄	Total
D	A	480	120	520	80	600
	Ā	320	80	300	100	400
	Total	800	200	820	180	1000
D̄	A	40	160	40	160	200
	Ā	160	640	540	260	800
	Total	200	800	580	420	1000
Total		1000	1000	1400	600	2000

Similar, though opposite, effects are at play with the association
between A and C:A and C are negatively related in not-D and positively
related in D, so that combining these two strata brings the mutual
dependence to zero.   ∎

A high correlation/association is just a pointer to the user, researcher or
manager alike, to look at what is behind. The data on their own cannot prove
any causal relations, especially when no timing is involved, as is the case in
all our case study problems. A causal relation can be established only with
a mechanism explaining the process in question theoretically, to which the
data may or may not add credibility.

2.3 Feature Space and Data Scatter

2.3.1 Data Matrix

A quantitative data table is usually referred to as a data matrix. Its rows correspond to entities and columns to features. Moreover, in most clustering
computations, all metadata are left aside so that a feature and entity are represented by the corresponding column and row only, under the assumption
that the labels of entities and variables do not change.

A data table with mixed scales such as Table 2.1 will be transformed to a
quantitative format. According to the rules described in the next section, this
is achieved by pre-processing each of the categories into a dummy variable
by assigning 1 to an entity that falls in it and 0 otherwise.

Example 2.4: Pre-Processing Colleges Data

Let us convert the Colleges data in Table 2.1 to the quantitative format.
The binary feature EL is converted by substituting Yes by 1 and No by 0.

TABLE 2.13

Quantitative Representation of the Colleges Data as an 8×7
Entity-to-Attribute Matrix

Entity	Stud	Acad	NS	EL	MSc	BSc	Certif.
1	3800	437	2	0	1	0	0
2	5880	360	3	0	1	0	0
3	4780	380	3	0	0	1	0
4	3680	279	2	1	1	0	0
5	5140	223	3	1	0	1	0
6	2420	169	2	1	0	1	0
7	4780	302	4	1	0	0	1
8	5440	580	5	1	0	0	1

A somewhat more complex transformation is performed at the three categories of feature Course type: each is assigned with a corresponding zero/one vector so that the original column Course type is converted into three columns (see Table 2.13). ∎

A data matrix row corresponding to an entity $i \in I$ constitutes what is called an *M-dimensional point* or *vector* $y_i = (y_{i1}, \ldots, y_{iM})$ whose components are the row entries. For instance, Colleges data in Table 2.13 is a 8×7 matrix, and the first row in it constitutes vector $y_1 = (3800, 437, 2, 0, 0, 1, 0)$ each component of which corresponds to a specific feature and, thus, cannot change its position without changing the feature's position in the feature list.

Similarly, a data matrix column corresponds to a feature or category with its elements corresponding to different entities. This is an *N-dimensional vector*.

Matrix and vector terminology is not just a fancy language but part of a well-developed part of mathematics, linear algebra, which is used throughout data analysis approaches.

2.3.2 Feature Space: Distance and Inner Product

Any *M-dimensional vector* $y = (y_1, \ldots, y_M)$ pertains to the corresponding combination of feature values. Thus, the set of all *M-dimensional vectors* y is referred to as the feature space. This space is provided with interrelated distance and similarity measures.

The distance between two *M-dimensional vectors*, $x = (x_1, x_2, \ldots, x_M)$ and $y = (y_1, y_2, \ldots, y_M)$, will be defined as the sum of the component-wise differences squared:

$$d(x, y) = (x_1 - y_1)^2 + (x_2 - y_2)^2 + \cdots + (x_M - y_M)^2 \qquad (2.17)$$

This difference-based quadratic measure is what the mathematicians call the squared Euclidean distance. It generalizes the basic property of plane

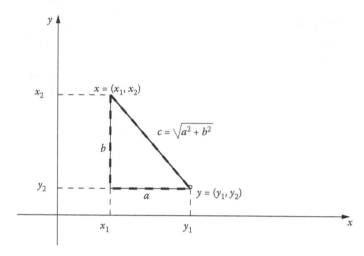

FIGURE 2.4
Interval between points *x* and *y* is the hypotenuse of the highlighted triangle, which explains the distance between *x* and *y*.

geometry, the so-called Pythagoras' theorem illustrated in Figure 2.4. Indeed, distance $d(x, y)$ in it is c^2 and c is the Euclidean distance between x and y.

Example 2.5: Distance between Entities

Let us consider three colleges, 1 and 2 from Science, and one, 7 from Arts, as row-points of the matrix in Table 2.13 presented in the upper half of Table 2.14. The mutual distances between them are calculated in the lower half. The differences in the first two variables, Stud and Acad, predefine the result, however, different the other features are, because their scales prevail. In this way, we get a counter-intuitive conclusion that a Science

TABLE 2.14

Computation of Distances between Three Colleges According to Table 2.13

Entity	Stud	Acad	NS	EL	MSc	BSc	Certif.	Distance
1	3800	437	2	0	1	0	0	
2	5880	360	3	0	1	0	0	
7	4780	302	4	1	0	0	1	
Distance								**Total**
$d(1,2)$	4,326,400	5929	1	0	0	0	0	4,332,330
$d(1,7)$	960,400	18,225	4	1	1	0	1	978,632
$d(2,7)$	1,210,000	3364	1	1	1	0	1	1,213,368

Note: Squared differences of values in the upper part are in the lower part of the matrix, column-wise; they are summed in the column "Distance" on the right.

college is closer to an Arts college than to the other Science colleges because $d(1,7) = 976,632 < d(1,2) = 4,332,330$. Therefore, feature scales must be rescaled to give greater weights to the other variables. ∎

The concept of M-dimensional feature space comprises not only all M-series of reals, $y = (y_1, \ldots, y_M)$, but also two mathematical operations with them: the component-wise summation defined by the rule $x + y = (x_1 + y_1, \ldots, x_M + y_M)$ and multiplication of a vector by a number defined as $\lambda x = (\lambda x_1, \ldots, \lambda x_M)$ for any real λ. These operations naturally generalize the operations with real numbers and have similar properties.

With such operations, variance s_v^2 also can be expressed as the distance, between column $v \in V$ and column vector $c_v e$, whose components are all equal to the column's average, c_v, divided by N, $s_v^2 = d(x_v, c_v e)/N$. Vector $c_v e$ is the result of multiplication of vector $e = (1, \ldots, 1)$, whose components are all unities, by c_v.

Another important operation is the so-called inner, or scalar, or dot, product. For any two M-dimensional vectors x, y, their inner product is a number denoted by $<x, y>$ and defined as the sum of component-wise products, $<x, y> = x_1 y_1 + x_2 y_2 + \cdots + x_M y_M$.

The inner product and distance are closely related. It is not difficult to see, just from the definitions, that for any vectors/points x, y: $d(x, 0) = <x, x> = \sum_{v \in V} x_v^2$ and $d(y, 0) = <y, y>$ and, moreover, $d(x, y) = <x - y, x - y>$. The symbol 0 refers here to a vector with all components equal to zero. The distance $d(y, 0)$ will be referred to as the scatter of y. The square root of the scatter $d(y, 0)$ is referred to as the (Euclidean) norm of y and denoted by $||y|| = \sqrt{(y,y)} = \sqrt{\sum_{i \in I} y_i^2}$. It expresses the length of y.

In general, for any M-dimensional x, y, the following equation holds:

$$d(x, y) = <x - y, x - y> = <x, x> + <y, y> - 2<x, y>$$
$$= d(0, x) + d(0, y) - 2<x, y> \qquad (2.18)$$

This equation becomes especially simple when $<x, y> = 0$. In this case, vectors x, y are referred to as mutually *orthogonal*. When x and y are mutually orthogonal, $d(0, x - y) = d(0, x + y) = d(0, x) + d(0, y)$, that is, the scatters of $x - y$ and $x + y$ are equal to each other and the sum of scatters of x and y. This is a multidimensional analogue to the Pythagoras theorem and the base for decompositions of the data scatter employed in many statistical theories including the theory for clustering presented in Chapter 7.

The inner product of two vectors has a simple geometric interpretation (see Figure 2.1 on p. 50), $<x, y> = ||x|| ||y|| \cos \alpha$ where α is the "angle" between x and y (at 0). This conforms to the concept of orthogonality above: vectors are orthogonal when the angle between them is a right angle.

2.3.3 Data Scatter

The summary scatter of all row-vectors in data matrix Y is referred to as the data scatter of Y and denoted by

$$T(Y) = \sum_{i \in I} d(0, y_i) = \sum_{i \in I} \sum_{v \in V} y_{iv}^2 \qquad (2.19)$$

Equation 2.19 means that $T(Y)$ is the total of all Y entries squared.

An important characteristic of feature $v \in V$ is its *contribution to the data scatter* defined as

$$T_v = \sum_{i \in I} y_{iv}^2 \qquad (2.20)$$

the distance of the N-dimensional column from the zero column. Data scatter is obviously the sum of contributions of all variables, $T(Y) = \sum_{v \in V} T_v$.

If feature v is centered, then its contribution to the data scatter is proportional to its variance:

$$T_v = N s_v^2 \qquad (2.21)$$

Indeed, $c_v = 0$ since v is centered. Thus, $s_v^2 = \sum_{i \in I}(y_{iv} - 0)^2 / N = T_v / N$.

The relative contribution $T_v / T(Y)$ is a characteristic which plays an important role in data standardization issues as explained in the next section.

2.4 Pre-Processing and Standardizing Mixed Data

The data pre-processing stage is to transform the raw entity-to-feature table into a quantitative matrix for further analysis. To this end, all categorical data are converted to a numerical format by using a dummy zero–one variable for each category. Then variables are standardized by shifting their origins and rescaling. This operation can be clearly substantiated from a classical statistics perspective, typically, by assuming that entities have been randomly and independently sampled from an underlying Gaussian distribution. In data analysis, a different reasoning should be applied [140]. By shifting all the origins to feature means, entities become scattered around the center of gravity so that clusters can be more easily "seen" from that point. In other words, the new origin plays a role of the background with which all the data entries are compared. With feature rescaling, feature scales should become better balanced according to the principle of equal importance of each feature brought into the data table.

To implement these general principles, we use a three-stage procedure. The stages are: (1) enveloping non-numeric categories, (2) standardization, and (3) rescaling, as follows:

1. *Quantitatively enveloping categories*: This stage is to convert a mixed scale data table into a quantitative matrix by treating every category as a separate dummy variable coded by 1 or 0 depending on whether an entity falls into the category or not. Binary features are coded similarly except that no additional columns are created. Quantitative features are left as they are. The converted data table will be denoted by $X = (x_{iv})$, $i \in I, v \in V$.

2. *Standardization*: This stage aims at transforming feature-columns of the data matrix by shifting their origins to a_v and rescaling them by b_v, $v \in V$, thus to create standardized matrix $Y = (y_{iv})$:

$$y_{iv} = \frac{x_{iv} - a_v}{b_v}, \quad i \in I, v \in V \tag{2.22}$$

In this text, the shift value a_v always will be the grand mean. In particular, the dummy variable corresponding to category $v \in V_l$ has its mean $c_v = p_v$, the proportion of entities falling in the category. This reflects the view that the shift should remove the "background norm." According to this view, the grand mean plays that role in a situation in which no other "norm" is known.

The scale factor b_v can be either the standard deviation or range or other quantity reflecting the variable's spread. In particular, for a category $v \in V_l$, the standard deviation can be either $\sqrt{p_v(1 - p_v)}$ (Bernoulli distribution) or $\sqrt{p_v}$ (Poisson distribution), see p. 47. The range of a dummy is always 1.

Using the standard deviation is popular in data analysis probably because it is used in the classical statistics relying on the theory of Gaussian distribution which is characterized by the mean and standard deviation. Thus standardized, contributions of all features to data scatter become equal to each other because of the proportionality of the contributions and standard deviations. At the first glance, this seems an attractive property guaranteeing equal contributions of all features to the results, an opinion to which the current author once also subscribed [135]. However, this is not so. Two different factors contribute to the value of standard deviation: the feature scale and the shape of its distribution. As shown in Section 2.1.2, within the same range scale the standard deviation may greatly vary from the minimum, at the peak unimodal distribution, to the maximum, at the peak bimodal distribution. By standardizing with the standard deviations, we deliberately bias data in favor of the unimodal distributions, although obviously it is the bimodal distribution that should contribute to clustering most. This is why the range, not the standard deviation, is our preferred option for the scaling factor b_v. In the case when there can be outliers in the data, which may highly affect the range, another more stable range-like scaling factor can be chosen, such as the difference between the percentiles, that does not much depend on the distribution shape. The range-based scaling option has been supported experimentally in [131,200].

3. *Rescaling*: This stage rescales those column-features v, that come from the same categorical variable l, back by further dividing y_{iv} with supplementary rescaling coefficients b'_v to restore the original weighting of raw variables. The major assumption in clustering is that all raw variables are supposed to be of equal weight. Having its categories enveloped, the "weight" of a nominal variable l becomes equal to the summary contribution $T_l = \sum_{v \in V_l} T_v$ to the data scatter where V_l is the set of categories belonging to l. Therefore, to restore the original "equal weighting" of l, the total contribution T_l must be related to $|V_l|$, which is achieved by taking $b'_v = \sqrt{|V_l|}$ for $v \in V_l$.

For a quantitative $v \in V$, b'_v is, typically, unity.

Sometimes, there can be available an expert evaluation of the relative weights of the original variables l. If such is the case, rescaling coefficients b'_v should be redefined with the square roots of the expert supplied relative weights. This option may be applied to both quantitative and non-quantitative features.

Note that two of the three steps above refer to categorical features.

Example 2.6: Effects of Different Scaling Options

Table 2.15 presents the Colleges data in Table 2.13 standardized according to the most popular transformation of feature scales, the so-called z-scoring, when the scales are shifted to their mean values and then

TABLE 2.15

Std Standardized College Matrix

Item	Stud	Acad	NS	EL	MSc	BSc	Certif.
1	−0.6	0.7	−0.9	−1.2	1.2	−0.7	−0.5
2	1.2	0.1	0.0	−1.2	1.2	−0.7	−0.5
3	0.3	0.3	0.0	−1.2	−0.7	1.2	−0.5
4	−0.7	−0.5	−0.9	0.7	1.2	−0.7	−0.5
5	0.6	−0.9	0.0	0.7	−0.7	1.2	−0.5
6	−1.8	−1.3	−0.9	0.7	−0.7	1.2	−0.5
7	0.3	−0.3	0.9	0.7	−0.7	−0.7	1.6
8	0.8	1.8	1.9	0.7	−0.7	−0.7	1.6
Mean	4490	341.3	3.0	0.6	0.4	0.4	0.3
Std	1123.9	129.3	1.1	0.5	0.5	0.5	0.5
Cntr (%)	14.3	14.3	14.3	14.3	14.3	14.3	14.3

Note: Mean is grand mean, Std the standard deviation, and Cntr the relative contribution of a feature in the standardized data matrix.

TABLE 2.16

Range Standardized Colleges Matrix

Item	Stud	Acad	NS	EL	MSc	BSc	Certif.
1	−0.2	0.2	−0.3	−0.6	0.6	−0.4	−0.3
2	0.4	0.0	0.0	−0.6	0.6	−0.4	−0.3
3	0.1	0.1	0.0	−0.6	−0.4	0.6	−0.3
4	−0.2	−0.2	−0.3	0.4	0.6	−0.4	−0.3
5	0.2	−0.3	0.0	0.4	−0.4	0.6	−0.3
6	−0.6	−0.4	−0.3	0.4	−0.4	0.6	−0.3
7	0.1	−0.1	0.3	0.4	−0.4	−0.4	0.7
8	0.3	0.6	0.7	0.4	−0.4	−0.4	0.7
Mean	4490	341.3	3.0	0.6	0.4	0.4	0.3
Range	3460	411	3.00	1.00	1.73	1.73	1.73
Cntr (%)	7.8	7.3	9.4	19.9	19.9	19.9	15.9

Note: Mean is grand mean, Range the range, and Cntr the relative contribution of a variable.

normalized by the standard deviations. Table 2.16 presents the same data matrix range standardized. All feature contributions are different in this table except for those of NS, MSc, and BSc which are the same. Why the same? Because they have the same variance $p(1 - p)$ corresponding to p or $1 - p$ equal to 3/8.

We can see how overrated the summary contribution of the nominal variable Course type becomes: three dummy columns on the right in Table 2.16 take into account 55.7% of the data scatter and thus highly affect any further results. A further rescaling of these three variables by the $\sqrt{3}$ decreases their total contribution three times. This should make up for the thrice enveloping of the single variable Course type. Table 2.17 presents the results of this operation applied to data in Table 2.16.

Note that the total Course type's contribution per cent has not decreased as much as one would expect: it is now about a half, yet not a one-third, of the contribution in the previous table!

Figure 2.5 shows how important the scaling can be for clustering results. It displays mutual locations of the eight Colleges on the plane of the first two principal components of the data in Table 2.13 at different scaling factors: (a) left top: no scaling at all; (b) right top: scaling by the standard deviations, see Table 2.15; (c) left bottom: scaling by ranges; (d) right bottom: scaling by ranges with the follow-up rescaling of the three dummy variables for categories of Course type by $\sqrt{3}$ taking into account that they come from the same three-category nominal feature; the scale shifting parameter is always the variable's mean.

The left top scatter plot displays no relation to Subject. It is a challenge for a clustering algorithm to properly identify the Subject classes with this standardization (Table 2.15). On the contrary, the Subject pattern is clearly displayed on the bottom right, and any reasonable clustering algorithm should capture them at this standardization.

TABLE 2.17

Range Standardized Colleges Matrix with the Additionally Rescaled Nominal Feature Attributes

Item	Stud	Acad	NS	EL	MSc	BSc	Certif.
1	−0.20	0.23	−0.33	−0.63	0.36	−0.22	−0.14
2	0.40	0.05	0.00	−0.63	0.36	−0.22	−0.14
3	0.08	0.09	0.00	−0.63	−0.22	0.36	−0.14
4	−0.23	−0.15	−0.33	0.38	0.36	−0.22	−0.14
5	0.19	−0.29	0.00	0.38	−0.22	0.36	−0.14
6	−0.60	−0.42	−0.33	0.38	−0.22	0.36	−0.14
7	0.08	−0.10	0.33	0.38	−0.22	−0.22	0.43
8	0.27	0.58	0.67	0.38	−0.22	−0.22	0.43
Mean	4490	341.3	3.0	0.6	0.4	0.4	0.3
Range	3460	411	3.00	1.00	1.73	1.73	1.73
Cntr (%)	12.42	11.66	14.95	31.54	10.51	10.51	8.41

Note: Mean is grand mean, Range the range, and Cntr the relative contribution of a variable.

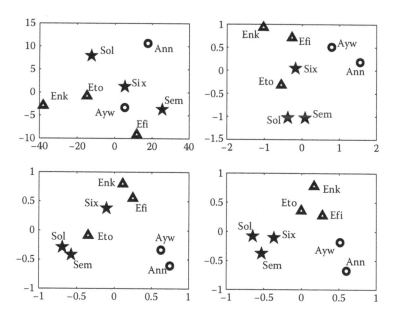

FIGURE 2.5
Colleges on the plane of the two first principal components at four different standardization options: no scaling (top left), scaling with standard deviations (top right), range scaling (bottom left) and range scaling with the follow-up rescaling (bottom right). Different subjects are labeled by different symbols: a pentagram for Science, a triangle for Engineering and a circle for Arts.

We can clearly see in Figure 2.5 that, in spite of the unidimensional nature of the transformation (2.22), its combination of shifts and scales can be quite powerful in changing the topology of the data structure. ∎

Example 2.7: Relative Feature Weighting under Standard Deviations and Ranges May Differ

For the Market town data, with $N = 45$ and $n = 12$, the summary feature characteristics are shown in Table 2.18.

As proven above, the standard deviations in Table 2.18 are at least twice as small as the ranges, which is true for all data tables. However, the ratio of the range and the standard deviation may differ for different features reaching as much as $3/0.6 = 5$ for DIY. Therefore, using standard deviations and ranges for scaling in Equation 2.22 may lead to differences in relative scales between the variables and, thus, to different clustering results, as well as at Colleges data. ∎

Are there any regularities in the effects of data standardization (and rescaling) on the data scatter and feature contributions to it? Not many. But there are items that should be mentioned:

Effect of shifting to the average. With shifting to the averages, the feature contributions and variances become proportional to each other, $T_v = Ns_v^2$, for all $v \in V$.

Effects of scaling of categories. Let us take a look at the effect of scaling and rescaling coefficients on categories. The contribution of a binary attribute v, standardized according to Equation 2.22, becomes $T_v = Np_v(1 - p_v)/(b_v)^2$ where p_v is the relative frequency of category v. This can be either $p_v(1 - p_v)$ or 1 or $1 - p_v$ depending on whether $b_v = 1$ (range) or $b_v = \sqrt{p_v(1 - p_v)}$ (Bernoulli's standard deviation) or $b_v = \sqrt{p_v}$ (Poisson's standard deviation), respectively. These can give some guidance in rescaling of binary categories: the first option should be taken when both zeros and ones are equally important, the second when the distribution does not matter, and the third when it is only unities that do matter.

TABLE 2.18

Summary Characteristics of the Market Town Data

	P	PS	Do	Ho	Ba	Su	Pe	DIY	SP	PO	CAB	FM
Mean	7351.4	3.0	1.4	0.4	4.3	1.9	2.0	0.2	0.5	2.6	0.6	0.2
Std	6193.2	2.7	1.3	0.6	4.4	1.7	1.6	0.6	0.6	2.1	0.6	0.4
Range	21,761.0	12.0	4.0	2.0	19.0	7.0	7.0	3.0	2.0	8.0	2.0	1.0
Cntr (%)	8.5	5.4	11.1	8.8	5.6	6.3	5.7	4.2	10.3	7.3	9.7	17.1

Note: Mean is grand mean, Std the standard deviation, Range the range, and Cntr the relative contribution of a variable.

Contributions of binary and two-category features made the same. An important issue faced by the user is how to treat a categorical feature with two categories such as gender (Male/Female) or voting pattern (Democrat/Republican) or belongingness to a group (Yes/No). The three-step procedure of standardization makes the issue irrelevant: there is no difference between a two-category feature and either of its binary representations.

Indeed, let x be a two-category feature assigning each entity $i \in I$ with a category "eks1" or "eks2" whose relative frequencies are p_1 and p_2 such that $p_1 + p_2 = 1$. Denote by $y1$ a binary feature corresponding to category "eks1" so that $y1_i = 1$ if $x_i = $ eks1 and $y1_i = 0$ if $x_i = $ eks2. Define a binary feature $y2$ corresponding to category "eks2" in an analogous way. Obviously, $y1$ and $y2$ complement each other so that their sum makes a vector whose all components are unities.

In the first stage of the standardization procedure, the user decides whether x is to be converted to a binary feature or left as a categorical one. In the former case, x is converted to a dummy feature, say, column $y1$; and in the latter case, x is converted into a two-column submatrix consisting of columns $y1$ and $y2$. Since the averages of $y1$ and $y2$ are p_1 and p_2, respectively, after shifting column $y1$'s entries become $1 - p_1$, for 1 and $-p_1$, for 0. Respective entries of $y2$ are shifted to $-p2$ and $1 - p2$, which can be expressed through $p_1 = 1 - p_2$ as $p_1 - 1$ and p_1. That means that $y_2 = -y_1$ after the shifting: the two columns become identical, up to the sign. That implies that all their square contribution characteristics become the same, including the total contributions to the data scatter so that the total contribution of the two-column submatrix is twice greater than the contribution of a single column $y1$, whichever scaling option is accepted. However, further rescaling the two-column submatrix by the recommended $\sqrt{2}$ restores the balance of contributions: the two-column submatrix will contribute as much as a single column.

Total contributions of categorical features. The total contribution of nominal variable l is

$$T_l = N \sum_{v \in V_l} p_v (1 - p_v)/(b_v)^2 \tag{2.23}$$

Depending on the choice of scaling coefficients b_v, this can be

1. $T_l = N(1 - \sum_{v \in V_l} p_v^2)$ if $b_v = 1$ (range normalization)
2. $T_l = N|V_l|$ if $b_v = \sqrt{p_v(1 - p_v)}$ (Bernoulli's normalization)
3. $T_l = N(|V_l| - 1)$ if $b_v = \sqrt{p_v}$ (Poisson's normalization)

where $|V_l|$ is the number of categories of l. The quantity on the top is the Gini coefficient of the distribution of v (2.3). As explained above, option 1 corresponds to the case at which 0 and 1 are equally important to the user. Option 3 should be used when only unities matter, and option 2, when the distribution of zeros and ones is of no importance.

The square roots of these should be used for further rescaling the categories stemming from the same nominal variable l to adjust their total impact on the data scatter.

2.5 Similarity Data

2.5.1 General

Unlike a feature over an entity set, a similarity index assigns a measure of similarity between entities from the same set I, in this text, so that the data can be represented by a square matrix A of similarity values a_{ij} for all $i, j \in I$. A characteristic property of the similarity data is that all the a_{ij} values are comparable across the entire matrix A, and moreover, it usually makes sense to compute the average similarity value for any subset of A's entries.

Here is a list of some, potentially overlapping, sources of similarity data:

1. *Experimental data.* This type of data emerges in Psychology and Marketing Research, among other domains, as the records of individual judgments on similarity, or dissimilarity, between various signals, images, situations or foods. Once being a major driving force for the development of data analysis techniques, this type of data currently is used in Computer Sciences for mostly illustrative purposes. In this text such a data set is the similarity evaluations between algebraic functions in Table 1.6 in the previous chapter.

2. *Contingency data.* This type of data comes from cross classifications of inputs and outputs in a big system. Consider a number of systems's elements, such as occupations, or countries, or stimuli. For each pair (i, j) of them, count the number of families in which the father's occupation is i, whereas the son's occupation is j (occupational mobility), or the amount of money generated in country i that has been spent in country j (inter-country transactions), or the number of times when stimulus i has been perceived as j (confusion table, like that in Table 1.10). Each of these tables summarizes a redistribution of some measurable amount between the system's elements. Therefore, its entries can be not only meaningfully compared with each other, but also meaningfully summed.

3. *Inner product, kernel, and affinity.* This is a similarity index produced by operating over the data collected in the entity-to-feature format. One of the simplest indexes is just the inner product of vectors in the feature space.

4. *Distance.* This is another form of a derived index-relating entities with each other. Unlike the similarity, though, the distance convention-ally is non-negative and symmetric. A defining property is that the smaller the value of a distance measure, the nearer the entities are; in contrast, the smaller the value of a similarity index, the farther away are the entities. A distance matrix is presented in Table 1.1 as a measure of difference between Primate species.

5. *Network.* This data type embraces interconnected sets of elements, such as websites and connections between them measured in various units, for example, the number of visits or time spent or the number of pages visited. It corresponds to the concept of (weighted) graph in discrete mathematics.

6. *Alignment.* When analyzing complex entities such as symbol sequences or sound signals or plane images, it can be difficult to develop a relevant feature set. Then a superposition, or an alignment, of the entities against each other can be a simpler task. A similarity index can involve a model of generation of the entities. For example, an alignment of protein sequences, under an assumption of a com-mon ancestor, can be accompanied by an estimate of the probability of accumulating the differences between the sequences that have been observed at the alignment. Models for signals may involve "time warping" for accounting for differences in time or speed.

7. *Subset similarity.* Given two subsets, A and B, typically as neighbor-hoods of two entities under consideration, the issue of scoring the similarity between them has received a lot of attention in the lit-erature. Unlike the alignment score, the similarity between subsets does not depend on specifics of the entities, such as their spatial or sequential or network topology. Moreover, the use of neighborhoods of entities rather than the entities themselves allows to represent the similarities between them in a much smoothed way (see examples in [143]). See more on subset similarities in Section 6.4.1.

In the following, items (2), (3), and (5) above will be considered in a greater detail.

2.5.2 Contingency and Redistribution Tables

Table $F = (f_{ij})$, $i \in I$, $j \in J$ will be referred to as a redistribution table if every entry expresses a quantity of the same matter in such a way that all of the entries can be meaningfully summed to a number expressing the total amount of the matter in the data. Examples of redistribution data tables are: (a) con-tingency tables that count numbers of co-occurred instances between two sets of categories; (b) mobility tables that count numbers of individual members of a group having changed their categories; and (c) trade transaction tables that show the money transferred from i to j during a specified period, for each $i, j \in I$.

This type of data can be of interest when processing massive information sources. The data itself can be untreatable within the time-memory constraints, but by counting co-occurrences of categories of interest in the data, that can be pre-processed into the redistribution data format and analyzed as such.

The nature of the redistribution data associates weights of row categories $i \in I, f_{i+} = \sum_{j \in J} f_{ij}$, and column categories $j \in J, f_{+j} = \sum_{i \in I} f_{ij}$, the total flow from i and that to j. The total redistribution volume is $f_{++} = \sum_{i \in I} \sum_{j \in J} f_{ij}$, which is the summary weight, $f_{++} = \sum_{j \in J} f_{+j} = \sum_{i \in I} f_{i+}$. Extending concepts introduced in Section 2.2.3 for contingency tables to the general redistribution data format, one can extend the definition of the relative Quetelet index as

$$q_{ij} = \frac{f_{ij} f_{++}}{f_{i+} f_{+j}} - 1 \qquad (2.24)$$

This index, in fact, compares the share of j in i's transaction, $p(j/i) = f_{ij}/f_{i+}$, with the share of j in the overall flow, $p(j) = f_{+j}/f_{++}$.

Indeed, it is easy to see that q_{ij} is the relative difference of the two, $q_{ij} = (p(j/i) - p(j))/p(j)$. Obviously, $q_{ij} = 0$ when there is no difference.

Transformation (2.24) is a standardization of the redistribution data which takes into account the data's properties. The standardization (2.24) is very close to the so-called normalization of Rao and Sabovala widely used in marketing research and, also, to the (marginal) cross-product ratio utilized in the analysis of contingency tables. Both can be expressed as p_{ij} transformed into $q_{ij} + 1$.

Example 2.8: Quetelet Coefficients for Confusion Data

Quetelet coefficients (2.24) for the digit-to-digit Confusion data are presented in Table 2.19. One can see from the table that the digits overwhelmingly correspond to themselves. However, there are also a few positive entries in the table outside of the diagonal. For instance, 7 is perceived as 1 with the frequency 87.9% greater than the average, and 3 and 9 can be confused both ways with higher frequencies than the average. ∎

Taking into account that the redistribution data entries can be meaningfully summed to the total volume available for redistribution, the distance between row entities in a contingency table is defined by weighting the columns by their "masses" p_{+l} as follows:

$$\chi(k, k') = \sum_{j \in J} p_{+j} (q_{kj} - q_{k'j})^2 \qquad (2.25)$$

TABLE 2.19

Relative Quetelet Coefficients for Confusion Data (%)

Stimulus	1	2	3	4	5	6	7	8	9	0
					Response					
1	512.7	−92.6	−92.7	−75.2	−95.8	−83.0	−31.8	−100.0	−95.9	−96.5
2	−90.2	728.9	−50.8	−95.5	−61.7	−46.7	−84.0	−69.6	−92.9	−84.1
3	−79.7	−69.3	612.1	−92.1	−80.9	−100.0	−54.5	−69.7	54.6	−86.8
4	4.1	−76.7	−95.8	725.9	−95.8	−87.6	−65.9	−92.7	−58.3	−100.0
5	−90.2	−72.5	−55.0	−84.2	610.7	−10.6	−92.0	−92.7	28.3	−87.6
6	−82.5	−85.2	−92.7	−87.6	2.9	615.8	−95.5	61.9	−88.8	−62.1
7	87.9	−95.8	−78.0	−76.3	−92.6	−100.0	658.6	−100.0	−95.9	−93.8
8	−92.3	−70.4	−70.7	−79.7	−80.9	−20.8	−87.5	502.7	−31.9	51.6
9	−82.5	−69.3	16.1	−48.1	−13.0	−87.6	−76.1	−14.3	459.3	−62.1
0	−87.4	−95.8	−92.7	−87.6	−92.6	−79.6	−71.6	−25.8	−78.6	621.1

This is equal to the so-called chi-squared distance defined between the row conditional profiles in the Correspondence factor analysis (see Section 7.1.5). A similar chi-squared distance can be defined for columns.

The concept of data scatter for contingency data is also modified by weighting rows and columns:

$$\chi^2(P) = \sum_{i \in I} \sum_{j \in J} p_{i+} p_{+j} q_{ij}^2 \qquad (2.26)$$

The notation reflects the fact that this value is closely connected with the Pearson chi-squared contingency coefficient, defined in Equation 2.13 above (thus to Q^2 (2.12) as well). Elementary algebraic manipulations show that $\chi^2(P) = \chi^2$.

Note that in this context the chi-squared index has nothing to do with the statistical independence concept for the analysis of which it was introduced by Pearson in 1901 [167]. In the context of this section, the index is just a weighted data scatter measure compatible with the specific properties of the redistribution and contingency data. In particular, it is not difficult to prove an analogue to a property of the conventional data scatter: $\chi^2(P)$ is the sum of chi-squared distances $\chi(k, 0)$ over all k.

Therefore, the concept of redistribution data leads to one more perspective at Pearson's chi-squared index. It is the redistribution data scatter after the data have been standardized into Quetelet indexes. For Table 1.10 transformed here into Table 2.19, the chi-squared value is equal to 4.21 which is 47% of 9, the maximum value that the index can reach at a 10 × 10 table.

In a more conventional perspective, one can view the Quetelet transformation as a measure of contrast between the observed distribution and the statistical independence as a background distribution. Indeed, the values

$\pi_{kk'} = p_{k+}p_{+k}$ can be considered as a measure of random interactions caused by the sheer "masses" p_{k+}, p_{+k} of the elements k, k' that therefore can be considered as a kind of background interaction (see [164,165]). Then the differences between the observed and background interactions would manifest the real structure of the phenomenon; it is these differences that a data analysis method should be applied to. The differences can be considered in various versions:

1. As they are:

$$d(k, k') = p_{kk'} - p_{k+}p_{+k} \qquad (2.27)$$

—this is advocated in the so-called modularity approach [164,165];

2. In a relative format:

$$q(k, k') = \frac{p_{kk'} - p_{k+}p_{+k}}{p_{k+}p_{+k}} \qquad (2.28)$$

—these are the relative Quetelet coefficients; and

3. In a relative format somewhat stretched in:

$$p(k, k') = \frac{p_{kk'} - p_{k+}p_{+k}}{\sqrt{p_{k+}p_{+k}}} \qquad (2.29)$$

—these indexes are not legitimate with the classical statistics approach, yet they can be seen in the definition of Pearson's chi-squared index—this is just the sum of squares of the $p(k, k')$.

The three formulas above can be considered different standardizations of contingency and redistribution data.

2.5.3 Affinity and Kernel Data

Affinity data are similarities between entities in an entity-to-feature table that are computed as inner products of the corresponding vectors in the data matrix. When the data are pre-procesed as described in Section 2.4, by shifting to grand means with a follow-up scaling and rescaling, it is only natural to pick up the row-to-row inner product $a_{ij} = <y_i, y_j> = \sum_{v \in V} y_{iv}y_{jv}$ (note, the inner product, not the correlation coefficient) as a good similarity measure. Indeed, each feature v contributes product $y_{iv}y_{jv}$ to the total similarity, which points to the mutual locations of entities i and j on the axis v with respect to grand mean c_v because the data standardization translates c_v to the scale's origin, 0. The product is positive when both entities are either larger than $c_v = 0$ or smaller than $c_v = 0$. It is negative when i and j are on different sides from $c_v = 0$. These correspond to our intuition. Moreover, the closer y_{iv} and

y_{jv} to 0, the smaller the product; the further away is either of them from 0, the larger the product. This property can be interpreted as supporting a major data-mining idea that the less ordinary a phenomenon, the more interesting it is: the ordinary in this case is just the average.

Example 2.9: Inner Product of Binary Data

Let us consider a binary data matrix such as Digit data in Table 1.9. The inner product of any two rows of this matrix is equal to the number of columns at which both rows have a unity entry. That means, that this similarity measure in fact just counts the number of attributes that are present at both of the entities. This is not a bad measure, especially when the numbers of attributes which are present at entities do not much differ. Let us denote V_i the set of attributes that are present at entity $i \in I$. When some entities may have hundreds of attributes in V_i while some other just dozens, this measure may suffer from the bias towards entities falling in greater numbers of features. To tackle this, various heuristics have been proposed of which probably the most popular is the Jackard similarity. The Jackard similarity index between two binary entities is the the number of common features related to the total number of features that are present on one or the other or both of the entities, $J = |V_i \cap V_j|/|V_i \cup V_j|, i, j \in I$.

If the data matrix is pre-processed by the centering columnwise as described in the previous section, the inner product leads to another attractive similarity measure. At a binary feature v, its average is but the proportion of entities at which feature v is present, $c_v = p_v$. After this is subtracted from every column v, the inner product can be expressed as:

$$a_{ij} = \sum_{v \in V} y_{iv} y_{jv} = |V_i \cap V_j| - |V_i| - |V_j| + t(i) + t(j) + \gamma \qquad (2.30)$$

where $t(i)$ (or $t(j)$) is the total frequency weight of features that are present at i (or, at j). The frequency weight of a feature v is defined here as $1 - p_v$; the more frequent is the feature, the smaller its frequency weight. The value of $\gamma = \sum_v p_v^2$ here is simply an averaging constant related to the Gini coefficient. The right-hand expression for a_{ij} follows from the fact that each of the pre-processed y_{iv} and y_{jv} can only be either $1 - p_v$ or $-p_v$.

This measure not only takes into account the sizes of the feature sets. It also involves evaluations of the information contents of all the binary features that are present at entities i or j. Amazingly, the measure (2.30) has never been explored or utilized in the data analysis community (see, e.g., a review in [26]).   ∎

An important property of the inner product matrix $A = YY^T$ is that it is positive semidefinite, that is, the quadratic form $x^T A x = \sum_{i,j \in I} a_{ij} x_i x_j$ cannot be negative at any x (see also Section 5.2). This property is a base to define a wide class of inner-product-like similarity measures referred to as kernels. A kernel $K(x, x')$ is a monotone non-increasing function of the distance between multivariate points x, x' generating positive semidefinite matrices on finite sets of multivariate points. This warrants that $K(x, x')$ is in fact the inner product in a high dimensional space generated by a non-linear transformation of the data features as defined by the eigenfunctions of the function $K(x, x')$ (see more details in [2,59,86,182,190]).

Arguably the most popular kernel function is the Gaussian kernel $G(x, y) = e^{\frac{-d(x,y)}{2\sigma^2}}$ where $d(x, y)$ is the squared Euclidean distance between x and y if $x \neq y$. The similarities produced by this kernel are usually referred to as the affinity data. The denominator $d = 2\sigma^2$ may greatly affect the structure of the affinity data. In our experiments, we define it so that to cover 99% of the interval of its values that are between 1 at $x = y$, and 0 at the infinite distance. This translates into a wish to see the exponent between 0 and -4.5, since $e^{-4.5}$ is approximately 0.01. After a feature has been normalized by its range, the expected value of the squared difference of its values should be about $1/4$, so that it should be multiplied by 18 to stretch it to 4.5. Therefore, we take denominator $d = M/18$ where M is the number of features.

The kernel affinity data format is becoming quite popular because of both practical and theoretical reasons. A practical reason is a huge success of the affinity data in advancing into such problems as the problem of image segmentation, especially when added with a distance threshold. The distance threshold sets the similarity between entities to 0 for all the distances greater than that [186]. Theoretical reasons relate to the hope that this concept may help in unifying three different data representations: distributions, entity-to-feature tables and entity-to-entity similarity matrices (see, e.g., [86]).

2.5.4 Network Data

A network, or graph, data set comprises two sets: (a) nodes to represent entities or objects, and (b) arcs connecting some pairs of nodes. Examples include:

i. *Telephone calls*: a set of individual units represented by nodes so that the presence of arc (i, j) means that one or more calls have been made from i to j.

ii. *Internet references*: a set of Web pages represented by nodes so that the presence of arc (i, j) means that page i has a reference to page j.

iii. *Citation network*: a set of research papers represented by nodes so that the presence of arc (i, j) means that paper i contains a reference to paper j.

Sometimes connections can be undirected, as, for example, a kin relation between individuals or context relation between articles. Then the network is referred to as undirected; an undirected ark is called an edge.

Two examples of networks are presented in Figure 2.6a and b. That in part (a) consists of two parts with no connections between them—these are the so-called (connected) components. To introduce the concept of component

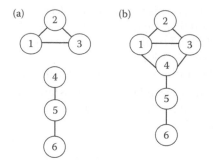

FIGURE 2.6
Two graphs on a set of six entities; that (a) consists of two components, (b) consists of one component.

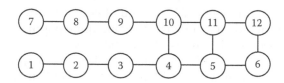

FIGURE 2.7
Cockroach graph: do six nodes on the right, 4, 5, 6, 10, 11, 12, form a cluster in the graph?

properly, one should use the concept of path: a path in a network is such a series of nodes $i_1 i_2 \cdots i_k$ that arc (i_j, i_{j+1}) belongs to the network for each $j = 1, 2, \ldots, k - 1$. Then a component is a part of the graph under consideration such that every two nodes in it are connected by a path and, moreover, no further node from the graph can be added without breaking the connectivity property.

The graph component is a simplest clustering concept; both of the components of graph in Figure 2.6a can be recognized as clusters. In contrast, graph in Figure 2.6b consists of one component only. Its part embracing nodes 1, 2, 3, and 4 seems deserving to be referred to as a cluster; the other two nodes are somewhat less related to that. A somewhat less pronounced cluster structure is presented in Figure 2.7.

A network on set I can be equivalently represented by a similarity matrix $A = (a_{ij})$ where $a_{ij} = 1$ whenever arc (i, j) belongs to the graph; otherwise, $a_{ij} = 0$. Likewise, any matrix $A = (a_{ij})$ with $i, j \in I$ can be represented by a network on I at which an arc (i, j) is present if and only if $a_{ij} \neq 0$; the value a_{ij} is referred to as the arc's weight, and the network is called weighted.

2.5.5 Similarity Data Pre-Processing

To help in sharpening the cluster structure hidden in the data, similarity data can be pre-processed with a normalization option. Three most effective, in our experience, options are considered below:

1. Removal of low similarities
2. Subtraction of background noise
3. Pseudo-Laplacian (Lapin) normalization

These are not universal. In some data structures none is working, whereas under some conditions one or another can help indeed. No general reasoning on this has been developed as yet; some advice will be given below.

2.5.5.1 Removal of Low Similarities: Thresholding

Given a similarity matrix $A = (a_{ij})$ of similarity scores a_{ij} between entities $i, j \in I$, specify a threshold level π within the score range such that any similarity score which is less than π can be considered insignificant, just a noise. Such a threshold can be extracted from experts in some domains. Let us now substitute all the a_{ij} that are less than π, $a_{ij} < \pi$, by zeros. The resulting similarity matrix A_{π} can be referred to as the result of π-thresholding of A. A further transformation would flatten all the remaining similarity values by changing them to unity to produce what is referred to as a threshold graph.

Consider, for example, matrix of similarity values between elementary functions in Table 1.6:

	1	2	3	4	5	6	7	8	9
1. e^x		6	1	1	2	3	2	2	2
2. lnx	6		1	1	2	2	4	4	3
3. $1/x$	1	1		6	1	1	1	1	1
4. $1/x^2$	1	1	6		1	1	1	1	1
5. x^2	2	2	1	1		6	5	5	5
6. x^3	3	2	1	1	6		4	3	2
7. $\sqrt{x}$	2	4	1	1	5	4		5	3
8. $\sqrt[3]{x}$	2	4	1	1	5	3	5		2
9. $|x|$	2	3	1	1	5	2	3	2	

The diagonal entries are left empty because the functions are not compared to themselves, as it frequently happens.

The result of 4-thresholding of this network is as follows:

	1	2	3	4	5	6	7	8	9
[1]		6	0	0	0	0	0	0	0
[2]	6		0	0	0	0	4	4	0
[3]	0	0		6	0	0	0	0	0
[4]	0	0	6		0	0	0	0	0
[5]	0	0	0	0		6	5	5	5
[6]	0	0	0	0	6		4	0	0
[7]	0	4	0	0	5	4		5	0
[8]	0	4	0	0	5	0	5		0
[9]	0	0	0	0	5	0	0	0	

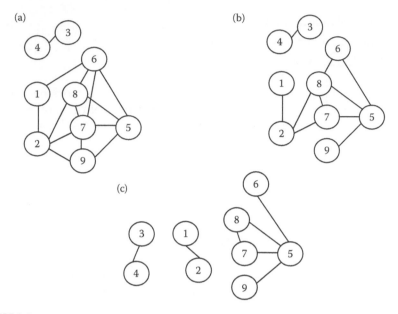

FIGURE 2.8
Threshold graphs for Elementary functions at $\pi = 2$ at part (a), at $\pi = 3$ at part (b), and at $\pi = 4$ at part (c).

Parts (a), (b), and (c) of Figure 2.8 demonstrate threshold graphs at different levels of thresholds corresponding to π equal to 2, 3 and 4, respectively.

One can see three components on 4-threshold graph in Figure 2.8c manifesting a cluster structure in the similarity matrix. Unfortunately, such a clear-cut structure of the graph does not happen too frequently in the real world data. This is why more complex methods for clustering graph data are needed sometimes.

2.5.5.2 Subtraction of Background Noise

There can be different background similarity patterns that should be subtracted to sharpen up the clusters hidden in data, of which two have been considered in the literature: (a) A constant "noise" level [135,140] and (b) Random interactions [164].

Given a similarity threshold value π, let us utilize it in a soft manner, by subtracting from all the similarity values in $A = (a_{ij})$ rather than by just removal of the smaller similarities as described above. Then the low values a_{ij}, those less than π, become negative: $a_{ij} - \pi < 0$, to make more difficult for i and j to be put together. Indeed, any reasonable clustering criterion would require a high similarity between elements of the same cluster; therefore, to overcome a negative similarity score, to put together these i and j, they must have very

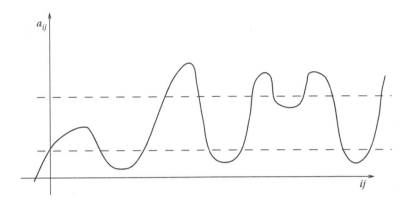

FIGURE 2.9
A pattern of similarity $a_{ij} = s_{ij} - a$ values depending on a subtracted threshold a.

high positive similarities with other elements of the cluster. Figure 2.9 demonstrates the effect of changing a positive similarity measure s_{ij} to $a'_{ij} = s_{ij} - a$ by subtracting a threshold $\pi = a > 0$: small similarities $s_{ij} < a$ can be transformed into negative similarities a'_{ij}. This can be irrelevant for some methods, such as the Single linkage (see Section 5.5), yet quite useful for some other methods such as Additive extraction of a cluster in Section 7.4.2.

Another interpretation of the soft threshold value: it is a similarity shift because by subtracting it from A, one changes the origin of the similarity scale from 0 to π.

The concept of random interactions involves a more specific modeling, a probabilistic interpretation of the similarities as emerging from some interactions between the entities. According to this interpretation, each entity i is assigned with a probability of interaction with other entities. A convenient definition sets the probability to be equal to the proportion of the summary similarity in ith row within the whole summary volume of the similarities (considered thus to be a kind of redistribution data). Then random interactions between two entities will occur with the probability that can be set as the product of their respective probabilities, which therefore should be subtracted from the similarity coefficients to clear up the nonrandom part of the similarity. This transformation will be referred to as the modularity standardization according to the terminology in [164].

2.5.5.3 Laplace Transformation

The Laplace transformation merges clustering into the classical mathematics realm. Conventionally, this transformation is considered as part of a clustering method, specifically, of the spectral clustering oriented at minimizing the normalized cut in a weighted graph [117,186]. Moreover, there are indications that this transformation and corresponding kernels are highly related [190].

Yet at the current state of the art, this transformation can be meaningfully applied to any symmetric similarity matrix, with no regard for a follow-up analysis, and therefore should be considered as part of the data pre-processing stage rather than part of any specific data analysis method [153].

Given a symmetric matrix A, its Laplace matrices are defined as follows. First, an $N \times N$ diagonal matrix D is computed, with its (i, i)-th entry equal to $d_i = a_{i+} = \sum_{j \in I} a_{ij}$, the sum of i's row of A. Then combinatorial Laplacian and normalized Laplacian are defined with equations $L = D - A$ and $L_n = D^{-1/2} L D^{-1/2}$, respectively. Both matrices are positive semidefinite and have zero as the minimum eigenvalue. The minimum non-zero eigenvalues and corresponding eigenvectors of the Laplacian matrices are utilized for relaxation of combinatorial partition problems [117,166,186] giving rise to what is referred to as the spectral clustering approach. The normalized Laplacian L_n's entries have the form $1 - (a_{ij})/(\sqrt{d_i d_j})$. Only L_n is used in this text because it leads to more straightforward mathematics [153] and, also, is more robust than L in some aspects [117]. Moreover, to align this transformation with the data recovery clustering models [140] in which maximum, rather than minimum, eigenvalues are prevalent, we bring forth the pseudo-inverse matrix L_n^+ for L_n that we refer to as the Laplace Pseudo-INverse transformation, Lapin for short. This is defined using the spectral decomposition of matrix L_n so that L_n^+ has the same eigenvectors as matrix L_n, but the corresponding eigenvalues are reciprocals of the eigenvalues of L_n. Of course, zero eigenvalues and their eigenvectors take no place in this. In this way, the minimum eigenvalue of L_n becomes the maximum eigenvalue of L_n^+. Moreover, the gap between that and the next eigenvalue increases too, so that the corresponding cluster can be easier to get extracted. For example, if three minimum eigenvalues of L_n are $0, 0.1, 0.2, 0.5$ then the three largest eigenvalues of L_n^+ are 10, 5, and 2, respectively, because $1/0.1 = 10$, $1/0.2 = 5$, and $1/0.5 = 2$.

To demonstrate the ability of the Lapin transformation in manifesting clusters according to human intuition, let us consider a two-dimensional set presented in the Table 2.20 and visualized in Figure 2.10.

This set has been generated as follows. Three 20×2 data matrices, $x1$, $x2$, and $x3$, were generated from Gaussian distribution $N(0, 1)$. Then matrix $x2$ was normed row-wise into b, so that each row in b is a 2D vector, whose squared entries total to unity, after which matrix c was defined as $c = x3 + 20b$. Its rows form a circular shape on the plane while rows of $x1$ fall into a ball in the circle's center as presented in Figure 2.10. Then $x1$ and c are merged together into a 40×2 matrix x, in which $x1$ takes the first 20 rows and c the next 20 rows.

The conventional data standardization methods would not much change the structure of Figure 2.10. Therefore, conventional clustering procedures, except for the Single linkage of course, would not be able to separate the ring, as a whole, from the ball. The affinity values illustrated in the left part of Table 2.21 would be high between points in the ball, and zero between not adjacent ring points. Yet Lapin transformation makes a perfect trick: all the Lapin similarities within the ball and ring are high and positive while all the

TABLE 2.20

Ball and Ring

Point	x	y	Point	x	y
1	−0.43	0.29	21	−19.38	−1.19
2	−1.67	−1.34	22	13.38	−17.75
3	0.13	0.71	23	−12.95	15.38
4	0.29	1.62	24	11.06	−16.48
5	−1.15	−0.69	25	−17.45	9.72
6	1.19	0.86	26	10.41	17.46
7	1.19	1.25	27	4.74	19.18
8	−0.04	−1.59	28	−17.69	9.62
9	0.33	−1.44	29	−18.91	−0.58
10	0.17	0.57	30	−1.87	19.55
11	−0.19	−0.40	31	−17.04	8.62
12	0.73	0.69	32	18.55	−8.74
13	−0.59	0.82	33	15.42	−10.46
14	2.18	0.71	34	19.14	−3.39
15	−0.14	1.29	35	7.88	−19.78
16	0.11	0.67	36	−19.75	−6.88
17	1.07	1.19	37	19.88	4.82
18	0.06	−1.20	38	−18.52	4.61
19	−0.10	−0.02	39	−1.09	19.74
20	−0.83	−0.16	40	−2.99	−18.86

Note: Coordinates of 40 points generated so that 20 fall in a ball around the origin, whereas the other 20 fall along a well-separated ring around that; see Figure 2.10.

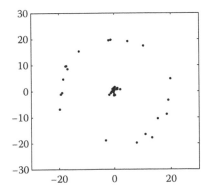

FIGURE 2.10
Two intuitively obvious clusters that are difficult to discern by using conventional approaches: the ball and ring; see Table 2.20.

TABLE 2.21

Lapin Transformation

Point	1	2	3	4	5	6	1	2	3	4	5	6
1	0.00	0.93	0.91	0.00	0.00	0.00	97.90	102.57	100.27	−52.220	−338.33	−188.52
2	0.93	0.00	0.97	0.00	0.00	0.00	102.57	107.48	105.06	−547.15	−354.50	−197.53
3	0.91	0.97	0.00	0.00	0.00	0.00	100.27	105.06	102.72.03	−534.89	−346.56	−193.11
4	0.00	0.00	0.00	0.00	0.00	0.00	−522.20	−547.15	−534.89	2938.20	1390.34	1060.72
5	0.00	0.00	0.00	0.00	0.00	0.00	−338.33	−354.50	−346.56	1390.34	2298.07	501.95
6	0.00	0.00	0.00	0.00	0.00	0.00	−188.52	−197.53	−193.11	1060.72	501.95	382.95

Note: Affinity values for three points taken from the ball and three from the ring (points 18–23, to be exact), in the left part of the table. The corresponding Lapin similarities between the points are on the right.

Lapin similarities between the ball and ring are high negative (see Table 2.21 on the right)! No clustering method is needed to separate the ball from the ring according to the Lapin transformation, just a 0 threshold would suffice.

This ability of Lapin transformation in transforming elongated structures into convex clusters has been a subject of mathematical scrutiny. An analogy with electricity circuits has been found. Roughly speaking, if w_{ij} measures the conductivity of the wire between nodes i and j in a "linear electricity network," then the corresponding element of a Lapin matrix expresses the "effective resistance" between i and j in the circuit [95]. Geometrically, this implies that two entities i and j, that are dissimilar according to the affinity A, may be highly related in L_n^+ if there are series of mutually linked entities connecting them according to A. Yet there can be cases of elongated structures at which Lapin transformation does not work at all [151].

Moreover, the Lapin transformation may fail sometimes—destroy the structure rather than to sharpen it [151]. At present, why and at what type of data this may happen is a mystery.

3

K-Means Clustering and Related Approaches

After reading this chapter, the reader will know about

1. Parallel and incremental K-Means
2. Instability of K-Means with regard to initial centroids
3. Approaches to the initial setting in K-Means
4. An anomalous cluster version of K-Means for incomplete clustering
5. An intelligent version of K-Means mitigating the issues of the choice of K and initial setting
6. A three-step version of K-Means involving feature weighting with Minkowski's distance
7. Popular extensions of K-Means
   a. Partitioning around medoids (PAM)
   b. Gaussian mixtures and expectation-maximization (EM) method
   c. Kohonen's self-organizing map (SOM)
   d. Fuzzy clustering
   e. Regression-wise K-Means

Key Concepts

Anomalous pattern A method for separating a cluster that is most distant from any reference point including the grand mean. The method is similar to K-Means at $K = 2$ except for the never-changing location of the reference point serving as the centroid of the other cluster.

Centroid A multidimensional vector minimizing the summary distance to cluster's elements. If the distance is Euclidean squared, the centroid is equal to the center's gravity center.

Expectation and maximization (EM) An algorithm of alternating maximization applied to the likelihood function for a mixture of distributions model. At each iteration, EM is performed according to the following steps: (1) expectation: given parameters of the mixture and individual density functions, find posterior probabilities for observations to belong to individual clusters; (2) maximization: given posterior probabilities g_{ik}, find parameters of the mixture maximizing the likelihood function.

Fuzzy clustering A clustering model at which entities $i \in I$ are assumed to belong to different clusters k with a degree of membership z_{ik} so that (1) $z_{ik} \geq 0$ and $\sum_{k=1}^{K} z_{ik} = 1$ for all $i \in I$ and $k = 1, \ldots, K$. Conventional "crisp" clusters can be considered a special case of fuzzy clusters in which z_{ik} may only be 0 or 1.

Gaussian distribution A popular probabilistic cluster model characterized by the modal point of the cluster mean vector and the feature-to-feature covariance matrix. The surfaces of equal density for a Gaussian distribution are ellipsoids.

iK-Means A version of K-Means, in which an initial set of centroids (seeds) is found with an iterated version of the Anomalous pattern algorithm.

Incremental K-Means A version of K-Means in which entities are dealt with one-by-one.

K-Means A major clustering method producing a partition of the entity set into non-overlapping clusters together with within-cluster centroids. It proceeds in iterations consisting of two steps each. One step updates clusters according to the minimum distance rule, the other step updates centroids as the centers of gravity of clusters. The method implements the so-called alternating minimization algorithm for the square error criterion. To initialize the computations, either a partition or a set of all K tentative centroids must be specified.

Medoid An entity in a cluster with the minimal summary distance to the other within cluster elements. It is used as a formalization of the concept of prototype in the PAM clustering method.

Minimum distance rule The rule which assigns each of the entities to its nearest centroid.

Minkowski metric clustering An extension of K-Means clustering criterion from the squares of errors to an arbitrary positive power (Minkowski's p). This makes the problem of finding centers harder than usual but allows, instead, to get rid of irrelevant features.

Mixture of distributions A probabilistic clustering model according to which each cluster can be represented by a unimodal probabilistic density function so that the population density is a probabilistic mixture of the individual cluster functions. This concept integrates a number of simpler models and algorithms for fitting them, including the K-Means clustering algorithm. The assumptions leading to K-Means can be overly restrictive, such as those leading to the compulsory z-score data standardization.

Reference point A vector in the variable space serving as the space origin. The anomalous pattern is sought starting from an entity which is the farthest from the reference point. The latter thus models the norm from which the Anomalous pattern deviates most.

Regression-wise clustering A clustering model in which a cluster is sought as a set of observations satisfying a regression function. Such a model can be fitted with an extension of K-Means in which prototypes are presented as regression functions.

Self-organizing map (SOM) A model for data visualization in the form of a 2D grid on plane in which entities are represented by grid nodes reflecting both their similarity and the grid topology.

Square error criterion The sum of summary distances from cluster elements to the cluster centroids, which is minimized by K-Means. The distance used is the Euclidean distance squared, which is compatible with the least-squares data recovery criterion.

Weighted K-Means A three-step version of K-Means including three groups of variables: cluster membership, centroids and feature weights. Especially effective when using Minkowski's distance at $p \neq 2$.

3.1 Conventional K-Means

3.1.1 Generic K-Means

K-Means is a most popular clustering technique. It is present, in various forms, in major statistical packages such as SPSS [66] and SAS [33,178], and datamining packages such as Clementine [27], iDA tool [174], and DBMiner [76].

The algorithm is appealing in many aspects. Conceptually, it may be considered a model for the cognitive process of making a data-conditioned typology. Also, it has nice mathematical properties. This method is computationally easy, fast, and memory-efficient. However, there are some problems too, especially with respect to the initial setting and stability of results, which will be dealt with in Section 3.2.

The cluster structure in K-Means is a partition of the entity set in K non-overlapping clusters represented by lists of entities and within cluster means of the variables. The means are aggregate representations of clusters and as such they are sometimes referred to as standard points or centroids or prototypes. These terms are considered synonymous in clustering. More formally, a cluster structure is represented by non-overlapping subsets $S_k \subset I$ and M-dimensional centroids $c_k = (c_{kv}), k = 1, \ldots, K$. Subsets S_k form partition $S = \{S_1, \ldots, S_K\}$ with a set of cluster centroids $c = \{c_1, \ldots, c_K\}$. More technical details can be found in Section 7.2.

Example 3.1: Centroids of Subject Clusters in Colleges Data

Let us consider the subject-based clusters in the Colleges data. The cluster structure is presented in Table 3.1 in such a way that the centroids are calculated twice, once for the raw data in Table 2.13 and the second time,

TABLE 3.1

Means of the Variables in Table 3.2 within $K = 3$ Subject-Based Clusters, Real (Upper Row) and Standardized (Lower Row)

Cl.	List	St (f1)	Ac (f2)	NS (f3)	EL (f4)	B (f5)	M (f6)	C (f7)
					Mean			
1	1, 2, 3	4820	392	2.67	0	0.67	0.33	0
		0.095	0.124	−0.111	−0.625	0.168	−0.024	−0.144
2	4, 5, 6	3740	224	2.33	1	0.33	0.67	0
		−0.215	−0.286	−0.222	0.375	−0.024	0.168	−0.144
3	7, 8	5110	441	4.50	1	0.00	0.00	1
		0.179	0.243	0.500	0.375	−0.216	−0.216	0.433

TABLE 3.2

Range Standardized Colleges Matrix with the Additionally Rescaled Nominal Feature Attributes Copied from Table 2.17

	St	Ac	NS	EL	M	B	C
1	−0.20	0.23	−0.33	−0.63	0.36	−0.22	−0.14
2	0.40	0.05	0.00	−0.63	0.36	−0.22	−0.14
3	0.08	0.09	0.00	−0.63	−0.22	0.36	−0.14
4	−0.23	−0.15	−0.33	0.38	0.36	−0.22	−0.14
5	0.19	−0.29	0.00	0.38	−0.22	0.36	−0.14
6	−0.60	−0.42	−0.33	0.38	−0.22	0.36	−0.14
7	0.08	−0.10	0.33	0.38	−0.22	−0.22	0.43
8	0.27	0.58	0.67	0.38	−0.22	−0.22	0.43

with the standardized data in Table 3.2, which is a copy of Table 2.17 of the previous chapter. ∎

Given K M-dimensional vectors c_k as cluster centroids, the algorithm updates cluster lists S_k according to the so-called *Minimum distance rule*. This rule assigns entities to their nearest centroids. Specifically, for each entity $i \in I$, its distances to all centroids are calculated, and the entity is assigned to the nearest centroid. When there are several nearest centroids, the assignment is taken among them arbitrarily. In other words, S_k is made of all such $i \in I$ that $d(i, c_k)$ is minimum over all k.

The Minimum distance rule is popular in data analysis and can be found in many approaches such as Voronoi diagrams and vector learning quantization.

In general, some centroids may be assigned with no entity at all with this rule.

Having cluster lists updated with the Minimum distance rule, the algorithm updates centroids as gravity centers of the cluster lists S_k; the gravity center coordinates are defined as within cluster averages, that is, updated

TABLE 3.3

Distances between the Eight Standardized College Entities and Centroids

Centroid	Entity, Row Point from Table 3.2							
	1	2	3	4	5	6	7	8
c_1	**0.22**	**0.19**	**0.31**	1.31	1.49	2.12	1.76	2.36
c_2	1.58	1.84	1.36	**0.33**	**0.29**	**0.25**	0.95	2.30
c_3	2.50	2.01	1.95	1.69	1.20	2.40	**0.15**	**0.15**

Note: Within column minima are highlighted in bold.

centroids are defined as $c_k = c(S_k), k = 1, \ldots, K$, where $c(S)$ is a vector whose components are averages of features over S.

Then the process is reiterated until clusters do not change.

Recall that the distance referred to is Euclidean squared distance defined, for any M-dimensional $x = (x_v)$ and $y = (y_v)$ as $d(x, y) = (x_1 - y_1)^2 + \cdots + (x_M - y_M)^2$.

Example 3.2: Minimum Distance Rule at Subject Cluster Centroids in Colleges Data

Let us apply the Minimum distance rule to entities in Table 3.2, given the standardized centroids c_k in Table 3.1. The matrix of distances between the standardized eight row points in Table 3.2 and three centroids in Table 3.1 is in Table 3.3.

The table, as expected, shows that points 1,2,3 are nearest to centroid c_1, 4,5,6 to c_2, and 7, 8 to c_3. This means that the rule does not change clusters. These clusters will have the same centroids. Thus, no further calculations can change the clusters: the subject-based partition is to be accepted as the result. ∎

Let us now explicitly formulate the algorithm of parallel K-Means. This procedure is frequently referred to as batch K-Means.

BATCH K-MEANS

0. *Data pre-processing.* Transform data into a standardized quantitative matrix Y. This can be done according to the three step procedure described in Section 2.4.

1. *Initial setting.* Choose the number of clusters, K, and tentative centroids $c_1, c_2, \ldots, c_K$, frequently referred to as seeds. Assume initial cluster lists S_k empty.

2. *Clusters update.* Given K centroids, determine clusters S'_k ($k = 1, \ldots, K$) with the Minimum distance rule.

---

3. *Stop-condition.* Check whether $S' = S$. If yes, end with clustering $S = \{S_k\}$, $c = \{c_k\}$. Otherwise, change S for S'.

4. *Centroids update.* Given clusters S_k, calculate within cluster means c_k ($k = 1, \ldots, K$) and go to Step 2.

This algorithm usually converges fast, depending on the initial setting. Location of the initial seeds may affect not only the speed of convergence but also, more importantly, the final results as well. Let us give examples of how the initial setting may affect results.

Example 3.3: A Successful Application of K-Means

Let us apply K-Means to the same Colleges data in Table 3.2, this time starting with entities 2, 5, and 7 as tentative centroids (Step 1). To perform Step 2, the matrix of entity-to-centroid distances is computed (see Table 3.4 in which within column minima are boldfaced). The Minimum distance rule produces three cluster lists, $S_1 = \{1, 2, 3\}$, $S_2 = \{4, 5, 6\}$, and $S_3 = \{7, 8\}$. These coincide with the subject-based clusters and produce within-cluster means (Step 4) already calculated in Table 3.1. Since these differ from the original tentative centroids (entities 2, 5, and 7), the algorithm returns to Step 2 of assigning clusters around the updated centroids. We do not do this here since the operation has been done already with distances in Table 3.3, which produced the same subject-based lists according to the Minimum distance rule. The process thus stops.   ∎

Example 3.4: An Unsuccessful Run of K-Means with Different Initial Seeds

Let us take entities 1, 2, and 3 as the initial centroids (assuming the same data in Table 3.2). The Minimum distance rule, according to entity-to-centroid distances in Table 3.5, leads to cluster lists $S_1 = \{1, 4\}$, $S_2 = \{2\}$, and $S_3 = \{3, 5, 6, 7, 8\}$. With the centroids updated at Step 4 as means of these clusters, a new application of Step 3 leads to slightly changed cluster

TABLE 3.4

Distances between Entities 2, 5, 7 as Seeds and the Standardized Colleges Entities

				Row-Point				
Centroid	1	2	3	4	5	6	7	8
2	**0.51**	**0.00**	**0.77**	1.55	1.82	2.99	1.90	2.41
5	2.20	1.82	1.16	**0.97**	**0.00**	**0.75**	0.83	1.87
7	2.30	1.90	1.81	1.22	0.83	1.68	**0.00**	**0.61**

TABLE 3.5

Distances between the Standardized Colleges Entities and Entities 1, 2, 3 as Seeds

Centroid	Row-Point							
	1	2	3	4	5	6	7	8
1	**0.00**	0.51	0.88	**1.15**	2.20	2.25	2.30	3.01
2	0.51	**0.00**	0.77	1.55	1.82	2.99	1.90	2.41
3	0.88	0.77	**0.00**	1.94	**1.16**	**1.84**	**1.81**	**2.38**

Note: Within column minima are highlighted in bold.

TABLE 3.6

Distances between the Standardized Colleges Entities and Entities 1, 4, 7 as Tentative Centroids

Centroid	Row-Point							
	1	2	3	4	5	6	7	8
1	**0.00**	**0.51**	**0.88**	1.15	2.20	2.25	2.30	3.01
4	1.15	1.55	1.94	**0.00**	0.97	**0.87**	1.22	2.46
7	2.30	1.90	1.81	1.22	**0.83**	1.68	**0.00**	**0.61**

Note: Within column minima are highlighted in bold.

lists $S_1 = \{1, 4, 6\}$, $S_2 = \{2\}$, and $S_3 = \{3, 5, 7, 8\}$. Their means calculated, it is not difficult to see that the Minimum distance rule does not change the clusters anymore. Thus the lists represent the final outcome, which differs from the subject-based solution. ∎

The poor performance in this example may be explained by the stupid choice of the initial centroids, all in the same subject. However, K-Means can lead to inconvenient results even if the initial setting fits well into clustering by subject.

Example 3.5: Unsuccessful K-Means with Subject-Based Initial Seeds

With the initial centroids at rows 1, 4, and 7, all of different subjects, the entity-to-centroid matrix in Table 3.6 leads to cluster lists $S_1 = \{1, 2, 3\}$, $S_2 = \{4, 6\}$, and $S_3 = \{5, 7, 8\}$ which do not change in the follow-up operations. These results put an Engineering college among the Arts colleges. Not a good outcome. ∎

3.1.2 Square Error Criterion

The instability of clustering results with respect to the initial setting leads to a natural question if there is anything objective in the method at all. Yes, there is.

It appears, there is a scoring function, an index, that is minimized by K-Means. To formulate the function, let us define the within cluster error. For a cluster S_k with centroid $c_k = (c_{kv})$, $v \in V$, its square error is defined as the summary distance from its elements to c_k:

$$W(S_k, c_k) = \sum_{i \in S_k} d(y_i, c_k) = \sum_{i \in S_k} \sum_{v \in V} (y_{iv} - c_{kv})^2 \qquad (3.1)$$

The square error criterion is the sum of these values over all clusters:

$$W(S, c) = \sum_{k=1}^{K} W(S_k, c_k) = \sum_{k=1}^{K} \sum_{i \in S_k} d(y_i, c_k) \qquad (3.2)$$

Criterion $W(S, c)$ (3.2) depends on two groups of arguments: cluster lists S_k and centroids c_k. Criteria of this type are frequently optimized with the so-called alternating minimization algorithm. This algorithm consists of a series of iterations. At each of the iterations, $W(S, c)$ is, first, minimized over S, given c, and, second, minimized over c, given the resulting S. In this way, at each iteration a set S is transformed into a set S'. The calculations stop when c is stabilized, that is, $S' = S$.

Statement 3.1. Batch K-Means is the alternating minimization algorithm for the summary square-error criterion (3.2) starting from seeds $c = \{c_1, \ldots, c_K\}$ specified in step 1.

Proof: Equation

$$W(S, c) = \sum_{k=1}^{K} \sum_{i \in S_k} d(i, c_k)$$

from Equation 3.2, implies that, given $c = \{c_1, \ldots, c_K\}$, the Minimum distance rule minimizes $W(S, c)$ over S. Let us now turn to the problem of minimizing $W(S, c)$ over c, given S. It is obvious that minimizing $W(S, c)$ over c can be done by minimizing each $W(S_k, c_k)$ (3.1) over c_k independently for $k = 1, \ldots, K$. Criterion $W(S_k, c_k)$ is a quadratic function of c_k and, thus, can be optimized with just first-order optimality conditions that the derivatives of $W(S_k, c_k)$ over c_{kv} must be equal to zero for all $v \in V$. These derivatives are equal to $F(c_{kv}) = -2 \sum_{i \in S_k} (y_{iv} - c_{kv})$, $k = 1, \ldots, K; v \in V$. The condition $F(c_{kv}) = 0$ obviously leads to $c_{kv} = \sum_{i \in S_k} y_{iv} / |S_k|$, which proves that the optimal centroids must be within cluster gravity centers. This proves the statement, q.e.d.

Square-error criterion (3.2) is the sum of distances from entities to their cluster centroids. This can be reformulated as the sum of within cluster variances

$\sigma^2_{kv} = \sum_{i \in S_k} (y_{iv} - c_{kv})^2 / N_k$ weighted by the cluster cardinalities:

$$W(S,c) = \sum_{k=1}^{K} \sum_{i \in S_k} \sum_{v \in V} (y_{iv} - c_{kv})^2 = \sum_{v \in V} \sum_{k=1}^{K} N_k \sigma^2_{kv} \qquad (3.3)$$

Statement 3.1 implies, among other things, that K-Means converges in a finite number of steps because the set of all partitions S over a finite I is finite and $W(S,c)$ is decreased at each change of c or S. Moreover, as experiments show, K-Means typically does not move far away from the initial setting of c. Considered from the perspective of minimization of criterion (3.2), this leads to the conventional strategy of repeatedly applying the algorithm starting from various randomly generated sets of prototypes to reach as deep a minimum of Equation 3.2 as possible. This strategy may fail especially if the feature set is large because in this case random settings cannot cover the space of solutions in a reasonable time.

Yet, there is a different perspective, of typology making, in which the criterion is considered not as something that must be minimized at any cost but rather a beacon for direction. In this perspective, the algorithm is a model for developing a typology represented by the prototypes. The prototypes should come from an external source such as the advice of experts, leaving to data analysis only their adjustment to real data. In such a situation, the property that the final prototypes are not far away from the original ones is more of an advantage than not. What is important in this perspective, though, is defining an appropriate, rather than random, initial setting.

The data recovery framework is consistent with this perspective. Indeed, the model underlying K-Means is based on a somewhat simplistic assumption that entities can be represented by their cluster's centroids, up to residuals. This model, according to Section 7.2.1, leads to an equation involving K-Means criterion $W(S,c)$ (3.2) and the data scatter $T(Y)$:

$$T(Y) = B(S,c) + W(S,c) \qquad (3.4)$$

where

$$B(S,c) = \sum_{k=1}^{K} N_k c^2_{kv} \qquad (3.5)$$

Criterion $B(S,c)$ measures the part of the data scatter taken into account by the cluster structure. Therefore, data scatter $T(Y)$ is decomposed in two parts: the part $B(S,c)$ explained by the cluster structure (S,c), and the unexplained part $W(S,c)$. The larger the explained part, the better the match between clustering (S,c) and data.

Example 3.6: Explained Part of the Data Scatter

The explained part of the data scatter, $B(S,c)$, is equal to 43.7% of the data scatter $T(Y)$ for partition $\{\{1,4,6\}, \{2\}, \{3,5,7,8\}\}$, found with entities 1,2,3 as initial centroids. The score is 58.9% for partition $\{\{1,2,3\}, \{4,6\}, \{5,7,8\}\}$, found with entities 1,4,7 as initial centroids. The score is maximum, 64.0%, for the subject-based partition $\{\{1,2,3\}, \{4,5,6\}, \{7,8\}\}$, which is thus superior. ∎

An advice for selecting the number of clusters and tentative centroids at Step 1 will be given in Section 3.2.

3.1.3 Incremental Versions of K-Means

Incremental versions of K-Means are those at which Step 4, with its Minimum distance rule, is executed not for all of the entities but for one of them only. There can be two principal reasons for doing so:

R1 The user is not able to operate with the entire data set and takes entities in one by one, because of either the nature of the data generation process or the massive data sets. The former cause is typical when clustering is done in real time as, for instance, in an on-line application. Under conventional assumptions of probabilistic sampling of the entities, the convergence of the algorithm was proven in [124].

R2 The user operates with the entire data set, but wants to smooth the action of the algorithm so that no drastic changes in the cluster centroids may occur. To do this, the user may specify an order of the entities and run entities one-by-one in this order for a number of times. (Each of the runs through the data set is referred to as an "epoch" in the neural network community.) The result of this may differ from that of Batch K-Means because of different computations. This computation can be especially effective if the order of entities is not constant but depends on their contributions to the criterion optimized by the algorithm. In particular, each entity $i \in I$ can be assigned value d_i, the minimum of distances from i to centroids $c_1, \ldots, c_K$, so that i minimizing d_i is considered first.

When an entity y_i joins cluster S_k whose cardinality is N_k, the centroid c_k changes to c'_k to follow the within cluster average values:

$$c'_t = \frac{N_k}{N_k + 1} c_k + \frac{1}{N_k + 1} y_i$$

When y_i moves out of cluster S_k, the formula remains valid if all the three pluses in it are changed for minuses. By introducing the variable z_i which is equal to +1 when y_i joins the cluster and -1 when it moves out of it, the formula becomes

$$c'_t = \frac{N_k}{N_k + z_i} c_k + \frac{z_i}{N_k + z_i} y_i \tag{3.6}$$

Accordingly, the distances from other entities change to $d(y_j, c'_k)$.

Because of the incremental setting, the stopping rule of the parallel version (reaching a stationary state) may be not necessarily applicable here. In case R1, the natural stopping rule is to end when there are no new entities observed. In the case R2, the process of running through the entities one-by-one stops when all entities remain in their clusters. Also, the process may stop when a pre-specified number of runs (epochs) is reached.

This gives rise to the following version of K-Means.

INCREMENTAL K-MEANS: ONE ENTITY AT A TIME

1. *Initial setting.* Choose the number of clusters, K, and tentative centroids, $c_1, c_2, \ldots, c_K$.

2. *Getting an entity.* Observe an entity $i \in I$ coming either randomly (setting R1) or according to a pre-specified or dynamically changing order (setting R2).

3. *Cluster update.* Apply Minimum distance rule to determine to what cluster list S_k ($k = 1, \ldots, K$) entity i should be assigned.

4. *Centroid update.* Update within cluster centroid c_k with formula (3.6). For the case in which y_i leaves cluster k' (in R2 option), $c_{k'}$ is also updated with Equation 3.6. Nothing is changed if y_i remains in its cluster. Then the stopping condition is checked as described above, and the process moves to observing the next entity (Step 2) or ends (Step 5).

5. *Output.* Output lists S_k and centroids c_k with accompanying interpretation aids (as advised in Section 6.3).

Example 3.7: A Smoothing Effect of the Incremental K-Means

Let us apply version R2 to the Colleges data with the entity order dynamically updated and $K = 3$, starting with entities 1, 4, and 7 as centroids. Minimum distances d_i to the centroids for the five remaining entities are presented in the first column of Table 3.7 along with the corresponding centroid (iteration 0). Since $d_2 = 0.51$ is minimum among them, entity 2 is put in cluster I whose centroid is changed accordingly. The next column, iteration 1, presents minimum distances to the updated centroids.

This time the minimum is at $d_8 = 0.61$, so entity 8 is put in its nearest cluster III and its center is recomputed. Column 2 presents the distances at iteration 2. Among remaining entities, 3, 5, and 6, the minimum distance is $d_3 = 0.70$, so 3 is added to its closest cluster I. Thus updated the centroid of cluster I leads to the change in minimum distances recorded at iteration 3. This time $d_6 = 0.087$ becomes minimum for the remaining entities 5 and 6 so that 6 joins cluster II and, in the next iteration, 5 follows it. Then the

TABLE 3.7

Minimum Distances between Standardized College Entities and
Dynamically Changed Centroids I, II, and III

	Iteration					
Entity	0	1	2	3	4	5
2	0.51/I	0.13/I	0.13/I	0.19/I	0.19/I	0.19/I
3	0.87/I	0.70/I	0.70/I	0.31/I	0.31/I	0.31/I
5	0.83/III	0.83/III	0.97/II	0.97/II	0.97/II	0.28/II
6	0.87/II	0.87/II	0.87/II	0.87/II	0.22/II	0.25/II
8	0.61/III	0.61/III	0.15/III	0.15/III	0.15/III	0.15/III

partition stabilizes: each entity is closer to its cluster centroid than to any
other. The final partition of the set of Colleges is the subject-based one.
We can see that this procedure smoothes the process indeed: starting from
the same centroids in Example 3.5, parallel K-Means leads to a different
partition which is not that good.   ∎

3.2 Choice of K and Initialization of K-Means

To initialize K-Means, one needs to specify:

1. The number of clusters, K
2. Initial centroids, $c_1, c_2, \ldots, c_K$

Each of these can be of an issue in practical computations. Both depend
on the user's expectations related to the level of resolution and typological
attitudes, which remain beyond the scope of the theory of K-Means. This is
why some claim these considerations are beyond the clustering discipline
as well. There have been, however, a number of approaches suggested for
specifying the number and location of initial centroids, which will be briefly
described in Section 6.2.1. Here we present, first, the most popular existing
approaches and, second, three approaches based on the idea that clusters
deviate from each other and from the norm.

3.2.1 Conventional Approaches to Initial Setting

Conventionally, either of two extremes is adhered to set initial centroids.
One view assumes no knowledge of the data and domain at all. The initial centroids are taken randomly then. The other approach, on the contrary,
relies on a user being an expert in the application domain and capable of
specifying initial centroids as hypothetical prototypes in the domain.

3.2.1.1 Random Selection of Centroids

According to this approach, K entities are randomly selected as the initial seeds (centroids) to run K-Means (either parallel or incremental). This is repeated pre-specified number, say 100, times after which the best solution according to criterion (3.2) is taken. This approach can be followed using any package containing K-Means. For example, SPSS takes the first K entities in the data set as the initial seeds. This can be repeated as many times as needed, each time with rows of the data matrix randomly permuted.

To choose the right K, one should follow this approach at different values of K, say, in the range from 2 to 15. Then that K is selected that leads to the best result. The issue here is that there is no "right" criterion for deciding what results are the best. The square-error criterion itself cannot be used for such a decision because it monotonically decreases when K grows.

Yet comparing clusterings found at different K may lead to insights on the cluster structure in data. In real world computations, the following phenomenon has been observed by the author and other researchers. When repeatedly proceeding from a larger K to $K-1$, the obtained $K-1$ clustering, typically, is rather similar to that found by merging two clusters in the K clustering, in spite of the fact that the K- and $(K-1)$-clusterings are found independently. However, in the process of decreasing K in this way, a critical value of K is reached such that $(K-1)$-clustering does not resemble K-clustering anymore. This value of K can be taken as that corresponding to the cluster structure.

3.2.1.2 Expert-Driven Selection of Centroids

The other approach relies on the opinion of an expert in the subject domain. An expert can point to a property that should be taken into account. For example, an expert may propose to distinguish numeral digits depending on the presence or absence of a closed drawing in them. This feature is present in 6 and absent from 1; therefore, these entities can be proposed as the initial seeds. By using another feature such as the presence or absence of a semi-closed drawing, the expert can further diversify the set of initial seeds. This approach would allow seeing how the conceptual types relate to the data.

However, in a common situation in which the user cannot make much sense of their data because the measured features are quite superficial, the expert vision may fail.

Example 3.8: K-Means at Iris Data

Table 3.8 presents results of the Batch K-Means applied to the Iris data on page 11 with $K = 3$ and specimens numbered 1, 51, and 101 taken as the initial centroids and cross classified with the prior three class partition. The clustering does separate genus *Setosa* but misplaces $14 + 3 = 17$ specimens between two other genera. This corresponds to the visual pattern in Figure 1.10, p. 25. ∎

TABLE 3.8

Cross Classification of 150 Iris Specimens According to K-Means Clustering and the Genera; Entries Show Count/Proportion

Cluster	Iris Genus			Total
	Setosa	Versicolor	Virginica	
S_1	50/0.333	0/0	0/0	50/0.333
S_2	0/0	47/0.313	14/0.093	61/0.407
S_3	0/0	3/0.020	36/0.240	39/0.260
Total	50/0.333	50/0.333	50/0.333	150/1.000

The current author advocates yet another approach at which the user should somewhat "lightly" explore the data structure to set initial centroids. Three data-driven approaches are described in the remainder of this section. They all rely on the idea that the centroids should represent somewhat anomalous patterns.

3.2.2 MaxMin for Producing Deviate Centroids

This approach is based on the following intuition. If there are cohesive clusters in the data, then entities within any cluster must be close to each other and rather far away from entities in other clusters. The following method, based on this intuition, has proved to work well in some real and simulated experiments.

MAXMIN

1. Take entities $y_{i'}$ and $y_{i''}$ maximizing the distance $d(y_i, y_j)$ over all $i, j \in I$ as c_1 and c_2 elements of the set C.

2. For each of the entities y_i, that have not been selected to the set c of initial seeds so far, calculate $d_c(y_i)$, the minimum of its distances to $c_t \in c$.

3. Find i^* maximizing $d_c(y_i)$ and check Stop-condition (see below). If it does not hold, add y_{i^*} to c and go to Step 2. Otherwise, end and output c as the set of initial seeds.

As the Stop-condition in MaxMin, either or all of the following pre-specified constraints can be utilized:

1. The number of seeds has reached a pre-specified threshold.

2. Distance $d_c(y_{i*})$ is larger than a pre-specified threshold such as $d = d(c_1, c_2)/3$.
3. There is a significant drop, such as 35%, in the value of $d_c(y_{i*})$ in comparison to that at the previous iteration.

Example 3.9: MaxMin for Selecting Intial Seeds

The table of entity-to-entity distances for Colleges is displayed in Table 3.9. The maximum distance here is 3.43, between An and En, which makes the two of them initial centroids according to MaxMin. The distances from other entities to these two are in Table 3.10; those minimal at the two are boldfaced. The maximum among them, the next MaxMin distance, is 2.41 between Se and An. The decrease here is <30% suggesting that this can represent a different cluster. Thus, we add Se to the list of candidate centroids and then take a look at the distances from other entities to these three (see Table 3.11). This time the MaxMin distance is 0.87 between Et and En. We might wish to stop the process at this stage since we expect only three meaningful clusters in Colleges data and, also, there is a significant drop, 64% of the previous MaxMin distance. It is useful to remember that such a clear-cut situation may not necessarily happen in other examples. The three seeds selected have been shown in previous examples to produce the subject-based clusters with K-Means. ∎

The issues related to this approach are typical in data analysis. First, it involves ad hoc thresholds which are not substantiated in terms of data and,

TABLE 3.9

Distances between Colleges from Table 3.2

	So	Se	Si	Et	Ef	En	Ay	An
So	0.00	0.51	0.88	1.15	2.20	2.25	2.30	3.01
Se	0.51	0.00	0.77	1.55	1.82	2.99	1.90	2.41
Si	0.88	0.77	0.00	1.94	1.16	1.84	1.81	2.38
Et	1.15	1.55	1.94	0.00	0.97	0.87	1.22	2.46
Ef	2.20	1.82	1.16	0.97	0.00	0.75	0.83	1.87
En	2.25	2.99	1.84	0.87	0.75	0.00	1.68	3.43
Ay	2.30	1.90	1.81	1.22	0.83	1.68	0.00	0.61
An	3.01	2.41	2.38	2.46	1.87	3.43	0.61	0.00

TABLE 3.10

Distances from College Entities to En and An

	So	Se	Si	Et	Ef	Ay
En	2.25	3.00	1.84	0.87	0.75	1.68
An	3.01	2.41	2.38	2.46	1.87	0.61

TABLE 3.11

Distances between Se, En, and Ak and Other Colleges Entities

	So	Si	Et	Ef	Ay
Se	**0.51**	**0.77**	1.55	1.82	1.90
En	2.25	1.84	**0.87**	**0.75**	1.68
Ak	3.01	2.38	2.46	1.87	**0.61**

Note: Within column minima are highlighted in bold.

thus, difficult to choose. Second, it can be computationally intensive when the number of entities N is large since finding the maximum distance at Step 1 involves computation of $O(N^2)$ distances. One more drawback of the approach is that some of the farthest away entities can be far away from all the others so that they would correspond to singletons and, therefore, should not be used as centroids at all. The other two approaches further on are better in this respect.

3.2.3 Anomalous Centroids with Anomalous Pattern

The Anomalous Pattern (AP) method provides an alternative to MaxMin for the initial setting, which is less intensive computationally and, also, reduces the number of ad hoc parameters. Moreover, the initial centroids here represent some "dense" parts of the data set rather than just standalone points.

The AP method involves the concept of a reference point which is chosen to represent an average or norm so that the clusters are to be as far away from that as possible. For example, an educator might choose, to represent a "normal student," a point which indicates good marks in tests and serious work in projects, and then look at those patterns of observed behavior that deviate from this. Or, a bank manager may set a customer having specific assets and backgrounds as a reference point, to look at customer types deviating from this. In engineering, a moving robotic device should be able to classify elements of the environment according to the robot's location, so that things that are closer to it are observed at a greater granularity than things that are farther away: the location is the reference point in this case. In many cases, the reference point can be reasonably set at the gravity center of the entire entity set.

With a reference point pre-specified, the user can compare entities with the point rather than with each other, which would drastically reduce computations. To find a cluster which is most distant from the reference point, a version of K-Means can be utilized [137,140]. According to this procedure, clusters are extracted one by one so that the only ad hoc choice is the reference point. Setting that would immediately lead to two seeds: the reference point itself and the entity which is the most distant from the reference point. Indeed the Anomalous Pattern cluster is found as the set of entities that are

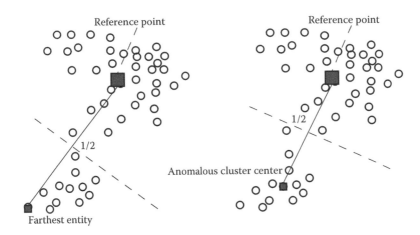

FIGURE 3.1
Extracting an "Anomalous Pattern" cluster with the reference point in the gravity center: the initial iteration is illustrated on the left and the final on the right.

closer to the cluster seed than to the reference point. This is done in a series of iterations. Each iteration starts with a seed c pre-specified. Then the set S of entities i whose distance to c is smaller than that to the reference point is determined, and its gravity center c' is computed. This c' is taken as the input c to the next iteration. The computation stops when $c' = c$. An exact formulation is given in Section 7.2.6.

The process is illustrated in Figure 3.1.

The Anomalous pattern method is a version of K-Means in which:

i. The number of clusters K is 2.

ii. Centroid of one of the clusters is 0, which is compulsory kept there through all the iterations.

iii. The initial centroid of the anomalous cluster is taken as an entity point which is the most distant from 0.

Property (iii) mitigates the issue of determining appropriate initial seeds, which would allow using Anomalous pattern algorithm for initializing K-Means.

Example 3.10: Anomalous Pattern in Market Towns

The Anomalous pattern method can be applied to the Market towns data in Table 1.1 assuming the grand mean as the reference point and scaling by range. The point farthest from 0, the tentative centroid at step 2, appears to be entity 35 (St Austell) whose distance from zero is 4.33, the maximum. Step 3 adds three more entities, 26, 29, and 44 (Newton Abbot, Penzance, and Truro), to the cluster. They are among the largest towns in the data, though there are some large towns like Falmouth that are out of

TABLE 3.12

Iterations in Finding an Anomalous Pattern in Market Towns Data

Iteration	List	#	Distance	Cntr	Cntr (%)
1	26, 29, 35, 44	4	2.98	11.92	28.3
2	4, 9, 25, 26, 29, 35, 41, 44	8	1.85	14.77	35.1

TABLE 3.13

Centroid of the Extracted Pattern of Market Towns

Centroid	P	PS	Do	Ho	Ba	SM	Pe	DIY	SP	PO	CAB	FM
Real	18,484	7.6	3.6	1.1	11.6	4.6	4.1	1.0	1.4	6.4	1.2	0.4
Stded	0.51	0.38	0.56	0.36	0.38	0.38	0.30	0.26	0.44	0.47	0.30	0.18
Over GM (%)	151	152	163	181	170	139	102	350	181	143	94	88
Related Cntr (%)	**167**	147	**154**	81	144	126	84	87	104	**163**	51	10

the list, thus being closer to 0 rather than to St Austell in the range standardized feature space. After one more iteration, the anomalous cluster stabilizes. The iterations are presented in Table 3.12. It should be noted that the scatter's cluster part (contribution) increases along the iterations as follows from the theory in Section 7.2.6: the decrease of the distance between centroid and zero is well compensated by the influx of entities. The final cluster consists of 8 entities and takes into account 35.13% of the data scatter. Its centroid is displayed in Table 3.13. As it frequently happens, the anomalous cluster here consists of better off towns, with all the standardized centroid values larger than the grand mean by 30–50% of the feature ranges. This probably relates to the fact that they comprise eight out of eleven towns which have a resident population greater than 10,000. The other three largest towns have not made it into the cluster because of insufficient services such as Hospitals and Farmers' markets. The fact that the scale of measurement of population is by far the largest in the original table does not affect the computation here because of the range standardized scales at which the total contribution of this feature is mediocre, about 8.5% only (see Table 2.18). It is rather a concerted action of all features associated with greater population which makes up the cluster. As the last line in Table 3.13 shows, the most important for cluster separation are the following features: Population, Post offices, and doctors, highlighted with the boldface. This analysis suggests a simple decision rule predicate to separate the cluster entities from the rest: "P is greater than 10,000 and Do is 3 or greater." ∎

3.2.4 Anomalous Centroids with Method Build

The Build algorithm [91] for selecting initial seeds proceeds in a manner resembling that of the iterated Anomalous pattern, yet the seeds here must belong to the data set while the number K is pre-specified. The computation starts by choosing an analogue to the grand mean among the entities, that is, such $i \in I$ that minimizes the total distance to the rest, $\sum_{j \in I} d(i, j)$, the medoid of set I [91], and takes it as the first seed c_1. Given a set C_m of m initial seeds, c_{m+1} is selected as follows ($1 \leq m < K$). For each $i \in I - C_m$, a cluster S_i is defined to consist of entities j that are closer to i than to C_m. The distance from j to C_m is taken as $D(j, C_m) = \min_{k=1}^{m} d(j, c_k)$, and S_i is defined as the set of all $j \in I - C_m$ for which $E_{ji} = D(j, C_m) - d(i, j) > 0$. The summary value $E_i = \sum_{j \in S_i} E_{ji}$ is used as a decisive characteristic of remoteness of i, along with its pack S_i, from C_m. The next seed c_{m+1} is defined as the most remote from C_m, that is, an entity i for which E_i is maximum over $i \in I - C_m$.

There is a certain similarity between selecting initial centroids in AP and initial seeds with Build. But there are certain differences as well:

1. K must be pre-specified in Build and not necessarily in AP.
2. The central point of the entire set I is taken as an initial seed in Build and is not in AP.
3. Adding a seed is based on different criteria in the two methods.

Example 3.11: Initial Seeds for Colleges Using Algorithm Build

Let us apply Build to the matrix of entity-to-entity distances for Colleges displayed in Table 3.14, which is replicated from Table 3.9.

Let us find three initial seeds. First, we determine that Ef is the medoid of the entire set I, because its total distance to the others, 9.60, is the minimum of total distances in the bottom line of Table 3.22. Thus, Ef is the first initial seed.

Then we build clusters S_i around all other entities $i \in I$. To build S_{So}, we take the distance between So and Ef, 2.20, and see that entities Se, Si, and Et have their distances to So smaller than that, which makes them So's cluster with $E_{So} = 4.06$. Similarly, S_{Se} is set to consist of the same entities,

TABLE 3.14

Distances between Colleges Data from Table 3.2

	So	Se	Si	Et	Ef	En	Ay	An
So	0.00	0.51	0.88	1.15	2.20	2.25	2.30	3.01
Se	0.51	0.00	0.77	1.55	1.82	2.99	1.90	2.41
Si	0.88	0.77	0.00	1.94	1.16	1.84	1.81	2.38
Et	1.15	1.55	1.94	0.00	0.97	0.87	1.22	2.46
Ef	2.20	1.82	1.16	0.97	0.00	0.75	0.83	1.87
En	2.25	2.99	1.84	0.87	0.75	0.00	1.68	3.43
Ay	2.30	1.90	1.81	1.22	0.83	1.68	0.00	0.61
An	3.01	2.41	2.38	2.46	1.87	3.43	0.61	0.00
Total	12.30	12.95	10.78	10.16	9.60	13.81	11.35	16.17

but its summary remoteness $E_{Se} = 2.98$ is smaller. Cluster S_{Si} consists of So and Se with even smaller $E_{Se} = 0.67$. Cluster S_{En} is empty and those of Ay and An contain just the other A college each contributing less than E_{So}. This makes So the next selected seed.

After the set of seeds has been updated by adding So, clusters S_i are to be recomputed on the remaining six entities. Clusters S_{Si} and S_{Et} are empty and the others are singletons. Of them, S_{An} consisting of Ay is the most remote, $E_{An} = 1.87 - 0.61 = 1.26$. This completes the set of initial seeds: Ef, So, and An. Note, these colleges are consistent with the different Subjects, Engineering, Science and Arts.   ∎

3.2.5 Choosing the Number of Clusters at the Post-Processing Stage

In the previous sections, we paid much attention to choosing a good initialization as a pre-processing step. Here, a review is given of attempts to select a right K number via post-processing of K-Means results at many random initializations. A number of different proposals in the literature can be categorized as follows [25]:

a. *Variance-based approach*: using intuitive or model-based functions of K-Means criterion (3.2) which should get extreme or "elbow" values at a correct K.

b. *Structural approach*: comparing within-cluster cohesion versus between-cluster separation at different K.

c. *Combining multiple clusterings*: choosing K according to stability of multiple clustering results at different K.

d. *Resampling approach*: choosing K according to the similarity of clustering results on randomly perturbed or sampled data.

Consider them in turn:

3.2.5.1 Variance-Based Approach

Let us denote the minimum of Equation 3.2 at a specified number of clusters K by W_K. Since W_K cannot be used for the purpose because it monotonically decreases when K increases, there have been other W_K-based indices proposed to estimate the K:

- A Fisher-wise criterion from [21] finds K maximizing $CH = ((T - WK)/(K - 1))/(WK/(N - K))$, where $T = \sum_{i \in I} \sum_{v \in V} y_{iv}^2$ is the data scatter. This criterion showed the best performance in the experiments by Milligan and Cooper [130], and was subsequently utilized by other authors.

- A "rule of thumb" by Hartigan [72] uses the intuition that when there are K^* well-separated clusters, then, for $K < K^*$, an optimal

$(K + 1)$-cluster partition should be a K-cluster partition with one of its clusters split in two, which would drastically decrease W_K because the split parts are well separated. On the contrary, W_K should not much change at $K \geq K^*$. Hartigan's statistic $H = (W_K/W_{K+1} - 1)(N - K - 1)$, where N is the number of entities, is computed while increasing K, so that the very first K at which H decreases to 10 or less is taken as the estimate of K^*. The Hartigan's rule indirectly, via a related criterion from [38], was supported by findings in [130]. Also, it did surprisingly well in experiments of Chiang and Mirkin [25], including the cases of synthetic data generated with overlapping non-spherical clusters.

- Gap statistic [205], a popular recommendation, compares the value of criterion (3.2) with its expectation under the uniform distribution and computes the logarithm of the average ratio of W_K values at the observed and multiple-generated random data. The estimate of K^* is the smallest K at which the difference between these ratios at K and $K + 1$ is greater than its standard deviation.

- Jump statistic [59] uses the average "distortion" of axis, $w_K = W_K/M$, where M is the number of features and W_K the value of K-Means square error criterion at a partition found with K-Means, to calculate "jumps" $j_K = w_K^{-M/2} - w_{K-1}^{-M/2}$ ($K = 1, 2, \ldots$). Defining $w_0 = 0$ and $w_1 = T/M$ where T is the data scatter, jumps can be defined for all reasonable values of $K = 0, 1, \ldots$. The maximum jump would correspond to the right number of clusters. This is supported with a mathematical derivation stating that if the data can be considered a standard sample from a mixture of Gaussian distributions and distances between centroids are great enough, then the maximum jump would indeed occur at K equal to the number of Gaussian components in the mixture [59].

  Experiments conducted by Chiang and Mirkin [25], with generated Gaussian clusters, show that Gap statistic tends to underestimate the number of clusters, whereas Jump statistic tends to do so only at clusters of small spreads, and it overestimates the number of larger spread clusters.

3.2.5.2 Within-Cluster Cohesion versus Between-Cluster Separation

A number of approaches utilize indexes comparing within-cluster distances with between-cluster distances: the greater the difference, the better the fit; many of them are mentioned in [130,141]. A well-balanced coefficient, the Silhouette width involving the difference between the within-cluster tightness and separation from the rest, promoted by Kaufman and Rousseeuw [91] and described in Section 6.2.1, has shown good performance in experiments. The largest average silhouette width, over different K, indicates the best number of clusters.

3.2.5.3 Combining Multiple Clusterings

The idea of combining results of multiple clusterings sporadically emerges here or there. The intuition is that results of applying the same algorithm at different settings or some different algorithms should be more similar to each other at the right K because a "wrong" K introduces more arbitrariness into the process of partitioning; this can be caught with an average characteristic of distances between partitions [104]. A different characteristic of similarity among multiple clusterings is the consensus entity-to-entity matrix whose (i,j)-th entry is the proportion of those clustering runs in which the entities $i, j \in I$ are in the same cluster (see Sections 5.4 and 7.5). This matrix typically is much better structured than the original data. An ideal situation is when the consensus matrix contains 0's and 1's only: this is the case when all the runs lead to the same clustering. This case can be caught by using characteristics of the distribution of consensus indices (the area under the cumulative distribution [156] or the difference between the expected and observed variances of the distribution, which is equal to the average distance between the multiple partitions [25]).

Let us consider the latter statement in more detail according to the analysis in Section 6.5.2.2. The average distance between m partitions equals $ad(K) = \sum_{u,w=1}^{m} M(S^u, S^w)/m^2$, where distance M is defined as the squared Euclidean distance between binary matrices of partitions S^u and S^w. A binary partition matrix is an entity-to-entity similarity matrix; its (i,j)-th entry is 1 if i and j belong to the same cluster, and 0, otherwise, so that consensus matrix $A(K)$ is the average of all m binary partition matrices. Denote the mean and the variance of matrix $A(K)$ by μ_K and σ_K^2, respectively. Then the average distance is proportional to $\mu_K(1 - \mu_K) - \sigma_K^2$, which also shows how close $A(K)$ is to being binary (see Statement 6.1 on page 249). Indeed, when $A(K)$ is binary, $\sigma_K^2 = \mu_K(1 - \mu_K)$, as is well known.

3.2.5.4 Resampling Methods

Resampling can be interpreted as using many randomly produced "copies" of the data for assessing statistical properties of a method in question. Among approaches for producing the random copies are

- Random sub-sampling in the data set
- Random splitting the data set into "training" and "testing" subsets
- Bootstrapping, that is, randomly sampling entities with replacement, usually to their original numbers
- Adding random noise to the data entries

All four have been tried for finding the number of clusters based on the intuition that different copies should lead to more similar results at the right number of clusters (for references, see [141]).

Of these perhaps most popular is approach by Dudoit and Fridlyand [39]. For each K, a number of the following operations is performed: the set is split into non-overlapping training and testing sets, after which both the training and testing part are partitioned in K parts; then a classifier is trained on the training set clusters and applied for predicting clusters on the testing set entities. The predicted partition of the testing set is compared with that found on the testing set by using an index of (dis)similarity between partitions. Then the average or median value of the index is compared with that found at randomly generated data sets: the greater the difference, the better the K. Yet on the same data generator, this setting was outperformed by a model-based statistic as reported in [123].

3.2.5.5 Data Structure or Granularity Level?

Other approaches into the problem probably can be structured along the conventional wisdom that the number of clusters is a matter of both data structure and the user's eye.

On the data side, two lines of development should be mentioned: (i) benchmark data for experimental verification and (ii) models of clusters. The data for experimental comparisons can be taken from real-world applications or generated artificially. There is nothing to report of research in the real data structures yet. As to the case of generated data, this involves choices of

1. Data sizes (for the two-three decades of computer-based developments the sizes grew from a hundred entities [130] to a few thousands by now).
2. Number of clusters (this has grown from 2–4 to a dozen or even couple of dozens currently).
3. Cluster sizes (here the fashion moves from equal-sized clusters to different and even random sizes; it looks that the difference in the numbers of points in clusters does not affect the results).
4. Cluster shapes (mostly spherical and elliptical have been considered so far).
5. Cluster intermix (from well-separated clusters to recognition of that as a most affecting factor).
6. Presence of irrelevant features (see Sections 3.4 and 7.2.7).
7. Data standardization (here the fashion moves from the conventional normalization by the standard deviation to normalization by a range-related value—such a shift should not frustrate anybody because normalizing by the standard deviation may distort the data structure: it downplays the role of features with multimodal distributions, which is counterintuitive as explained above in Section 2.4).

The approach related to modeling clusters is in a rather embryonic state too; a few attempts are referred to in [141].

Regarding the "user's eye" side, that is, the external perspective, one can distinguish between at least these lines of development:

- Cognitive classification structures
- Preferred granularity level
- Additional knowledge of the domain

Of the first of these items, one cannot deny that cluster algorithms do model such structures as typology (K-Means), taxonomy (agglomerative algorithms), or conceptual classification (classification trees). Yet no meaningful formalizations of the classification structures themselves have been developed so far to include, say, "real" types, "ideal" types and social "strata" within the same concept of typology. A deeper line of development should include a dual archetype structure potentially leading to formalization of the process of generation of useful features from the raw data such as sequence motifs or attribute subsets—a matter of future developments going far beyond the scope of this book.

Of the next item, the number of clusters itself characterizes the extent of granularity; the issue is in simultaneously utilizing other characteristics of granularity, such as the minimal size at which a set can be considered a cluster rather than outlier, and the proportion of the data variance taken into account by a cluster—these can be combined within a data recovery and spectral clustering frameworks. Yet there should be other more intuitive characteristics of granularity developed and combined with those above.

The knowledge of domain suggests a formalized view of a most important clustering criterion, quite well known to all practitioners in the field—consistency between clusters and other aspects of the phenomenon in question. Long considered as a purely intuitive matter, this emerges as a powerful device, first of all in bioinformatics studies. Papers such as [37,53] show how the knowledge of biomolecular functions embedded in the so-called Gene Ontology can be used to cut functionally meaningful clusters out of a hierarchical cluster structure. Another approach along the same line will be described in Section 6.5.4. The more knowledge of different aspects of real-world phenomena emerges, the greater importance of the consistency criterion in deciding of the right number of clusters.

3.3 Intelligent K-Means: Iterated Anomalous Pattern

When clusters in the feature space are well separated from each other or the cluster structure can be thought of as a set of differently contributing clusters, the clusters can be found with iterative application of Anomalous pattern that mitigates the need for pre-setting the number of clusters and their initial centroids. Moreover, this can be used as a procedure to meaningfully determine the number of clusters and initial seeds for K-Means. In this way, we come to an algorithm that can be referred to as an intelligent K-Means, because it relieves the user from the task of specifying the initial setting.

Some other potentially useful properties of the method relate to its flexibility with regard to dealing with outliers and the "swamp" of inexpressive, ordinary, entities around the grand mean.

IK-MEANS

0. *Setting.* Put $t = 1$ and I_t the original entity set. Specify a threshold r of resolution to discard all AP clusters whose cardinalities are less than r. The default setting is $r = 2$ so that only singletons are to be discarded here.

1. *Anomalous pattern.* Apply AP to I_t to find S_t and c_t. There can be either option taken: do Step 1 of AP (standardization of the data) at each t or only at $t = 1$. The latter is the recommended option as it is compatible with the theory in Section 7.2.6.

2. *Test.* If Stop-condition (see below) does not hold, put $I_t \leftarrow I_t - S_t$ and $t \leftarrow t + 1$ and go to Step 1.

3. *Removal of the small clusters.* Remove all of the found clusters that are smaller than a pre-specified *cluster discarding threshold* r. (Entities comprising singleton clusters should be checked for the errors in their data entries anyway.) Denote the number of remaining clusters by K and their centroids by $c_1, ..., c_K$.

4. *K-Means.* Do Batch (or Incremental) K-Means with $c_1, ..., c_K$ as the initial seeds.

The Stop-condition in this method can be any or all of the following:

1. *All clustered.* $S_t = I_t$ so that there are no unclustered entities left.

2. *Large cumulative contribution.* The total contribution of the first t clusters to the data scatter has reached a pre-specified threshold such as 50%.

3. *Small cluster contribution.* Contribution of t-th cluster is too small, say, compared to the order of average contribution of a single entity, $1/N$.

The first condition is natural if there are "natural" clusters that indeed differ in their contributions to the data scatter. The second and third conditions can be considered as imposing a finer granularity at the user's viewing of the data.

At step 4, K-Means can be applied to either the entire data set or to the set from which the smaller clusters have been removed. This may depend on the situation: in some problems, such as structuring of a set of settlements for better planning or monitoring, no entity should be left out of the consideration, whereas in other problems, such as developing a summary for text corpora, some weird texts should be left out of the coverage indeed.

If K is pre-specified, the iterated AP can be utilized to initialize K-Means by centroids of the largest K AP clusters.

Example 3.12: Iterated Anomalous Patterns in Market Towns

Applied to the Market towns data with Stop-condition 1, the iterated AP algorithm has produced 12 clusters of which five are singletons. Each of the singletons has a strange pattern of town facilities with no similarity to any other town in the list. For instance, entity 19 (Liskeard, 7044 residents) has an unusually large number of Hospitals (6) and CABs (2), which makes it a singleton cluster.

The seven non-singleton clusters are in Table 3.15, in the order of their extraction in the iterated AP. Centroids of the seven clusters are presented in Table 6.4 in Section 6.3.

The cluster structure does not much change when, according to the iK-Means algorithm, Batch K-Means is applied to the seven centroids (with the five singletons put back into the data). Moreover, similar results have been observed with clustering of the original list of about 1300 Market towns described by an expanded list of 18 characteristics of their development: the number of non-singleton clusters was the same, with their descriptions (see Tables 6.4 and 6.6 on p. 216) very similar.   ■

TABLE 3.15

Iterated AP Market Towns Clusters

Cluster	Size	Elements	Cntr (%)
1	8	4, 9, 25, 26, 29, 35, 41, 44	35.1
3	6	5, 8 , 12, 16, 21, 43	10.0
4	18	2, 6, 7, 10, 13, 14, 17, 22, 23, 24, 27, 30, 31, 33, 34, 37, 38, 40	18.6
5	2	3 , 32	2.4
6	2	1,11	1.6
8	2	39, 42	1.7
11	2	20, 45	1.2

Example 3.13: Intelligent K-Means on the Bribery Data

Let us apply iK-Means to the Bribery data in Table 1.13 on page 21. According to the prescriptions above, the data processing includes the following steps:

1. Data standardization. This is done by subtracting the feature averages (grand means) from all entries and then dividing them by the feature ranges. For a binary feature corresponding to a non-numerical category, this reduces to subtraction of the category proportion, p, from all the entries which in this way become either $1 - p$, for "yes," or $-p$, for "no."

2. Repeatedly performing AP clustering. Applying AP to the pre-processed data matrix with the reference point taken as the space origin 0 and never altered, 13 clusters have been produced as shown in Table 3.16. They explain 64% of the data variance.

3. Initial setting for K-Means. There are only five clusters that have more than three elements according to Table 3.16. This defines the number of clusters as well as the initial setting: the first elements of the five larger clusters, indexed as 5, 12, 4, 1, and 11, are taken as the initial centroids.

4. Performing K-Means. K-Means, with the five centroids from the previous step, produces five clusters presented in Table 3.17. They explain 45% of the data scatter. The reduction of the proportion of the explained data scatter is obviously caused by the reduced number of clusters. ■

TABLE 3.16

Clusters Found by the Iterated AP Algorithm in the Bribery Data

Cluster	Size	Elements	Contribution (%)
1	7	5,16,23,27,28,41,42	9.8
2	1	25	2.2
3	2	17,22	3.3
4	1	49	2.2
5	1	2	2.1
6	1	35	2.1
7	13	12,13,20,33,34,38,39,43 45,47,48,50,51	10.7
8	9	4,6,9,10,21,26,30,31,40	10.2
9	5	1,3,15,29,32	6.3
10	2	7,52	3.3
11	3	8,14,36	3.4
12	8	11,24,37,44,46,53,54,55	7.8
13	2	18,19	2.6

TABLE 3.17

Clusters Found by K-Means in the Entire Bribery Data Set from the Largest Clusters in Table 3.16

Cluster	#	Elements	Contribution (%)
1	8	5,16,23,25,27,28,41,42	10.0
2	19	7,8,12,13,14,20,33,34,35,36,38,39,43,45,47,48,50,51,52	9.8
3	10	4,6,9,10,21,22,26,30,31,40	10.0
4	7	1,3,15,17,29,32,49	7.0
5	11	2,11,18,19,24,37,44,46,53,54,55	8.1

3.4 Minkowski Metric K-Means and Feature Weighting

3.4.1 Minkowski Distance and Minkowski centers

In data analysis, Minkowski distance, or p-metric at $p > 0$, between vectors $x = (x_v)$ and $y = (y_v)$ is defined by formula

$$d(x,y) = \left(\sum_{v \in V} |x_v - y_v|^p \right)^{\frac{1}{p}}$$ (3.7)

Its p power,

$$d^p(x,y) = \sum_{v \in V} |x_v - y_v|^p$$

is equivalently used in Minkowski metric K-Means clustering. Just the distance in Equation 3.3 is changed for Minkowski distance so that K-Means criterion is specified as

$$W_p(S,c) = \sum_{k=1}^{K} \sum_{i \in S_k} d^p(y_i, c_k) = \sum_{k=1}^{K} \sum_{i \in S_k} \sum_{v \in V} |y_{iv} - c_{kv}|^p$$ (3.8)

Obviously, Minkowski K-Means criterion is the conventional K-Means criterion at $p = 2$.

The formula in Equation 3.8 suggests that the same Batch K-Means algorithm is applicable in this case. Just the Minimum distance rule applies with the distance being the p power of Minkowski p-metric rather than the squared Euclidean distance. Similarly, at the centroids update stage, the components of cluster k centroid, c_k, are not necessarily the within cluster averages. These are values, minimizing the Minkowski distances to within-cluster feature values. Specifically, given a series of reals $x_1, x_2, \ldots, x_n$, its Minkowski p-center

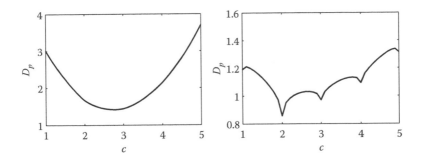

FIGURE 3.2
Graphs of Minkowski $D_p(c)$ function for the series $I = \{2,2,3,4,5,1\}$ at $p = 1.5$, on the left, and $p = 0.5$, on the right.

is c minimizing the summary distance

$$D_p(c) = \sum_{i=1}^{n} |x_i - c|^p \tag{3.9}$$

At $p = 1$, the Minkowski p-center is known to be the median, that is, the mid-term in the sorted list of x-values. At p tending to infinity, the p-center tends to be the mid-range, that is, the middle of the interval between the minimum and maximum of x-values. At $p > 1$ the function $D_p(c)$ is strictly convex, as illustrated on the left graph of Figure 3.2, and always has a unique minimum within the range of x-values. This minimum, the Minkowski p-center can be found with a steepest descent procedure such as described in Section 7.2.7. In contrast, at $p < 1$, the function $D_p(c)$ is strictly concave in the intervals between each consecutive x-values, so that its minimum can be reached only at a series value x_i, $i = 1, \ldots, n$, as illustrated in Figure 3.2, on the right. That is, the Minkowski p-center at $p < 1$ can be found by just taking the minimum of $D_p(x_i)$ for $i = 1, \ldots, n$.

Example 3.14: Minkowski p-Center Function at Different p

Consider a data set of six values $I = \{2,2,3,4,5,1\}$ and function $D_p(c)$ (3.9) at this set. Its graphs at $p = 1.5$ and $p = 0.5$ are presented in Figure 3.2, the left part and right part, respectively. ∎

Example 3.15: Minkowski p-Centers at Different p

Consider the BMI data set on p. 13 as presented in Table 1.8. Its Minkowski p-centers at different p are presented in Table 3.18. A feature of this data set is that the median (second row), the mean (third row) and the mid-range (last row) are in different relationships at different components: the mid-range is larger than the other two over "Height" while it is between

TABLE 3.18

Minkowski centers of BMI Data Set at Different
Values of Power p

Power p	Height (cm)	Weight (kg)
0.5	170.00	75.00
1	170.00	73.50
2	168.91	71.50
3	169.22	71.39
4	169.56	71.53
5	169.82	71.69
10	170.54	72.11
20	170.91	72.30
∞	171.00	72.50

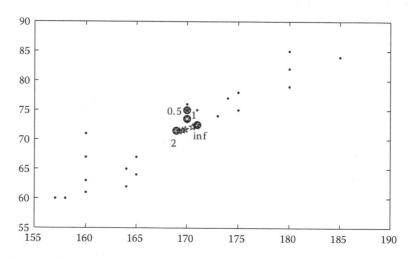

FIGURE 3.3

Minkowski p-centers for the BMI set at different p.

the two over "Weight." These center points are presented on the scatter-
plot of the set by stars of which those labeled by p value are circled, see
Figure 3.3.   ∎

3.4.2 Feature Weighting at Minkowski Metric K-Means

The issue of feature weighting has attracted considerable attention in the
literature since K-Means itself appears helpless at situations in which some
features do reflect a viable cluster structure while the others have nothing to
do with that. K-Means by itself cannot distinguish a relevant feature from an
irrelevant one (see, for example, [4,115] and references therein). However, the

format of Minkowski metric K-Means criterion allows its natural extension involving feature weights and a three-stage version of K-Means iterations. Each iteration of the Minkowski metric K-Means involves a separate optimization of clusters, centroids, and weights, in this sequence. The optimal weights are inversely proportional to the within-cluster variances so that noise features, those distributed randomly, would get zero or almost zero weights. A technical description of the method is given in Section 7.2.7.

Given a standardized data matrix $Y = (y_{iv})$, where $i \in I$ are entities and $v \in V$ features, the non-negative weights w_v, satisfying condition $\sum_{v \in V} w_v = 1$, serve as rescaling factors so that the criterion (3.8) is reformulated over the rescaled data $y'_{iv} = w_v y_{iv}$ and rescaled centroids $c'_{kv} = w_v c_{kv}$:

$$W_p(S, c, w) = \sum_{k=1}^{K} \sum_{i \in S_k} d^p(y'_i, c'_k) = \sum_{k=1}^{K} \sum_{i \in S_k} \sum_{v \in V} w_v^p |y_{iv} - c_{kv}|^p \qquad (3.10)$$

Criterion (3.10) holds to the generic form of K-Means criterion. This is in stark contrast to the other approaches to feature weighting: they change the criterion by supplementary items either added to or factored in the criterion (for a review, see [4]).

The alternating minimization of the criterion (3.10) over three groups of variables, corresponding to S, c, and w, respectively, leads to the following 3-step version of K-Means.

BATCH MINKOWSKI METRIC WEIGHTED

K-MEANS (MWK–MEANS)

0. *Data pre-processing.* Transform data into a standardized quantitative matrix Y. This can be done according to the three step procedure described in Section 2.4.

1. *Initial setting.* Choose the number of clusters, K, and tentative centroids $c_1, c_2, \ldots, c_K$. Assume initial weights all equal to each other, $w_v = 1/|V|$. Assume initial cluster lists S_k empty.

2. *Clusters update.* Given centroids c_k and weights w_v, determine clusters S'_k ($k = 1, \ldots, K$) with the Minimum distance rule applied at p power of Minkowski distance between the rescaled entities and centroids.

3. *Stop-condition.* Check whether $S' = S$. If yes, end with clustering $S = \{S_k\}$, $c = \{c_k\}$, $w = (w_v)$. Otherwise, change S for S'.

4. *Centroids update.* Given clusters S_k and weights w_v, calculate within cluster Minkowski centers $c'_{kv} = w_v c_{kv}$ ($k = 1, \ldots, K$, $v \in V$).

5. *Weights update.* Given clusters S_k and centroids c_k, compute weights using formula:

$$w_v = \frac{1}{\sum_{u \in V} (D_{vp}/D_{up})^{\frac{1}{p-1}}} \tag{3.11}$$

where $D_{vp} = \sum_{k=1}^{K} \sum_{i \in S_k} |y_{iv} - c_{kv}|^p$.

6. Go to 2.

The structure of the criterion and the algorithm suggests that the feature weights can be made cluster-specific with a corresponding adjustment of the formulas in step 5 of the MWK-Means algorithm. In this case, w_v would be defined by a corresponding modification of Equation 3.11 to the within-cluster case:

$$w_{kv} = \frac{1}{\sum_{u \in V} [D_{kvp}/D_{kup}]^{\frac{1}{p-1}}}$$

where k is the cluster's label and $D_{kvp} = \sum_{i \in S_k} |y_{iv} - c_{kv}|^p$.

Example 3.16: Clustering Iris Data Set with Variable Feature Weights

Let us refer to the Iris data set in Table 1.7. This set is characterized by four features (Sepal Length, Sepal Width, Petal Length, Petal Width) and comprises three subsets representing three Iris species, 50 specimens in each. Assuming that the number of clusters, 3, is pre-specified, apply the following intelligent version of MWK-Means: find Anomalous pattern clusters at Minkowski's distance, pick up the three largest of them, and apply Minkowski Weighted K-Means starting from the AP centroids. All the centers, including the center of the entire set are found as Minkowski centers according to the algorithm on p. 287. The best results, six misclassified entities can be found at different p values. Table 3.19 presents the results at $p = 1.3$ and $p = 2$.

One can see that in both versions the Sepal sizes are less important than those of Petal, although the difference between the two is not that large. Of the six misclassified entities, five—71, 78, 107, 120, and 134— are present in both cases, whereas entities 84 and 135 are case specific. Specific conditions at which both of the entities can be correctly clustered may exist.

When the feature weights are cluster specific, as described above, clustering results best matching the Iris taxa are found with iMWK-Means at $p = 1.2$. The five entities above are misclassified again, but all the rest are clustered correctly.

The cluster-specific feature weights [4] are presented in Table 3.20, at which rows correspond to clusters and columns to features, for both

TABLE 3.19

iMWK-Means Results at Different p

	Final Feature Weights				
p-Value	w1	w2	w3	w4	Misclassified Entities
1.3	0.1047	0.1097	0.4403	0.3452	71, 78, 84, 107, 120, 134
2.0	0.1839	0.1835	0.3211	0.3116	71, 78, 107, 120, 134, 135

TABLE 3.20

iMWK-Means Cluster-Specific Feature Weights in the Iris
Data Set at $p = 1.2$

Cluster	Sepal Length	Sepal Width	Petal Length	Petal Width
1	0.0022	0.0182	0.9339	0.0458
2	0.0000	0.0000	0.9913	0.0086
3	0.0000	0.0000	1.0000	0.0000
1	0.0228	0.1490	0.5944	0.2338
2	0.0508	0.0036	0.5898	0.3558
3	0.0233	0.0386	0.4662	0.4719

Note: Those on top are found with a Minkowski metric version of AP
clustering; those on the bottom are final.

anomalous patterns found first and those MWK-Means clusters to which
the process converges.

As one can see from Table 3.20, the AP algorithm relies almost entirely
on feature Petal length that takes all or almost all of the total feature
weight. Yet, in the final version, this is distributed between features of
Petal sizes leaving very little to the Sepal size features. This is well aligned
with the literature which tends to show these features, or their product,
as being the informative ones. Another thing which is worth mentioning
is that the error of five entities misclassified is the absolute minimum
achieved so far in both supervised classification and clustering applied to
Iris data set. ∎

Another example concerns the ability of MWK-Means to suppress noise
features.

**Example 3.17: Clustering a Noisy Version of Colleges Data Set with
Variable Feature Weights**

Consider the Colleges data in Table 3.2. The maximum across the table
is 0.67 and the minimum, −0.63. Let us generate seven random features
whose values are uniformly randomly distributed between these two (see
Table 3.21).

TABLE 3.21

Noise Features to Be Included into the Standardized Colleges Data Set in Table 3.2

Entity	f1	f2	f3	f4	f5	f6	f7
1	−0.079	0.064	−0.357	−0.282	−0.111	0.109	0.241
2	0.003	−0.548	−0.068	−0.367	0.226	0.559	−0.486
3	0.420	0.073	0.592	−0.074	0.121	−0.329	−0.051
4	0.621	0.036	−0.011	0.488	−0.263	−0.259	−0.120
5	0.102	−0.432	0.641	0.256	0.284	0.637	0.317
6	0.228	0.495	0.205	0.006	0.128	−0.422	−0.329
7	0.328	−0.065	0.098	−0.315	0.013	−0.066	−0.614
8	0.388	0.635	0.464	−0.173	−0.346	−0.226	0.061

When applying iMWK-Means to the 8×14 data comprising the College data in Table 3.2 supplemented with the data in Table 3.21, at any p between 1.1 and 3.6 inclusive, the resulting partition at $K = 3$ is always the three subject-based clusters. The noise features get almost zero weights. ∎

An issue of Minkowski clustering methods is in choosing the value of Minkowski exponent p. The value of p leading to the best clustering results is data dependent and can be learnt in a semi-supervised manner. Amorim and Mirkin [4] report of a successful experiment at which p has been learnt by comparing the results of clustering of the entity set with "supervised" cluster labels for 20% of the entities.

3.5 Extensions of K-Means Clustering

3.5.1 Clustering Criteria and Implementation

Traditionally, a cluster is understood as a subset of entities which are similar to each other and dissimilar from the rest. However, in real-world situations additional properties may be required which may complement and sometimes even contradict this view. Consider, for instance, a set of young men with only one variable, their height, measured. Depending on our perspective, we may be interested in the deviate clusters of those who are too short or too tall to carry out regular sailor duties, or rather in the central core cluster of those who normally fit them. The issue of definition of what is a central core cluster becomes less trivial when encountered in the setting of a web-based or other network. Defining what is a deviate cluster can be of issue in specific settings such as dealing with banking customers. Extending the men's height example to a mixed sex group, one might be interested in defining clusters of male and female individuals in terms of probabilistic normal distributions with differing means and variances, which is a traditional abstraction for height measure on humans. A different clustering goal invites a different formalization and different methods for clustering.

In Chapter 1, a number of user-defined goals for clustering have been described. These goals typically cannot be explicitly formalized and, thus, are implicit in clustering methods. The explicit goals are much more technical and down to earth. In particular, criteria optimized by K-Means and Ward methods are such that the found clusters:

1. Consist of entities that are most similar to each other and their centroids.

2. Are the most associated with features involved in the data.

3. Are the most anomalous and thus interesting.

These claims can be supported with materials in Sections 3.1.2, 6.3.4, and 3.3, respectively.

The goals above obviously could be pursued with variously defined concepts of distance (dissimilarity), centroid and correlation. Even within the data recovery approach, different approximation criteria would lead to differently defined distances and centroids within the goal 1 of the list above. If, for example, the quality of the data recovery is scored with the sum of moduli, rather than squares, of differences between the original and recovered data, then the corresponding distance between entities $x = (x_1, \ldots, x_m)$ and $y = (y_1, \ldots, y_M)$ is measured by the sum of moduli, $d = \sum_v |x_v - y_v|$, not the sum of squares. This Minkowski metric, at $p = 1$, is frequently used in data analysis and referred to as Manhattan distance or city-block metric. Within the data recovery approach with the least-moduli criterion, the concept of centroid would slightly change too: it is the medians, not the averages that would populate them! Both Manhattan distance and median-based centroids bring more stability to cluster solutions, especially with respect to outliers. However, the very same stability properties can make least-moduli-based data-recovery clustering less versatile with respect to the presence of mixed scale data [136].

One may use different distance and centroid concepts with computational schemes of K-Means and hierarchical clustering without any strictly defined model framework as, for instance, proposed in [91]. A popular method from that book, PAM (Partitioning around medoids), will be described in the next subsection.

There can be other cluster structures as well. In particular, the following are rather popular:

1. Cluster membership of an entity may not necessarily be confined to one cluster only but shared among several clusters (Fuzzy clustering).

2. Geometrically, cluster shapes may be not necessarily spherical as in the classic K-Means but may have shapes elongated along regression lines (Regression-wise clustering).

3. Data may come from a probabilistic distribution which is a mixture of unimodal distributions that are to be separated to represent different clusters (EM).

4. A visual representation of clusters on a two-dimensional screen can be explicitly embedded in clustering (SOM).

Further on in this section, we present extensions of K-Means clustering techniques that are oriented toward these goals.

3.5.2 Partitioning around Medoids

K-Means centroids are average points rather than individual entities, which may be considered too artificial in some clustering problems in which the user may wish to involve nothing artificial but only genuinely occurring entities. To hold on to this idea, let us change the concept of a cluster's centroid for that of a cluster's medoid. An entity of a cluster S, $i^* \in S$, will be referred to as its medoid if it minimizes the sum of distances to other elements of S, that is, $\sum_{j \in S} d(i^*, j) = \min_{i \in S} \sum_{j \in S} d(i, j)$. The symbol $d(i, j)$ is used here to denote a dissimilarity function, not necessarily squared Euclidean distance, between entities $i, j \in I$.

Having this concept in mind, the method of partitioning around medoids PAM from [91] can be formulated analogously to that of Batch K-Means. Our formulation slightly differs from the formulation in [91], though it is equivalent to the original, to make its resemblance to K-Means more visible.

PARTITIONING AROUND MEDOIDS PAM

1. *Initial setting.* Choose the number of clusters, K, and select K entities $c_1, c_2, \ldots, c_K \in I$, for example, with algorithm Build from Section 3.2.4. Assume initial cluster lists S_k empty.

2. *Clusters update.* Given K medoids $c_k \in I$, determine clusters S'_k ($k = 1, \ldots, K$) with the Minimum distance rule applied to dissimilarity $d(i, j), i, j \in I$.

3. *Stop-condition.* Check whether $S' = S$. If yes, end with clustering $S = \{S_k\}$, $c = (c_k)$. Otherwise, change S for S'.

4. *Medoids update.* Given clusters S_k, determine their medoids c_k ($k = 1, \ldots, K$) and go to Step 2.

TABLE 3.22

Distances between Colleges from Table 3.2

	So	Se	Si	Et	Ef	En	Ay	An
So	0.00	0.51	0.88	1.15	2.20	2.25	2.30	3.01
Se	0.51	0.00	0.77	1.55	1.82	2.99	1.90	2.41
Si	0.88	0.77	0.00	1.94	1.16	1.84	1.81	2.38
Et	1.15	1.55	1.94	0.00	0.97	0.87	1.22	2.46
Ef	2.20	1.82	1.16	0.97	0.00	0.75	0.83	1.87
En	2.25	2.99	1.84	0.87	0.75	0.00	1.68	3.43
Ay	2.30	1.90	1.81	1.22	0.83	1.68	0.00	0.61
An	3.01	2.41	2.38	2.46	1.87	3.43	0.61	0.00
Total	12.30	12.95	10.78	10.16	9.60	13.81	11.35	16.17

Example 3.18: Partitioning around Medoids for Colleges

Let us apply PAM to the matrix of entity-to-entity distances for Colleges displayed in Table 3.22, which is replicated from Table 3.9.

We take the three initial seeds found on p. 106 by using Build algorithm, Ef, So, and An.

With these seeds, the Minimum distance rule produces the subject-based clusters (Step 2). The Stop-condition sends us to Step 4, because these clusters differ from the initial, empty, clusters. At Step 4, clusters' medoids are selected: they are obviously Se in the Science cluster, En in the Engineering cluster and either Ay or An in the Arts cluster. With the set of medoids changed to Se, En, and An, we proceed to Step 2, and again apply the Minimum distance rule, which again leads us to the science-based clusters. This time the Stop-condition at Step 3 halts the process. ∎

3.5.3 Fuzzy Clustering

A fuzzy cluster is represented by its membership function $z = (z_i)$, $i \in I$, in which z_i ($0 \le z_i \le 1$) is interpreted as the degree of membership of entity i to the cluster. This extends the concept of a usual, hard (crisp) cluster, which can be considered a special case of the fuzzy cluster corresponding to membership z_i restricted to only 1 or 0 values. A set of fuzzy clusters $z_k = (z_{ik})$, $k = 1, \ldots, K$, forms a fuzzy partition if the total degree of membership for any entity $i \in I$ is 1, $\sum_k z_{ik} = 1$. This requirement follows the concept of a crisp partition in which any entity belongs to one and only one cluster so that for every i $z_{ik} = 0$ for all k but one. In a fuzzy partition, the full degree of membership of an entity is also 1, but it may be distributed among different clusters. This concept is especially easy to grasp if membership z_{ik} is considered as the probability of belongingness. However, in many cases fuzzy partitions have nothing to

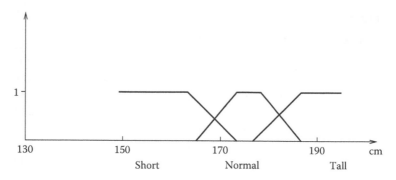

FIGURE 3.4
Fuzzy sets: Membership functions for the concepts of short, normal, and tall in men's height.

do with probabilities. For instance, dividing all people by their height may involve fuzzy categories "short," "medium," and "tall" with fuzzy meanings such as shown in Figure 3.4. Fuzzy clustering can be of interest in applications related with natural fuzziness of the cluster borders such as image analysis, robot planning, geography, and so on.

The following popular fuzzy version of Batch K-Means involves a fuzzy K-class partition and cluster centroids. The fuzzy partition is represented with an $N \times K$ membership matrix (z_{ik}) $(i \in I, k = 1, \ldots K)$ where z_{ik} is the degree of membership of entity i in cluster k satisfying conditions: $0 \leq z_{ik} \leq 1$ and $\sum_{k=1}^{K} z_{ik} = 1$ for every $i \in I$. With these conditions, one may think of the total membership of item i as a unity that can be differently distributed among centroids.

The criterion of goodness of fuzzy clustering is a modified version of the square-error criterion (3.2),

$$W_F(z, c) = \sum_{k=1}^{K} \sum_{i \in I} z_{ik}^{\alpha} d(y_i, c_k) \qquad (3.12)$$

where $\alpha > 1$ is a parameter affecting the shape of the membership function and d distance (2.17) (squared Euclidean distance); as usual, y_i is an entity point and c_k a centroid. In computations, typically, the value of α is put at 2.

Analogously to Batch K-Means, which is an alternating optimization technique, Fuzzy K-Means can be defined as the alternating minimization technique for criterion (3.12). The centroids are actually weighted averages of the entity points, while memberships are related to the distances between entities and centroids. More precisely, given centroids, $c_k = (c_{kv})$, the optimal membership values are determined as

$$z_{ik} = 1 / \sum_{k'=1}^{K} [d(y_i, c_k)/d(y_i, c_k')]^{\frac{2}{\alpha - 1}} \qquad (3.13)$$

Given membership values, centroids are determined as convex combinations of the entity points:

$$c_k = \sum_{i \in I} \lambda_{ik} y_i \tag{3.14}$$

where λ_{ik} is a convex combination coefficient defined as $\lambda_{ik} = z_{ik}^\alpha / \sum_{i' \in I} z_{i'k}^\alpha$. These formulas can be derived from the first-degree optimality conditions for criterion (3.12).

Thus, starting from a set of initial centroids and repeatedly applying formulas (3.13) and (3.14), a computational algorithm has been proven to converge to a local optimum of criterion (3.12).

Further improvements of the approach are reported in [85] and [57]. The Anomalous pattern method is applicable as a tool for initializing Fuzzy K-Means as well as crisp K-Means, leading to reasonable results. Nascimento and Franco [161] applied this method for segmentation of sea surface temperature maps; found fuzzy clusters closely follow the expert-identified regions of the so-called coastal upwelling that are relatively cold, and nutrient rich, water masses. In contrast, the conventional fuzzy K-Means, with user-defined K, under- or over-segments the images.

Criterion (3.12) as it stands cannot be associated with a data recovery model. An attempt to build a criterion fitting into the data recovery approach is made in [162].

3.5.4 Regression-Wise Clustering

In principle, centroids c_k can be defined in a space which is different from that of the entity points y_i. Such is the case of regression-wise clustering illustrated in Figure 3.5. Let us recall that a regression function $x_n = f(x_1, \ldots, x_{n-1})$ may relate a target feature, x_n, to (some of the) other features $x_1, \ldots, x_{n-1}$ as, for instance, the price of a product to its consumer value and production cost attributes. In regression-wise clustering, entities are grouped together according to the degree of their correspondence to a regression function rather than according to their closeness to the gravity center. That means that regression functions play the role of centroids in regression-wise clustering.

Let us consider a version of Batch K-Means for regression-wise clustering to involve linear regression functions relating standardized y_n to other variables, $y_1, \ldots, y_{n-1}$, in each cluster. Such a function is defined by the equation $y_n = a_1 y_1 + a_2 y_2 + \cdots + a_{n-1} y_{n-1} + a_0$ for some coefficients $a_0, a_1, \ldots, a_{n-1}$. These coefficients form a vector, $a = (a_0, a_1, \ldots, a_{n-1})$, which can be referred to as the regression-wise centroid. When a regression-wise centroid is given, its distance to an entity point $y_i = (y_{i1}, \ldots, y_{in})$ is defined as $r(i, a) = (y_{in} - a_1 y_{i1} - a_2 y_{i2} - \cdots - a_{n-1} y_{i,n-1} - a_0)^2$, the squared difference between the observed value of y_n and that calculated from the regression equation. To determine the regression-wise centroid $a(S)$, given a cluster list $S \subset I$, the standard technique

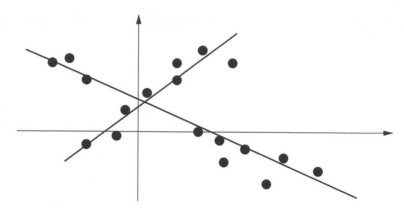

FIGURE 3.5
Regression-wise clusters: solid lines as centroids.

of multivariate linear regression analysis is applied, which is but minimizing the within-cluster summary residual $\sum_{i\in S} r(i, a)$ over all possible a.

Then Batch K-Means can be applied with the only changes being that: (1) centroids must be regression-wise centroids and (2) the entity-to-centroid distance must be $r(i, a)$.

3.5.5 Mixture of Distributions and EM Algorithm

Data of financial transactions or astronomic observations can be considered as a random sample from a (potentially) infinite population. In such cases, the data structure can be analyzed with probabilistic approaches of which arguably the most radical is the mixture of distributions approach.

According to this approach, each of the yet unknown clusters k is modeled by a density function $f(x, a_k)$ which represents a family of density functions over x defined up to a parameter vector a_k. A one-dimensional density function $f(x)$, for any small $dx > 0$, assigns probability $f(x)dx$ to the interval between x and $x + dx$; multidimensional density functions have similar interpretation.

Usually, the density $f(x, a_k)$ is considered unimodal (the mode corresponding to a cluster standard point), such as the normal, or Gaussian, density function defined by its mean vector μ_k and covariance matrix Σ_k:

$$f(x, a_k) = \left(2^M \pi^M |\Sigma_k|\right)^{-1/2} \exp\left\{-(x - \mu_k)^T \Sigma_k^{-1}(x - \mu_k)/2\right\} \tag{3.15}$$

The shape of Gaussian clusters is ellipsoidal because any surface at which $f(x, a_k)$ (3.15) is constant satisfies equation $(x - \mu_k)^T \Sigma_k^{-1}(x - \mu_k) = \text{const}$ defining an ellipsoid. The mean vector μ_k specifies the k-th cluster's location.

The mixture of distributions clustering model can be set as follows. The row points $y_1, \ldots, y_N$ are considered a random sample of $|V|$-dimensional observations from a population with density function $f(x)$ which is a mixture of individual cluster density functions $f(x, a_k)$ $(k = 1, \ldots, K)$ so that $f(x) = \sum_{k=1}^{K} p_k f(x, a_k)$ where $p_k \geq 0$ are the mixture probabilities, $\sum_k p_k = 1$. For $f(x, a_k)$ being the normal density, $a_k = (\mu_k, \Sigma_k)$ where μ_k is the mean and Σ_k the covariance matrix.

To estimate the individual cluster parameters, the main approach of mathematical statistics, the maximum likelihood, is applied. The approach is based on the postulate that really occurred events are those that are most likely. In its simplest version, the approach requires finding of the parameters p_k, a_k, $k = 1, \ldots, K$, by maximizing the logarithm of the likelihood of the observed data under the assumption that the data come from a mixture of distributions:

$$L = \log \left\{ \prod_{i=1}^{N} \sum_{k=1}^{K} p_k f(y_i, a_k) \right\}$$

To computationally handle the maximization problem, this criterion can be reformulated as

$$L = \sum_{i=1}^{N} \sum_{k=1}^{K} g_{ik} \log p_k + \sum_{i=1}^{N} \sum_{k=1}^{K} g_{ik} \log f(y_i, a_k) - \sum_{i=1}^{N} \sum_{k=1}^{K} g_{ik} \log g_{ik} \qquad (3.16)$$

where g_{ik} is the posterior density of class k, defined as

$$g_{ik} = \frac{p_k f(y_i, a_k)}{\sum_k p_k f(y_i, a_k)}$$

In this way, criterion L can be considered a function of two groups of variables: (1) p_k and a_k, and (2) g_{ik}, so that the method of alternating optimization can be applied. The alternating optimization algorithm for this criterion is referred to as the EM-algorithm since computations are performed as a sequence of the so-called Expectation (E) and Maximization (M) steps.

EM-ALGORITHM

Start: With any initial values of the parameters,

E-step: Given p_k, a_k, estimate g_{ik}.

M-step: Given g_{ik}, find p_k, a_k maximizing the log-likelihood function (3.16).

Halt: When the current parameter values approximately coincide with the previous ones.

If f is the Gaussian density function, then the optimal values of parameters, in M-step, can be found using the conventional formulas:

$$\mu_k = \sum_{i=1}^{N} g_{ik} y_i / g_k, \quad \Sigma_k = \sum_{i=1}^{N} g_{ik} (y_i - \mu_k)(y_i - \mu_k)^T / g_k$$

where $g_k = \sum_{i=1}^{N} g_{ik}$.

If the user needs an assignment of the observations to the classes, the posterior probabilities g_{ik} can be utilized: i is assigned to that k for which g_{ik} is the maximum. Also, ratios g_{ik}/g_k can be considered as fuzzy membership values.

The situation, in which all covariance matrices Σ_k are diagonal and have the same variance value σ^2 on the diagonal, corresponds to the assumption that all clusters have the same spherical distributions. This situation is of particular interest because the maximum likelihood criterion here is equivalent to $W(S,c)$ criterion of K-Means and, moreover, there is a certain homology between the EM and Straight K-Means algorithms.

Indeed, under the assumption that feature vectors corresponding to entities $x_1, \ldots, x_N$ are randomly and independently sampled from the population, with unknown assignment of the entities to clusters S_k, the likelihood function in this case has the following formula:

$$L(\{\mu_k, \sigma^2, S_k\}) = A \prod_{k=1}^{K} \prod_{i \in S_k} \sigma^{-M} \exp\{-(x_i - \mu_k)^T \sigma^{-2}(x_i - \mu_k)/2\} \qquad (3.17)$$

so that to maximize its logarithm, the following function is to be minimized:

$$l(\{\mu_k, \sigma^2, S_k\}) = \sum_{k=1}^{K} \sum_{i \in S_k} (x_i - \mu_k)^T (x_i - \mu_k)/\sigma^2 \qquad (3.18)$$

This function is but a theoretic counterpart to K-Means criterion $W(S, \mu) = \sum_{k=1}^{K} \sum_{i \in S_k} d(y_i, \mu_k)$ applied to vectors y_i obtained from x_i with z-scoring standardization (shifting scales to grand means and normalizing them by standard deviations).

Thus, the mixture model can be considered a probabilistic model behind the conventional K-Means method. Moreover, it can handle overlapping clusters of not necessarily spherical shapes (see Figure 3.6). Note, however, that the K-Means data recovery model assumes no restricting hypotheses on the mechanism of data generation. We also have seen how restricting is the requirement of data standardization by z-scoring, associated with the model.

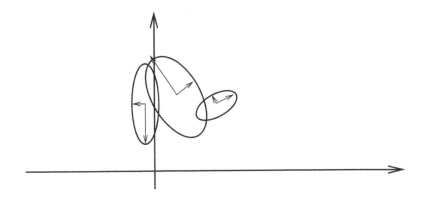

FIGURE 3.6
A system of ellipsoids corresponding to a mixture of three normal distributions with different means and covariance matrices.

Moreover, there are numerous computational issues related to the need in estimating much larger numbers of parameters in the mixture model.

The experimental findings in [25] show that K-Means is applicable at different cluster spreads and elongated, not necessaryly spheric, shapes. This supports the view that K-Means should be considered within the data recovery approach rather than the probabilistic one.

One of successful attempts in application of the mixture of distributions approach is described in [217]. The authors note that there is a tradeoff between the complexity of the probabilistic model and the number of clusters: a more complex model may fit to a smaller number of clusters. To select the better model, one can choose that one which gives the higher value of the likelihood criterion which can be approximately evaluated by the so-called Bayesian Information Criterion (BIC) equal, in this case, to

$$BIC = 2 \log p(X/p_k, a_k) - v_k \log N \tag{3.19}$$

where X is the observed data matrix, v_k, the number of parameters to be fitted and N the number of observations, that is, the rows in X. The BIC analysis has been demonstrated to be useful in accessing the number of clusters.

Further advances in mixture of distributions clustering are described in [8,223].

3.5.6 Kohonen Self-Organizing Maps

The Kohonen SOM is an approach to visualize a data cluster structure by mapping it onto a plane grid [97]. Typically, the grid is rectangular and its size is determined by the user-specified numbers of rows and columns, r and c, respectively, so that there are $r \times c$ nodes on the grid. Each of the grid nodes, e_t $(t = 1, \ldots, rc)$, is one-to-one associated with the so-called model, or reference, vector m_t which is of the same dimension as the entity points $y_i, i \in I$. Initially, vectors m_t are to be set in the data space either randomly or according to an assumption of the data structure such as, for instance, K-Means centroids. Given vectors m_t, entity points y_i are partitioned according to a version of the Minimum distance rule into sets I_t. For each t^*, I_{t^*} consists of those y_i whose distance to m_{t^*} is minimum over all $t = 1, \ldots, rc$.

Besides, a neighborhood structure is assigned to the grid. In a typical case, the neighborhood of node e_t is set E_t of all nodes on the grid whose path distance from e_t on the grid is smaller than a pre-selected threshold value.

Historically, all SOM algorithms worked in an incremental manner as neural networks, but later on, after some theoretical investigation, parallel versions appeared, such as the following.

BATCH SOM

1. *Initial setting.* Select r and c for the grid and initialize model vectors (seeds) m_t ($t = 1, \ldots, rc$) in the entity space.

2. *Neighborhood update.* For each grid node e_t, define its grid neighborhood E_t and collect the list I_u of entities most resembling the model m_u for each $e_u \in E_t$.

3. *Seeds update.* For each node e_t, define new m_t as the average of all entities y_i with $i \in I_u$ for some $e_u \in E_t$.

4. *Stop-condition.* Halt if new m_ts are close to the previous ones (or after a pre-specified number of iterations). Otherwise go to 2.

As one can see, the process much resembles that of Batch K-Means, with the model vectors similar to centroids, except for two items:

1. The number of model vectors is large and has nothing to do with the number of clusters, which are to be determined visually in the end as clusters on the grid.

2. The averaging goes along the grid neighborhood, not that of the entity space.

These features provide for less restricted visual mapping of the data structures to the grid. On the other hand, the interpretation of results here remains more of an intuition rather than instruction. A framework relating SOM and EM approaches is proposed in [18]. One can notice that the popularity of SOM is in a steady decline, probably because it offers visualization without interpretation.

3.6 Overall Assessment

K-Means is a good method because it

1. Models typology building.
2. Computationally effective both in memory and time.
3. Can be utilized incrementally, "on-line."
4. Straightforwardly associates feature salience weights with feature scales. Moreover, in a three-step version it allows for adjustment of the feature weights according to their relation to the clusters being built.
5. Applicable to both quantitative and categorical data and mixed data, provided that care has been taken of the relative feature scaling.
6. Provides a number of interpretation aids including cluster prototypes as well as features and entities most contributing to the cluster specificity, as described in Chapter 6.

K-Means issues:

1. Simple convex spherical shape of clusters.
2. No advice for choosing the number of clusters and locations of the initial seeds.
3. The final clusters are usually close to the initial seeds so that the results much depend on initialization.

The issues above are not necessarily shortcomings but rather properties that could and should be taken care of. To address the issue 1, the feature set should be chosen carefully according to the goals of the analysis. Moreover, the option of cluster-dependent weights such as described at MWK-Means, much abates the issue. To address the issues 2 and 3, the number and location of the initial seeds should be selected based on the conceptual understanding of the substantive domain or preliminary data structure analysis such as that at the AP clustering approach.

There can be some advantages in the issues as well. The issue 3 of solutions being close to the initial centroids may be not as bad when centroids have been conceptually substantiated. The issue 1, the convexity of clusters, provides for a possibility of deriving simple conjunctive descriptions of them (see Section 6.4).

A clustering algorithm should present the user with a comfortable set of options. For example, the iK-Means can be supplemented with options for removal of either (1) "anomalous" or (2) "intermediate" or (3) "trivial" entities. The options can be implemented within the same framework so that: (1) the "deviant" entities would be taken as the contents of small Anomalous pattern clusters; (2) "intermediate" entities can be defined as those that are far away from their cluster centroids; (3) "trivial" entities would be those nearest to the grand mean.

There is a great number of clustering methods that have been and are being developed. In Section 3.5, a number of different clustering methods that can be treated as extensions of K-Means clustering have been presented. These methods are selected because of their popularity. There are rather straight-forward extensions of K-Means among them, such as Partitioning around medoids, in which the concept of centroid is specified to be necessarily an observed entity, and Fuzzy K-Means, in which entity memberships can be spread over several clusters. Less straightforward extensions are represented by the regression-wise clustering, in which centroids are regression functions, and self-organizing maps SOM that combine the Minimum distance rule with visualization of the cluster structure over a grid. The probabilistic mixture of distributions is a far reaching extension, falling within the classical statistics paradigm.

4

Least-Squares Hierarchical Clustering

After reading this chapter, the reader will know about

1. Cluster hierarchies as mathematical structures
2. Ward distance as a least-squares hierarchical clustering criterion
3. Agglomerative and divisive Ward clustering
4. Divisive projection clustering
5. Visualization of hierarchical clusters with heighted tree diagrams and box charts
6. Decomposition of the data scatter involving both Ward and K-Means criteria
7. Contributions of individual splits to: (i) the data scatter, (ii) feature variances and covariances, and (iii) individual entries
8. Extensions of Ward clustering to dissimilarity, similarity and contingency data

Key Concepts

Agglomerative clustering Any method of hierarchical clustering that works bottom up, by merging two nearest clusters at each step.

Aggregation Transformation of a contingency table into a smaller size table by aggregating its row and column categories with summing up corresponding entries. Can be done with Ward clustering extended to contingency tables.

Binary tree A rooted tree in which every internal node has two children.

Box chart A visual representation of a cluster up-hierarchy involving a triple partition of a rectangular box corresponding to each split. The middle part is proportional to the contribution of the split and the other two to contributions of resulting clusters.

Cluster hierarchy A rooted binary tree whose leaves correspond to individual entities, and internal nodes to clusters obtained by either merging of its children or by splitting its parents. The former is part of agglomerative clustering, and the latter, of divisive clustering.

Conceptual clustering Any divisive clustering method that uses only a single feature at each splitting step. The purity and category utility scoring functions are closely related to Ward clustering criterion.

Contribution of a split Part of the data scatter that is explained by a split and equal to the Ward distance between the split parts. Features most contributing to a split can be used in taxonomic analysis. Split contributions to covariances between features and individual entities can also be considered.

Divisive clustering Any method of hierarchical clustering that works from top to bottom, by splitting a cluster in two distant parts, starting from the universal cluster containing all entities.

Heighted tree A visual representation of a cluster hierarchy by a tree diagram in which nodes correspond to clusters and are positioned along a vertical axis in such a way that the height of a parent node is always greater than the heights of its child nodes.

Hierarchical clustering An approach to clustering based on representation of data as a hierarchy of clusters nested over set-theoretic inclusion. Usually, hierarchical clustering is a tool for partitioning, though there are cases, such as that of the evolutionary tree, in which the hierarchy reflects the substance of a phenomenon.

Node cluster For a node in a cluster hierarchy, the set of all entities corresponding to the leaves descending from the node.

Split For an internal node in a cluster hierarchy, a partition of its node cluster in two parts, corresponding to the children node clusters.

Split base vector An N-dimensional vector defined for each internal node of a cluster hierarchy. Its components correspond to entities and take either of three values: 0, for entities outside of the node cluster, a value proportional to N_1 for entities in one split part, and $-N_2$ in the other split part, where N_2 and N_1 are cardinalities of the respective parts.

Ward clustering A method of hierarchical clustering involving Ward distance between clusters. Ward distance is maximized in Ward divisive clustering and minimized in Ward agglomerative clustering. Ward clustering accords with the least squares data recovery approach.

Ward distance A measure of dissimilarity between clusters, equal to the squared Euclidean distance between cluster centroids weighted by the product of cluster sizes related to the sum of cluster sizes. This is equal to the difference between values of K-Means square-error criterion at (i) a partition in which the two clusters are present separately, and (ii) same partition in which the two clusters are merged together.

4.1 Hierarchical Cluster Structures

Hierarchical clustering is devoted to representing data in the form of a hierarchy over the entity set. Sometimes features are also put into the same or separate hierarchy (hierarchical biclustering).

The concept of hierarchy here applies to what is called a *rooted binary tree* and presented as shown in Figure 4.1a. Unlike in Biology, in Computer Science, a hierarchy is drawn in such a way that its root comes on top, and leaves—they are labeled by the entities—on bottom. Accordingly, an interior node such as E in Figure 4.1a has children immediately connected to it from below, leaf 4 and node B in this case, to which it is the parent. The set of all the leaves descending from a node forms its cluster, which is {4, 5, 6} for the node E.

There are two approaches to building a cluster hierarchy:

a. *Agglomerative clustering*: this builds a hierarchy in the bottom-up fashion by starting from smaller clusters and sequentially merging them into "parental nodes," and

b. *Divisive clustering*: this builds a hierarchy top-to-bottom by splitting greater clusters into smaller ones starting from the entire data set. This full set corresponds to the root.

The cluster hierarchy trees are binary to reflect the fact that agglomerative clusters are obtained by merging just two clusters, whereas divisive clusters are split in two parts only.

To have an opportunity to look at the substance of these processes, one needs concepts to express the intermediate results, such as that resulting from an agglomeration process, Figure 4.1c and that resulting from a divisive process, Figure 4.1b. These are referred to, respectively, as a down hierarchy

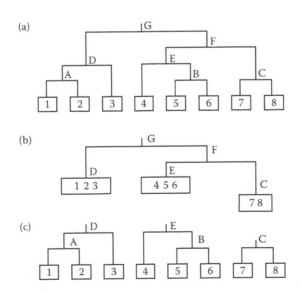

FIGURE 4.1
Binary cluster hierarchies: a full hierarchy (a), an up-hierarchy (b), and down-hierarchy (c).

and an up hierarchy (see later Section 7.3.1). Each interior node of the tree corresponds to a split of the parental cluster and can be represented by a three-valued split base vector. Components of this vector one-to-one correspond to the leaves, that is, entities in the set being hierarchically clustered. For the hierarchy in Figure 4.1a, these vectors are presented by columns of the Table 4.1. To define the split base vector for a node, say E, all the components corresponding to leaves outside of its cluster are set to zero. The cluster's leaves are assigned either of two values each depending on which part of the split a leaf belongs to. The two values are sizes of the split parts, and an element of a split part is assigned the size of the other part. Moreover, all elements of one of the split parts are assigned with a minus sign. This makes the total of the vector to be equal to 0. For the node E, its split parts are {4} and {5,6} of sizes 1 and 2, respectively. Therefore, the former is assigned with +2, and the latter with −1 (see column E in Table 4.1). Analogously, all the elements of split part {1,2,3} of node G are assigned with 5, whereas all elements of its counterpart, {4,5,6,7,8}, are assigned with −3. This arrangement warrants that each of the vectors corresponding to interior nodes is centered: its mean is equal to 0. Moreover, these centered vectors are mutually orthogonal. Taking into account that their number is $N − 1$, the number of interior node in a cluster hierarchy, we come to the conclusion that these vectors form an orthogonal base of the set of all N-dimensional centered vectors. Indeed, the dimension of the space of centered N-dimensional vectors is $N − 1$: it is a subspace of the N-dimensional space of all the N-dimensional vectors that is orthogonally complementary to the one-dimensional subspace of bisector vectors $\alpha(1, 1, \ldots, 1)$ where α is any real. This explains the usage of the term "split base vector."

TABLE 4.1

Split Base Vectors

Node Leaf	A	B	C	D	E	F	G
1	1			1			5
2	−1			1			5
3				−2			5
4					2	2	−3
5		1			−1	2	−3
6		−1			−1	2	−3
7			1			−3	−3
8			−1			−3	−3

Note: Here are unnormed three-valued split base vectors corresponding to the interior nodes of the hierarchy in Figure 4.1. The zeros are not shown, for the sake of convenience.

Example 4.1: Vectors Corresponding to the Interior Nodes of a Hierarchy Are Mutually Orthogonal

Let us demonstrate that all the column-vectors in Table 4.1 are mutually orthogonal, that is, the inner product of any two columns in the table is zero. This is quite obvious for those vectors that correspond to different, non-overlapping, branches of the tree, like say D and E. Since they do not overlap, at least one of the corresponding components in the columns is zero for each of the eight leaves, 1–8, to guarantee that the product is 0, too: $<D, E> = 1 * 0 + 1 * 0 - 2 * 0 + 0 * 2 - 0 * 1 - 0 * 1 + 0 * 0 + 0 * 0 = 0$.

Take now two vectors corresponding to nodes in the same branch of the tree, like F and B. Here is another catch: one node cluster must be part of just one of the split parts in the larger node cluster. In this example, cluster of B, {5,6}, is part of cluster of F, {4,5,6,7,8}, falling within only one of its split parts, E. This implies that all the components of F-vector corresponding to the non-zero components of B-vector are the same! Indeed, in this example, they are all equal to 2. That is, all the nonzero components of B-vector are summed together, under the common factor 2, to produce 0 according to the definition. As mentioned above, the definition of split base vectors implies that each of them sums to 0. In our example, this leads to $< B, F >= 0 * 0 + 0 * 0 + 0 * 0 + 0 * 2 + 1 * 2 - 1 * 2 - 0 * 3 - 0 * 3 = 2 * (1 - 1) = 2 * 0 = 0$. This logics applies to each other split base vector pair, which completes the analysis. ∎

4.2 Agglomeration: Ward Algorithm

The agglomerative approach in clustering builds a cluster hierarchy by merging two clusters at a time, starting from singletons (one-entity clusters) or other pre-specified clusters. Thus, each non-singleton cluster in the hierarchy is the union of two smaller clusters, which can be drawn like a genealogical tree. The smaller clusters are called children of the merged cluster which is referred to as their parent. The singletons are referred to as terminal nodes or leaves, and the combined universal cluster consisting of the entire entity set is referred to as the root of the tree. All other clusters are referred to as nodes of the tree.

The singletons and their successive mergers at every intermediate step form what is called a cluster down-hierarchy, until the root is reached, at which point a full cluster hierarchy emerges.

Besides the tree topology, some metric information is usually also specified: each cluster-node is accompanied with a positive number referred to as its height. A heighted tree is drawn in such a way that each cluster is represented with a node whose height is reflected in its position over the vertical axis. The heights then should satisfy the natural monotonicity requirement: the parent's height is greater than its children's heights.

At each step of an agglomerative clustering algorithm, a set of already formed clusters **S** is considered along with the matrix of distances between maximal clusters. Then two nearest maximal clusters are merged and the newly formed cluster is supplied with its height and distances to other clusters. The process ends, typically, when all clusters have been merged into the universal cluster consisting of the set I of all entities under consideration.

Agglomerative algorithms differ depending on between-cluster distance measures used in them. Especially popular are the so-called single linkage, full linkage and group average criteria. The distance between clusters is defined as the minimum or maximum distance between cluster elements in the single linkage and full linkage methods, respectively. The group average criterion takes the distance between cluster centroids as the between-cluster distance. Quite a broad set of agglomerative algorithms has been defined by Lance and Williams in terms of a formula for dynamic recalculation of the distances between clusters being merged and other clusters into the distances from the merged cluster (see, for instance, [83,109,135]). Lance and Williams formula covers all interesting algorithms proposed in the literature so far and much more.

Here we concentrate on a weighted group average criterion first proposed by Ward [211].

Specifically, for clusters S_{w1} and S_{w2} whose cardinalities are N_{w1} and N_{w2} and centroids c_{w1} and c_{w2}, respectively, Ward distance is defined as

$$dw(S_{w1}, S_{w2}) = \frac{N_{w1}N_{w2}}{N_{w1} + N_{w2}}d(c_{w1}, c_{w2}) \tag{4.1}$$

where $d(c_{w1}, c_{w2})$ is the squared Euclidean distance between c_{w1} and c_{w2}.

To describe the intuition behind this criterion, let us consider a partition S on I and two of its classes S_{w1} and S_{w2} and ask ourselves the following question: how the square error of S, $W(S,c)$, would change if these two classes are merged together?

To answer the question, let us consider the partition that differs from S only in that respect that classes S_{w1} and S_{w2} are changed in it for the union $S_{w1} \cup S_{w2}$ and denote it by $S(w1, w2)$. Note that the combined cluster's centroid can be expressed through centroids of the original classes as $c_{w1 \cup w2} = (N_{w1}c_{w1} + N_{w2}c_{w2})/(N_{w1} + N_{w2})$. Then calculate the difference between the square error criterion values at the two partitions, $W(S(w1, w2), c(w1, w2)) - W(S, c)$, where $c(w1, w2)$ stands for the set of centroids in $S(w1, w2)$. The difference is equal to the Ward distance between S_{w1} and S_{w2}, as proven in Section 7.3:

$$dw(S_{w1}, S_{w2}) = W(S(w1, w2), c(w1, w2)) - W(S, c) \tag{4.2}$$

Because of the additive nature of the square error criterion (3.2), all items on the right of Equation 4.2 are self-subtracted except for those related to S_{w1}

and S_{w2} so that the following equation holds

$$dw(S_{w1}, S_{w2}) = W(S_{w1} \cup S_{w2}, c_{w1 \cup w2}) - W(S_{w1}, c_{w1}) - W(S_{w2}, c_{w2}) \qquad (4.3)$$

where, for any cluster S_k with centroid c_k, $W(S_k, c_k)$ is the summary distance (3.1) from elements of S_k to c_k ($k = w1, w2, w1 \cup w2$).

The latter equation can be rewritten as

$$W(S_{w1} \cup S_{w2}, c_{w1 \cup w2}) = W(S_{w1}, c_{w1}) + W(S_{w2}, c_{w2}) + dw(S_{w1}, S_{w2}) \qquad (4.4)$$

which shows that the summary square error of the merged cluster is the sum of square errors of the original clusters and Ward distance between them.

Since all expressions on the right side in Equation 4.4 are positive, the square error $W(S_{w1} \cup S_{w2}, c_{w1 \cup w2})$ of the merged cluster is always greater than that of either of the constituent clusters, which allows using the cluster square error as the height function in visualizing a cluster hierarchy.

Equation 4.2 justifies the use of Ward distance if one wants to keep the within cluster variance as small as possible at each of the agglomerative steps. The following presents the Ward agglomeration algorithm.

WARD AGGLOMERATIVE ALGORITHM

1. *Initial setting.* The set of maximal clusters is all the singletons, their cardinalities being unity, heights zero, themselves being centroids.

2. *Cluster update.* Two clusters, S_{w1} and S_{w2}, that are closest to each other (being at the minimum Ward distance) among the maximal clusters, are merged together forming their parent cluster $S_{w1 \cup w2} = S_{w1} \cup S_{w2}$. The merged cluster's cardinality is defined as $N_{w1 \cup w2} = N_{w1} + N_{w2}$, centroid as $c_{w1 \cup w2} = (N_{w1}c_{w1} + N_{w2}c_{w2})/N_{w1 \cup w2}$ and its height as $h(w1 \cup w2) = h(w1) + h(w2) + dw(S_{w1}, S_{w2})$.

3. *Distance update.* Put $S_{w1 \cup w2}$ into and remove S_{w1} and S_{w2} from the set of maximal clusters. Define Ward distances between the new cluster $S_{w1 \cup w2}$ and other maximal clusters S_t.

4. *Repeat.* If the number of maximal clusters is larger than 1, go to Step 2. Otherwise, output the cluster merger tree along with leaves labeled by the entities.

Ward agglomeration starts with singletons whose variance is zero and produces an increase in criterion (3.2) that is as small as possible, at each agglomeration step. This justifies the use of Ward agglomeration results by practitioners to get a reasonable initial setting for K-Means. Two

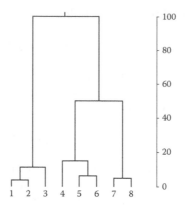

FIGURE 4.2
Cluster tree over Colleges built with Ward clustering algorithm; the node heights are scaled in per cent to the height of the entire entity set.

methods supplement each other in that clusters are carefully built with Ward agglomeration, and K-Means allows overcoming the inflexibility of the agglomeration process over individual entities by reshuffling them. There is an issue with this strategy though: Ward agglomeration, unlike K-Means, is a computationally intensive method, not applicable to large sets of entities.

The height of the new cluster is defined as its square error according to Equation 4.4. Since the heights of merged clusters include the sums of heights of their children, the heights of the nodes grow fast, with an "exponential" speed. This can be used to address the issue of determining what number of clusters is "relevant" to the data by cutting the hierarchical tree at the layer separating long edges from shorter ones if such a layer exists (see, e.g., Figure 4.2 whose three tight clusters can be seen as hanging on longer edges), see [45], pp. 76–77, for heuristic rules on this matter. For an update review, see [141].

Example 4.2: Agglomerative Clustering of Colleges

Let us apply the Ward algorithm to the pre-processed and standard-ized Colleges data in Table 2.13 presented in the right-bottom display of Figure 2.5. The algorithm starts with the matrix of Ward distances between all singletons, that is a matrix of entity-to-entity Euclidean distances squared and divided by 2, as obviously follows from Equation 4.1. The Ward distance matrix is presented in Table 4.2.

Minimum non-diagonal value in the matrix of Table 4.2 is $dw(1,2) = 0.25$ with $dw(7,8) = 0.30$ and $d(5,6) = 0.37$ as the second and third runners-up, respectively. These are to start agglomerations according to Ward algorithm: clusters $\{1,2\}$ $\{7,8\}$, and $\{5,6\}$ whose heights are 0.25, 0.30, and 0.37, respectively, shown in Figure 4.2 as percentages of the data scatter $T(Y) = \sum_{i,v} y_{iv}^2$ which is the height of the maximum cluster comprising all the entities as proven in Section 7.3. Further mergers are also shown in Figure 4.2 with their heights.

TABLE 4.2

Matrix of Ward Distances between Eight Entities in Table 3.2

Entity	1	2	3	4	5	6	7	8
1	0.00	0.25	0.44	0.57	1.10	1.13	1.15	1.51
2	0.25	0.00	0.39	0.78	0.91	1.50	0.95	1.21
3	0.44	0.39	0.00	0.97	0.58	0.92	0.91	1.19
4	0.57	0.78	0.97	0.00	0.49	0.44	0.61	1.23
5	1.10	0.91	0.58	0.49	0.00	0.37	0.41	0.94
6	1.13	1.50	0.92	0.44	0.37	0.00	0.84	1.71
7	1.15	0.95	0.91	0.61	0.41	0.84	0.00	0.30
8	1.51	1.21	1.19	1.23	0.94	1.71	0.30	0.00

The subject classes hold on the tree for about 35% of its height, then the Arts cluster merges with that of Engineering as should be expected from the bottom-right display in Figure 2.5. The hierarchy drastically changes if a different feature scaling system is applied. For instance, with the standard deviation-based standardization, the Arts two colleges do not constitute a single cluster but are separately merged within the other clusters. This does not change even with the follow-up rescaling of categories of Course type by dividing them over $\sqrt{3}$. ∎

4.3 Least-Squares Divisive Clustering

4.3.1 Ward Criterion and Distance

A divisive algorithm builds a cluster hierarchy from top to bottom, each time by splitting a cluster in two, starting from the entire set. Such an algorithm will be referred to as a Ward-like divisive clustering algorithm if the splitting steps maximize Ward distance between split parts. Let us denote a cluster by S_w, its split parts by S_{w1} and S_{w2}, so that $S_w = S_{w1} \cup S_{w2}$ and consider Equation 4.4 which is applicable here: it decomposes the square error, that is, the summary distance between elements and the centroid of S_w into the sum of the square errors of the split parts and the Ward distance between them:

$$W(S_w, c_w) = W(S_{w1}, c_{w1}) + W(S_{w2}, c_{w2}) + dw(c_{w1}, c_{w2}) \qquad (4.5)$$

where the indexed c refers to the centroid of the corresponding cluster.

In the process of divisions, a divisive clustering algorithm builds what is referred to as a cluster up hierarchy, which is a binary tree rooted at the universal cluster I such that its leaves are not necessarily singletons. One may think of a cluster up-hierarchy as a cluster hierarchy halfway through construction from top to bottom, in contrast to cluster down-hierarchies that are cluster hierarchies built halfway through bottom up.

Thus, a Ward-like divisive clustering algorithm goes like this.

WARD-LIKE DIVISIVE CLUSTERING

1. *Start.* Put $S_w \leftarrow I$ and draw the tree root as a node corresponding to S_w at the height of $W(S_w, c_w)$.

2. *Splitting.* Split S_w in two parts, S_{w1} and S_{w2}, to maximize Ward distance $wd(S_{w1}, S_{w2})$.

3. *Drawing attributes.* In the drawing, add two children nodes corresponding to S_{w1} and S_{w2} at the parent node corresponding to S_w, their heights being their square errors.

4. *Cluster set's update.* Set $S_w \leftarrow S_{w'}$ where $S_{w'}$ is the node of maximum height among the leaves of the current up-hierarchy.

5. *Halt.* Check the stopping condition (see below). If it holds, halt and output the hierarchy and interpretation aids described in Section 6.3.5; otherwise, go to 2.

There are two issues of divisive clustering:

i. What cluster to split first and
ii. When to stop the splitting process

that are to be decided when organizing the clustering process.

These are not of an issue with the least squares approach. The cluster and split contributions are to be used for decision making on them. One should try all leaf splits to choose that maximizing the contribution, that is, Ward distance between split parts. As this can be computationally intensive, one may use a proxy criterion, just maximum of leaf cluster contributions. Specifically, that cluster S is to be split first, as a proxy, whose contribution, that is, the total square error value, $W(S, c) = \sum_{i \in S} d(i, c) = \sum_{i \in S} \sum_{v \in V} (y_{iv} - c_v)^2$, where c is centroid of S, is maximum over the candidate clusters.

The stopping condition for the process of divisions should take into account either or all of the following:

1. *The number of final clusters.* The number of terminal nodes (leaves) has reached a pre-specified threshold.

2. *Cluster height.* The height $W(S_w, c_w)$ of S_w has decreased to a pre-specified threshold such as the average contribution of a single entity, $T(Y)/N$, or a pre-specified proportion, say 5%, of the data scatter.

3. *Contribution to data scatter.* The total contribution of the current cluster hierarchy, that is, the sum of Ward distances between split parts in it, has reached a pre-specified threshold such as 50% of the data scatter.

Each of these effectively specifies the number of final clusters. Other approaches to choosing the number of clusters are considered in Section 3.2.5.

It should be noted that the drawn representation of an up-hierarchy may follow the format of the binary rooted tree graph utilized for representing results of an agglomerative method. Or, one may use a different visualization format. In particular, we propose to dwell on the property of cluster hierarchies that all the least-squares contributions are totaling to 100% of the data scatter. Therefore, a divisive cluster up-hierarchy can be represented with a box chart such as in Figure 4.4 on p. 145. At such a box chart each splitting is presented with a partition of a corresponding box in three sections. The section in the middle corresponds to the split itself, whereas those on the right and left correspond to split clusters. The section areas are proportional to their contributions: the split section area follows the Ward distance between the clusters, and areas corresponding to the clusters obtained are proportional to the cluster errors, that is, to the summary distances of the cluster's entities to their centroids, $W(S_w, c_w) = \sum_{i \in S_w} d(y_i, c_w)$, for any $S_w \in \mathbf{S}$.

The box chart concept is similar to that of the pie chart except for the fact that the pie chart sections are of the same type whereas there are two types of section in the box chart, those corresponding to splits and those to split clusters.

Developing a good splitting algorithm at Step 2 of Ward-like divisive clustering is of an issue. Four different procedures are described in Section 7.3.4:

1. Alternating minimization: Bisecting K-Means, or 2-splitting
2. Spectral approach: Principal direction division PDDP
3. One-feature splitting: Conceptual clustering
4. Separation of just one cluster

These four will be covered in the remainder of this section.

4.3.2 Bisecting K-Means: 2-Splitting

This method uses K-Means at $K = 2$ for splitting, and as usual, the most difficult issue here is the initial setting. It is advisable to first find all AP clusters and then use centroids of the two largest of them to start the computation.

Example 4.3: 2-Means Splitting of Colleges from the Farthest Entities

Let us apply the Ward-like divisive clustering method to the Colleges data in Table 2.13 range standardized with the follow-up rescaling the dummy variables corresponding to the three categories of Course type. The method with 2-Means splitting may produce a rather poor structure if the most distant entities, 6 and 8 according to the distance matrix in Table 4.2, would be taken as the initial seeds. In this case, the very first division would produce tentative clusters {1, 3, 4, 5, 6} and {2, 7, 8} because 2 is closer to 8 than to 6, as easily seen in Table 4.2. This partition breaks the subject-based classes. Unfortunately, no further K-Means iterations can change it. Similar results will be obtained by taking 1 and 8 as the initial seeds. This shows how vulnerable can be results found with the rule for initial seed setting at the two farthest entities. ■

4.3.3 Splitting by Separation

This method utilizes the same criterion of maximization of Ward distance between split parts S_1 and S_2 in a cluster S expressed as

$$dw(S_1, S_2) = \frac{N_1}{N_2} d(c_1, c)$$

in Equation 7.55, where c is centroid of S, c_1 of S_1, and N_1, N_2 are cardinalities of S_1 and S_2, respectively. This criterion is somewhat easier to utilize, because it may start from selecting just c_1, to initialize the splitting process. Moreover, it takes into account S_1 only, making the other part just the difference $S_2 = S - S_1$.

Example 4.4: A Tree and Box-Chart for Divisive Clustering of Colleges with Splitting by Separation

Splitting by Separation starting from the entity 8, that is the farthest from the origin, applied to Colleges data, produces the tree presented in Figure 4.3. This tree also leads to the three subject areas, yet it differs from the tree found with the agglomerative Ward method not only in the order of the Subjects (the Arts cluster goes first here) but also in the node heights.

Figure 4.4 illustrates the splitting process in the Ward-like divisive clustering with a box chart. The first split separates Arts two colleges, 7 and 8, from the rest. Contributions are calculated according to the decomposition (4.5). The split itself contributes to the data scatter 34.2%, the Arts cluster 5.1%, and the rest 60.7%, which is reflected in the areas occupied by the vertically split parts in Figure 4.4. The second split (horizontal lines across the right-hold part of the box) produces the Science cluster, with entities 1, 2, and 3, contributing 12.1% to the data scatter, and the Engineering cluster, with entities 4, 5, and 6, contributing 14.5%; the split itself contributes 34.1%. Consider $1/N = 1/8 = 12.5\%$, which is the average contribution of an entity, as a division stopping criterion. Then the process would halt at this point. A box chart in Figure 4.4 illustrates

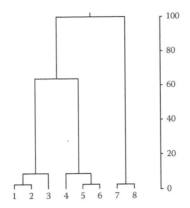

FIGURE 4.3
Cluster tree of Colleges built with Splitting by separation; the node heights are scaled as percentages of the pre-processed data scatter.

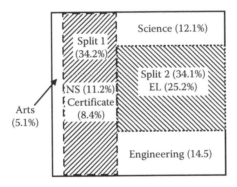

FIGURE 4.4
A box chart to illustrate clustering by separating; splitting has been halted at split contributions to the data scatter that are smaller than the average contribution of an individual entity.

the process. Sections corresponding to splits are shadowed and those corresponding to clusters are left blank. The most contributing features are printed in split sections along with their relative contributions. The thinner the area of a cluster, the closer its elements to the centroid and thus to each other. ∎

Example 4.5: Evolutionary Tree for Gene Profile Data and Mapping Gene Histories

Take agglomerative and divisive Ward clustering, the latter with the Bisecting K-Means, that is, 2-Means splitting, at every step, and apply them to the Gene profiles data in Table 1.3. The goal is to cluster genomes, which are columns of the table. The methods lead to almost identical trees shown in Figure 4.5a and b, respectively; the height of each split reflects its contribution to the data scatter as described in Section 7.3.2. The rows here are features. It should be noted that the first two lines in which all entries are unities do not affect the results of the computation at all, because their contributions to the data scatter are zero. Similarly, in the bacterial cluster appearing after the first split on the right, the next nine COGs (rows 3–11) also become redundant because they have constant values throughout this cluster.

The only difference between the two trees is the position of species *b* within the bacterial cluster *dcrbjqv*: *b* belongs to the left split part in tree (a), and to the right split part in tree (b). The first two splits below the root reflect the divergence of bacteria (the cluster on the right), then eukaryota (the leaf *y*) and archaea. All splits in these trees are compatible with the available biology knowledge. This is not by chance, but because of a targeted selection of the COGs: out of the original 1700 COGs considered in [142], more than one-third are at odds with the major divisions between bacteria, archaea, and eukaryota; probably, because of the extensive loss and horizontal transfer events during the evolution. Because of these and similar processes, the results obtained with a variety of tree-building algorithms on the full data are incompatible with the tree found with more robust data, such as similarities between their ribosomal proteins.

The COGs which contribute the most to the splits seem to be biologically relevant in the sense that they tend to be involved in functional systems and processes that are unique to the corresponding cluster. For

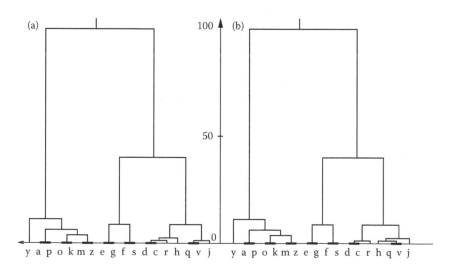

FIGURE 4.5
Evolutionary trees built with: (a) agglomerative Ward algorithm and (b) divisive Ward-like algorithm involving Bisecting K-Means.

instance, COG3073, COG3115, COG3107, and COG2853 make the maximum contribution, of 13% each, to the split of bacteria in both trees. The respective proteins are unique to the bacterial cluster *egfs*; they are either bacteria-specific cell wall components or regulators of expression.

Curiously, the divisive Ward-like algorithm with splitting by Separation produces a different tree in which a subset of bacteria *efgsj* splits off first. In contrast to the Colleges data set analyzed above, the procedure incorrectly determines the starting divergence here. ∎

4.3.4 Principal Direction Partitioning

This method, originally proposed by Boley [19] as a heuristic, can be considered in the framework of the least squares divisive clustering as a method within the spectral approach to clustering. The least-squares splitting criterion in the format of Rayleigh quotient (7.53) leads to the spectral approach as a viable approximation strategy. Indeed, the constraint that a solution vector must be a three-valued split base vector to correspond to a split can be relaxed to admit any vector as a solution. This would lead to the optimal solution being the first principal component of the fragment of the data matrix related to the subset being split. After the principal direction has been found, there can be different approaches to using it for producing the best split. Specifically, all the entities under consideration are sorted in the ascending order of the principal direction vector. Then this order is split in two fragments. Four approaches to splitting described in Section 4.3.4 are: PDDP-Sign, PDDP-Projection, PDDP-RQ, and PDDP-MD. The PDDP-Sign, originally proposed in [19], makes the split parts according to sign of the principal vector components: those positive go to one part and those negative go to the other. This does not necessarily lead to the best approximating split base vector, which is what PDDP-Projection does. The PDDP-RQ version takes that of the splits that maximizes the Rayleigh quotient (7.53). Yet the PDDP-MD approach makes a cut at the deepest minimum of the Parzen's density function built over the principal direction vector [204].

The difference between PDDP-Sign and PDDP-Projection is pinpointed in the next example.

Example 4.6: Sign Is Not Necessarily the Best Projection

Consider a 10-dimensional vector y in Table 4.3. It is centered and normed by range. The PDDP-Sign would lead to a split base vector equal to ϕ_1 in Table 4.3. Yet a split base vector can be found that is nearer to y than ϕ_1—just take a look at ϕ_2 differently dividing the set. Vector ϕ_1 corresponds to partition at which the 10-strong set is divided in clusters of 6 and 4 elements, whereas ϕ_2, to 2 and 8 entities in the split parts. The latter's distance from y is slightly less than that of the former. ∎

TABLE 4.3

Approximate Split Base Vectors

Entity	y	ϕ_1	ϕ_2
1	0.5009	0.2582	0.6325
2	0.3650	0.2582	0.6325
3	0.0974	0.2582	−0.1581
4	0.0708	0.2582	−0.1581
5	0.0198	0.2582	−0.1581
6	0.0059	0.2582	−0.1581
7	−0.1552	−0.3873	−0.1581
8	−0.1634	−0.3873	−0.1581
9	−0.2422	−0.3873	−0.1581
10	−0.4991	−0.3873	−0.1581
Distance from y		0.3893	0.3885

Note: Two different results according to sign and projection approximations of y.

The following example gives another illustration of shortcomings of the PDDP-Sign approach.

Example 4.7: Principal Direction Divisive Partitioning PDDP in Three Versions

Consider the Colleges standardized data set as presented in Table 2.17 copied here as Table 4.4.

The principal direction vector is the first singular vector scoring the entities (see Table 4.5). It has four negative components and four positive ones, thus leading to the corresponding split according to the PDDP-Sign version by [19]. This split is not consistent with the subject areas of the

TABLE 4.4

The Standardized Colleges Data Set Replicated from Table 2.17

	St	Ac	NS	EL	M	B	C
1	−0.20	0.23	−0.33	−0.63	0.36	−0.22	−0.14
2	0.40	0.05	0.00	−0.63	0.36	−0.22	−0.14
3	0.08	0.09	0.00	−0.63	−0.22	0.36	−0.14
4	−0.23	−0.15	−0.33	0.38	0.36	−0.22	−0.14
5	0.19	−0.29	0.00	0.38	−0.22	0.36	−0.14
6	−0.60	−0.42	−0.33	0.38	−0.22	0.36	−0.14
7	0.08	−0.10	0.33	0.38	−0.22	−0.22	0.43
8	0.27	0.58	0.67	0.38	−0.22	−0.22	0.43

TABLE 4.5

PDDP Results for Three Versions

Entity	1	2	3	4	6	5	7	8
PD	−0.518	−0.426	−0.294	−0.005	0.132	0.222	0.409	0.479
Sign	−	−	−	−	+	+	+	+
Projection								
A	0.892	0.459	0.193	0.243	0.378	0.548	0.975	
B	−0.456	−0.456	−0.456	0.274	0.274	0.274	0.274	0.274
C	−	−	−	+	+	+	+	+
RQ								
A	0.907	1.722	2.291	1.956	1.894	2.034	1.462	
B	−0.456	−0.456	−0.456	0.274	0.274	0.274	0.274	0.274
C	−	−	−	+	+	+	+	+

Note: Sign, Projection, and Rayleigh Quotient. For the latter two versions rows are: A—values of the criterion at split fragments (to be minimized at Projection and maximized at RQ); B—the corresponding split vector; C—the corresponding split.

Colleges. In contrast, both Projection and RQ versions lead to the splitting of the three Science colleges from the rest.

The next splits in these also follow the pattern of the hierarchy on Figure 4.2. ∎

Example 4.8: PDDP-MD: Looking at Minima of the Parzen Density

Let us build a Parzen density function for the scoring vector in the first line of Table 4.5. Its standard deviation is $\sigma = 0.378$ and the number of scores is 8, which leads to the Parzen window value mentioned in Section 4.3.4, $h = \sigma(4/(3 * 8))^{1/5} = 0.264$. Then each score z_i gives rise to the function $f(x, z_i) = \exp{-(x - z_i)^2/(2h^2))}/\sqrt{2\pi}$. The density function $f(x)$ found by averaging of all $f(x, z_i)$, $i = 1, 2, \ldots, 8$ is presented in Figure 4.6.

As one can see, the density has no minima at all, though it has an inflection at about −0.10, which does separate the first three, Science, colleges from the rest. Perhaps this result reflects the fact that there are too few entities here to warrant a success: the probability-based concept does need more points to be relevant. ∎

4.3.5 Beating the Noise by Randomness

In spite of the unsuccessful attempt in the previous section, method PDDP-MD overall performs quite well, as experiments reported in [204] clearly demonstrate. To make the method more flexible, Kovaleva and Mirkin [100] proposed extending it to projections of the data set to, not necessarily principal, but rather random directions. The method by Kovaleva and Mirkin [100]

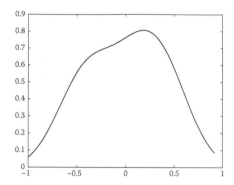

FIGURE 4.6
Graph of the Parzen density function for the scoring vector in the first line of Table 4.5.

takes as many random directions as there are variables in the data set. The split goes along the deepest minimum found. The process stops when there are no mimima at all the leaf clusters.

Table 4.6 reports of some results in comparing the original PDDP-MD method and its random projections version, 2-splitting-MDR. The table reports of an experiment at data sets generated as Gaussian clusters. The number of entities was set to be $N = 1500$, and the number of features, $M = 15$. The number of clusters has been set at three different levels: 9, 15, and 25. The cluster centers have been generated uniformly random within a cube between -2.5 and 2.5. The cluster cardinalities have been kept uniformly random, subject to a constraint that the number of entities in a cluster must be at least

TABLE 4.6

Minimum of Parzen Density Function Results at the Principal and Random Directions, with No Noise Points, on the Left, and with 300 Random Points Added to the Set, on the Right

Number of Clusters	Version	No Noise		20% Noise	
		K	ARI	K	ARI
9	PDDP-MD	12.3	0.96	10.3	0.90
	2-split-MDR	1.5	0.07	11.7	0.89
15	PDDP-MD	16.5	0.68	9.2	0.36
	2-split-MDR	1.2	0.01	14.7	0.85
25	PDDP-MD	9.5	0.13	5.3	0.06
	2-split-MDR	1.0	0.00	11.8	0.35

50 so that the density functions make sense. The covariance matrices have been taken diagonal, with the standard deviations generated uniformly random within interval (0,0.25) so that at larger numbers of clusters they are intermixed, to an extent. After the 1500 data points have been generated, 300 uniformly random points have been generated additionally within the hyper-rectangle defined by the minima and maxima of the original data generation.

At each of the numbers of clusters, a hundred data sets have been generated and several divisive clustering methods, with an innate stopping condition, have been applied to each [100]. The Table presents only two methods: (a) PDDP-MD, a version of the PDDP method utilizing the minimum of the Parzen density function for both choosing the cluster to split and stopping the process (when no within leaf cluster minima have been observed) [204]; (b) 2-split-MDR, Bisecting K-Means with a similar stopping criterion applied to $M = 15$ random projections so that the process halts when a third of projections or more report no minima of the density function in them [100].

The quality of the found cluster hierarchies is evaluated by using two criteria: (i) the number of leaf clusters obtained and (ii) Adjusted Rand index (ARI) value between the generated and found partitions (see Equation 6.17, p. 245). Table 4.6 presents the average performance values; the corresponding standard deviations are omitted because they are rather low.

The table shows that when no noise is present and the number of clusters is relatively small, 15 or less, PDDP-MD works quite well, whereas 2-split-MDR tends to shrink the numbers of clusters. However, in the presence of noise points, 2-split-MDR works better at the larger numbers of clusters.

4.3.6 Gower's Controversy

Gower [63] provided an example demonstrating a peculiarity of Ward distance as a clustering criterion to reflect the fact that factor $N_{w1}N_{w2}/(N_{w1} + N_{w2})$ in Equations 4.1 or 7.50 favors the equal distribution of entities between split parts of a cluster and, thus, the criterion may fail to immediately separate singleton outliers.

To be more specific, let us refer to the following example (see Figure 4.7).

Let c_1, c_2, and c_3 be centroids of groups containing N_1, N_2, and N_3 entities, respectively, and all three located on a straight line so that c_2 lies between c_1 and c_3. The difference in distances would imply that, in a cluster hierarchy, c_1 and c_2 should be merged first with c_3 joining next, or, with splitting, c_3 would diverge first. However, this is not necessarily so with Ward distances.

FIGURE 4.7
Three clusters with respect to Ward clustering: which is the first to go, S_3 or S_1?

Indeed, let the third group be quite small and consist, for instance, just of one entity. Let us further assume that the Euclidean squared distance d between c_2 and c_3 is 20 times as great as between c_2 and c_1. This means that $N_3 = 1$ and $d(c_2, c_3)/d(c_1, c_2) = 20$. Then the ratio of Ward distances $q = dw(c_2, c_3)/dw(c_1, c_2)$ will be equal to $q = 20(N_1 + N_2)/(N_1 N_2 + 1)$. Depending on the value of q either of the mergers can be of favor according to the Ward criterion. If $q \geq 1$ then $dw(c_1, c_2) \leq d(c_2, c_3)$ and the intuitively obvious merger of S_1 and S_2 should go first. If, however, $q < 1$ then $dw(c_2, c_3)$ is smaller and S_2, S_3 are to be joined first. Obviously, q can be <1, for example, when $N_1 = N_2 = 50$, thus leading to first merging the more distant clusters with centroids c_2 and c_3.

Similarly, in Ward-like divisive clustering with this setting, the first split must be done to separate S_1, not the distant cluster S_3, from the rest.

This argumentation perhaps has contributed to the practice of using the unweighted distance between centroids $d(c_{w1}, c_{w2})$ as an ad hoc criterion in hierarchical clustering, the so-called UPGMA method [45,191]. In our view, such a conclusion, however, would probably be an overstretching application of a mathematical property. In fact, with local search algorithms, which are the only ones currently available, the property may not work at all.

Let us consider, for instance, Ward-like divisive clustering with 2-Means splitting. It appears, the controversy does not show up in it. Indeed, one starts with the entities farthest from each other, c_1 and c_3, as initial seeds. Then, with the Minimum distance rule, all points in S_2 are put in the closest cluster S_1 and never removed, as the centroid of the merged cluster $S_1 \cup S_2$ is always closer to c_1 than c_3. This way, 2-Means will produce the intuitively correct separation of the farthest singleton from the rest.

Similarly, in Splitting by separation, c_3 is selected as the initial seed of the cluster to be separated. Then adding an element from the middle cluster will produce d/d' larger than the right-hand side expression in Equation 7.56, thereby halting the splitting process. For instance, with $N_1 = N_2 = 50$ in the example above, $d/d' = 3.89$, while $(N_1 N_2 + N_2)/(N_1 N_2 - N_1) = 2.02$. Actually, the latter becomes larger than the former only after more than half of the entities from the middle cluster have been added to the distant singleton, which is impossible with local search heuristics.

This is an example of the situation in which the square error criterion would have led to a wrong partition if this was not prevented by constraints associated with the local search nature of 2-Means and Splitting by separation procedures.

4.4 Conceptual Clustering

This method of divisive clustering has been proposed by Sonquist et al. [195] back in 60-es, then revitalized by Michalsky [127], but still has not found much popularity. The method builds a cluster up-hierarchy by sequentially splitting clusters, each time using a single attribute rather than all of them as in the previous section. Such is the hierarchy of Digits in Figure 1.5 built by splitting the entire set according to the presence or absence of the bottom edge e7 and then splitting Yes entities according to the presence or absence of e5 and e1. The scoring functions used here are impurity function and category utility function, see p. 297.

Example 4.9: Conceptual Clustering of Digits

The classification tree of the Digit data in Figure 1.5 on p. 16 has been produced with the process above, assuming all the features are binary nominal. Let us take, for instance, partition $S = \{S_1, S_2\}$ of I according to attribute e2 which is present at S_1 comprising 4, 5, 6, 8, 9, and 0, and is absent at S_2 comprising 1, 2, 3, and 7. Cross classification of S and e7 in Table 4.7 yields $\Delta(e7, S) = 0.053$.

To see what this has to do with the setting in which K-Means complementary criterion applies, let us pre-process the Digit data matrix by subtracting the averages within each column (see Table 4.8); note that the scaling coefficients are all unity here.

However, the data in Table 4.8 is not exactly the data matrix Y considered theoretically because both Y and X must have 14 columns after enveloping each of the 14 categories reflected in Table 1.9. Columns corresponding to the category "e1 is absent" in all features $i = 1, 2, \ldots, 7$ are not included in Table 4.8, because they provide no additional information.

The data scatter of this matrix is the summary column variance times $N = 10$, which is 13.1. However, to get the data scatter in the lefthand side of Equation 7.20, this must be doubled to 26.2 to reflect the "missing half" of the virtual data matrix Y.

Let us now calculate within class averages c_{kv} of each of the variables, $k = 1, 2, v = e1, \ldots, e7$, and take contributions $N_k c_{kv}^2$ summed up over clusters S_1 and S_2. This is done in Table 4.9, the last line in which contains contributions of all features to the explained part of the data scatter.

The last item, 0.267, is the contribution of e7. Has it anything to do with the reported value of impurity function $\Delta(e7, S) = 0.053$? Yes, it does. There are two factors that make these two quantities different. First, to get the contribution from Δ it must be multiplied by $N = 10$ leading to $10\Delta(e7, S) = 0.533$. Second, this is the contribution to the data

TABLE 4.7

Cross Tabulation of S (or, e2) against e7

e7	S_1	S_2	Total
e7 $= 1$	5	2	7
e7 $= 0$	1	2	3
Total	6	4	10

TABLE 4.8

Data in Table 1.9 1/0 Coded with the Follow-Up Centring of the Columns

e1	e2	e3	e4	e5	e6	e7
−0.8	−0.6	0.2	−0.7	−0.4	0.1	−0.7
0.2	−0.6	0.2	0.3	0.6	−0.9	0.3
0.2	−0.6	0.2	0.3	−0.4	0.1	0.3
−0.8	0.4	0.2	0.3	−0.4	0.1	−0.7
0.2	0.4	−0.8	0.3	−0.4	0.1	0.3
0.2	0.4	−0.8	0.3	0.6	0.1	0.3
0.2	−0.6	0.2	−0.7	−0.4	0.1	−0.7
0.2	0.4	0.2	0.3	0.6	0.1	0.3
0.2	0.4	0.2	0.3	−0.4	0.1	0.3
0.2	0.4	0.2	−0.7	0.6	0.1	0.3

TABLE 4.9

Feature Contributions to Digit Classes Defined by e2

e2	e1	e2	e3	e4	e5	e6	e7
e2 = 1	0.007	0.960	0.107	0.107	0.060	0.060	0.107
e2 = 0	0.010	1.440	0.160	0.160	0.090	0.090	0.160
Total	0.017	2.400	0.267	0.267	0.150	0.150	0.267

scatter of matrix Y obtained after enveloping all 14 categories which has not been done in Table 4.8, thus, not taken into account in the contribution 0.267. After the contribution is properly doubled, the quantities do coincide.

Similar calculations made for the other six attributes, e1, e2, e3, ..., e6, lead to the total $\sum_{l=1}^{7} \Delta(el, S) = 0.703$ and, thus, to $u(S) = 0.352$ according to Statement 7.5 on p. 297 since $M = 2$. The part of the data scatter taken into account by partition S is the total of $\Delta(S, el)$ over $l = 1, \ldots, 7$ times $N = 10$, according to Equation 7.29, that is, 7.03% or 26.8% of the scatter 26.2.

The evaluations at the first splitting step of the total Digit set actually involve all pairwise contingency coefficients $G^2(l, l') = \Delta(l, l')$ $(l, l' = 1, \ldots, 7)$ displayed in Table 4.10. According to this data, the maximum summary contribution is supplied by the S made according to e7; it is equal to 9.63 which is 36.8% of the total data scatter.

Thus, the first split is to be done according to e7. The second split, according to e5, contributes 3.90, and the third split, according to e1, 3.33, so that the resulting four-class partition, $S = \{\{1, 4, 7\}, \{3, 5, 9\}, \{6, 8, 0\}, \{2\}\}$, contributes $9.63 + 3.90 + 3.33 = 16.87 = 64.4\%$ to the total data scatter. The next partition step would contribute <10% of the data scatter, that is, less than an average entity, which may be considered a signal to stop the splitting process.

TABLE 4.10

Pairwise Contingency Coefficients

Target	e1	e2	e3	e4	e5	e6	e7
e1	0.320	0.003	0.020	0.015	0.053	0.009	0.187
e2	0.005	0.480	0.080	0.061	0.030	0.080	0.061
e3	0.020	0.053	0.320	0.034	0.003	0.009	0.034
e4	0.020	0.053	0.045	0.420	0.003	0.020	0.115
e5	0.080	0.030	0.005	0.004	0.480	0.080	0.137
e6	0.005	0.030	0.005	0.009	0.030	0.180	0.009
e7	0.245	0.053	0.045	0.115	0.120	0.020	0.420
Total	0.695	0.703	0.520	0.658	0.720	0.398	0.963

Note: In each column, e*l* values of Δ for all variables are given under the assumption that partition S is made according to e*l* $(l = 1, \ldots, 7)$.

One should note that the category utility function $u(S)$, see p. 297, after the first split is equal to $9.63/2 = 4.81$, and after the second split, to $(9.63 + 3.90)/3 = 13.53/3 = 4.51$. The decrease means that calculations must be stopped after the very first split, according to the category utility function, which is not an action of our preference. ∎

Example 4.10: Relation between Conceptual and Ward Clustering of Gene Profiles

The divisive tree of species according to gene profiles in Figure 4.5b can be used for analysis of the conceptual clustering category utility score criterion (7.54), which is equivalent to the ratio of the explained part of the data scatter over the number of clusters. Indeed, the first split contributes 50.3% to the data scatter, which makes the category utility function u be equal to $50.3/2 = 25.15$. The next split, of the bacterial cluster, adds 21.2%, making the total contribution $50.3 + 21.2 = 71.5\%$, which decreases the utility function to $u = 71.5/3 = 23.83$. This would force the division process to stop at just two clusters, which shows that the normalizing value K might be overly stringent.

This observation holds not only at the general divisive clustering results in Figure 4.5b but at the conceptual clustering results as well. Why? Because each of the two splits, although found with the multidimensional search, also can be done monothetically, with one feature only: the first split at COG0290 or COG1405 (lines 4, 17 in Table 1.3) and the second one at COG3073 or COG3107 (lines 22, 28 Table 1.3). These splits must be optimal because they have been selected in the much less restrictive multidimensional splitting process. ∎

4.5 Extensions of Ward Clustering

4.5.1 Agglomerative Clustering with Dissimilarity Data

Given a dissimilarity matrix $D = (d_{ij})$, $i, j \in I$, the Ward algorithm can be reformulated as follows.

WARD DISSIMILARITY AGGLOMERATION

1. *Initial setting.* All the entities are considered as singleton clusters so that the between-entity dissimilarities are between clusters. All singleton heights are set to be 0.

2. *Agglomeration rule.* Two candidate clusters, S_{w1} and S_{w2}, that are nearest to each other (that is, being at the minimum Ward distance) are merged together forming their parent cluster $S_{w1 \cup w2} = S_{w1} \cup S_{w2}$, and the merged cluster's height is defined as the sum of the children's heights plus the distance between them.

3. *Distance.* If the newly formed cluster $S_{w1 \cup w2}$ coincides with the entire entity set, go to Step 4. Otherwise, remove S_{w1} and S_{w1} from the set of candidate clusters and define distances between the new cluster $S_{w1 \cup w2}$ and other clusters S_k as follows:

$$dw_{w1 \cup w2, k} = \frac{N_{w1} + N_k}{N^+} dw_{w1, k} + \frac{N_{w2} + N_k}{N^+} dw_{w2, k} - \frac{N_k}{N^+} dw_{w1, w2} \quad (4.6)$$

where $N^+ = N_{w1 \cup w2} + N_k$. The other distances remain unvaried. Then, having the number of candidate clusters reduced by one, go to Step 2.

4. *Output.* Output upper part of the cluster tree according to the height function.

It can be proven that distance (4.6) is equal to Ward distance between the merged cluster and other clusters when D is a matrix of Euclidean squared distances. In fact, Formula (4.6) allows the calculation and update of Ward distances without calculation of cluster centroids.

Agglomeration step 2 remains computationally intensive. However, the computation can be shortened because of properties of the Ward distance [159].

4.5.2 Hierarchical Clustering for Contingency Data

The contingency, or redistribution, data format has been introduced in Section 2.2.3. Due to the fact that contingency data are not only measured in the same scale but also measure different parts of the data flow and thus can be summed to the total number of observations, they can be processed with a greater extent of comparability than the ordinary entity-to-feature data.

To introduce the concepts needed, let us consider a contingency table $P = (p_{tu})$, $t \in T$, $u \in U$, whose entries have been divided by the total flow p_{++}, which means that $p_{++} = 1$. The marginals p_{t+} and p_{+u}, which are just within-row and within-column totals, will be referred to as the weights of rows $t \in T$ and columns $u \in U$.

For any subset of rows $S \subseteq T$, the conditional probability of a column u can be defined as $p(u/S) = p_{Su}/p_{S+}$ where p_{Su} is the sum of frequencies p_{tu} over all $t \in S$ and p_{S+} the summary frequency of rows $t \in S$.

Example 4.11: Aggregating Confusion Data

For the Confusion data in Table 1.10, the matrix of relative frequencies is in Table 4.11. For example, for $S = \{1, 4, 7\}$ and $u = 3$, $p_{S3} = 0.001 + 0.000 + 0.002 = 0.003$ and $p_{S+} = 0.100 + 0.100 + 0.100 = 0.300$ so that $p(3/S) = 0.003/0.300 = 0.010$. Analogously, for $u = 1$, $p_{S1} = 0.088 + 0.015 + 0.027 = 0.130$ and $p(1/S) = 0.130/0.100 = 1.30$. ∎

The row set S will be characterized by its profile, the vector of conditional probabilities $g(S) = (p(u/S))$, $u \in U$. Then, the chi-squared distance between any two non-overlapping row sets, S_1 and S_2, is defined as

$$\chi(g(S_1), g(S_2)) = \sum_{u \in U} (p(u/S_1) - p(u/S_2))^2/p_{+u} \tag{4.7}$$

TABLE 4.11

Confusion Contingency Data

Entity	Feature										Total
	1	2	3	4	5	6	7	8	9	0	
1	0.088	0.001	0.001	0.002	0.000	0.002	0.006	0	0.000	0.000	0.100
2	0.001	0.078	0.005	0.000	0.004	0.005	0.001	0.003	0.001	0.002	0.100
3	0.003	0.003	0.068	0.100	0.002	0.000	0.004	0.003	0.015	0.002	0.100
4	0.015	0.002	0.000	0.073	0.000	0.001	0.003	0.001	0.004	0	0.100
5	0.001	0.003	0.004	0.001	0.067	0.008	0.001	0.001	0.013	0.001	0.100
6	0.002	0.001	0.001	0.001	0.010	0.063	0.000	0.016	0.001	0.004	0.100
7	0.027	0.000	0.002	0.002	0.001	0	0.067	0	0.000	0.001	0.100
8	0.001	0.003	0.003	0.002	0.002	0.007	0.001	0.058	0.007	0.017	0.100
9	0.002	0.003	0.011	0.005	0.008	0.001	0.002	0.008	0.055	0.004	0.100
0	0.002	0.000	0.001	0.001	0.001	0.002	0.002	0.007	0.002	0.082	0.100
Total	0.143	0.095	0.096	0.089	0.094	0.088	0.088	0.096	0.098	0.113	1.000

Using this concept, Ward's agglomeration algorithm applies to contingency data exactly as it has been defined in Section 4.2 except that the Ward distance is modified here to adapt to the situation when both rows and columns are weighted:

$$w(S_{h1}, S_{h2}) = \frac{p_{S_{h1}} + p_{S_{h2}}}{p_{S_{h1} \cup S_{h2}}} \chi(g(S_{h1}), g(S_{h2})) \tag{4.8}$$

This definition differs from the standard definition in the following three aspects:

1. Profiles are taken as cluster centroids.
2. Chi-squared distance is taken instead of the Euclidean squared.
3. Marginal frequencies are used instead of cardinalities.

These formulas are derived in Section 7.4.4 from a data recovery model relating Quetelet coefficients in the original data table and that aggregated according to the clustering. It appears that they express the decrement of the Pearson chi-squared contingency coefficient under the aggregation.

Example 4.12: Clustering and Aggregation of Confusion Data

The drawing in Figure 4.8 shows the hierarchy of Digits row clusters found with the agglomerative clustering algorithm which uses the chi-squared distance (4.7). Curiously, the same topology, with slightly changed heights, emerges when the data table is aggregated over rows and columns simultaneously to minimize the decrement of $\chi^2(F, F)$ of the aggregated table.

The aggregate confusion rates and Quetelet coefficient data corresponding to the four-class partition, $S = \{\{1, 4, 7\}, \{3, 5, 9\}, \{6, 8, 0\}, \{2\}\}$, are on the right in Table 4.12. ∎

4.6 Overall Assessment

Advantages of hierarchical clustering:

1. Visualizes the structure of similarities in a convenient form
2. Models taxonomic classifications
3. Provides a bunch of interpretation aids at the level of entities, variables and variable covariances

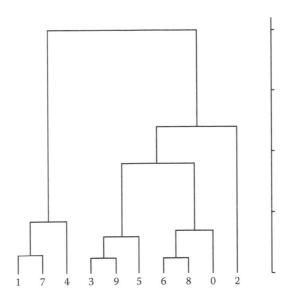

FIGURE 4.8
The hierarchy found with the modified Ward algorithm for Confusion data.

TABLE 4.12

Confusion

I	2827	33	96	44	I	1.95	−0.88	−0.89	−0.95
II	33	783	90	94	II	−0.90	7.29	−0.69	−0.68
III	203	85	2432	280	III	−0.79	−0.70	1.81	−0.69
IV	134	46	263	2557	IV	−0.84	−0.84	−0.70	1.86

Note: Four cluster aggregate Confusion data and corresponding Quetelet coefficients.

Less attractive features of the approach:

1. Massive computations related to finding minimum distances at each step, which is especially time consuming in agglomerative algorithms.
2. Rigidity: Having a splitting or merging step done, no possibility to change it afterwards.

Computations in agglomerative clustering can be drastically reduced if the minima from previous computations are kept.

5

Similarity Clustering: Uniform, Modularity,
Additive, Spectral, Consensus,
and Single Linkage

After reading through this chapter, the reader will know about the most popular similarity clustering approaches including

1. Sources of similarity data including networks and graphs, as well as kernel, or affinity, transformations
2. Summary clustering with background noise subtracted, including
   a. Uniform clustering in which a noise threshold, or the similarity shift, value is subtracted
   b. Modularity clustering in which random interaction noise is subtracted
3. Additive clustering
4. Spectral clustering including Laplacian, and pseudo-inverse Laplacian, transformation of similarity data
5. Consensus clustering including
   a. Consensus ensemble data recovery clustering
   b. Combined consensus data recovery clustering
   c. Concordant partition and a drawback of its criterion
6. Graph-theoretic clustering including connected components, Maximum Spanning Tree and Single Linkage clustering

---

Key Concepts

Additive cluster A subset of the entity set with a positive intensity weight assigned to it. A set of additive clusters generates a similarity matrix so that every two entities are assigned with a similarity score which is equal to the sum of the intensity weights of those clusters that contain both of the entities.

Additive clustering The additive cluster analysis is oriented at solving the following problem: given a similarity matrix, find a small number of additive clusters such that their generated summary similarity matrix is as close as possible to the given matrix.

Affinity data Similarity data obtained from an entity-to-feature data matrix with the so-called affinity, or kernel, function, which usually monotonically increases if the Euclidean distance between entities decreases. Such is Gaussian kernel $G(x, y) = \exp(-d(x, y)/s)$ where $s > 0$ and $d(x, y)$ is the squared Euclidean distance. This function has one more desirable property: the Gaussian kernel similarity matrices are semidefinite positive.

Combined consensus clustering Given a number of partitions on the same entity set, a combined consensus partition is defined according to the principle that its clusters can be predicted from those given ones as accurately as possible.

Concordant partition Given a number of partitions on the same entity set, a combined consensus partition is defined as that minimizing the summary distance to the given partitions. The concordant partition fails a natural test and is not advisable to use.

Connected component Part of a graph or network comprising a maximal subset of vertices that can be connected to each other by a path involving the graph's edges only.

Consensus clustering An approach to clustering that is gaining popularity as a clustering post-processing stage. Given a number of clusterings found with different parameter or algorithm settings, consensus clustering should be as close to the clusterings as possible.

Cut and normalized cut Given a split of a data set in two clusters, the cut is the sum of all the between cluster similarities. The normalized cut relates this to individual cluster volumes. The volume of a cluster is the total of all the similarities of its elements and all the entities. A good split minimizes either cut or normalized cut. At non-negative similarities, the cut criterion leads to a silly, highly unbalanced, split. This, however, can be cured with a transformation of the similarity data, such as the uniform or modularity or Laplacian one.

Distance between partitions This is the number of entity pairs which are discordant between the two given partitions. A pair i, j is discordant if these two fall in a same class in one of the partitions and in different classes in the other partition. The distance is complementary to the popular Rand index designed to score similarity between partitions.

Ensemble consensus clustering Given a number of partitions on the same entity set, an ensemble consensus partition is defined according to the principle that its clusters can be used to predict those given ones as accurately as possible.

Graph A formalization of the concept of network involving two sets: a set of nodes, or vertices, and a set of edges or arcs connecting pairs of nodes.

Laplacian A standardization of the similarity data involving division of its elements by the square roots of the products of the corresponding row and column summary similaritiy values. This allows applying the spectral approach to the normalized cut criterion.

Minimum spanning tree A tree structure within a graph such that its summary edge weights are minimal. This concept is a mathematical backbone for the method of single link clustering.

Modularity transformation A transformation of similarity data to apply the minimum cut criterion. The so-called random interactions, proportional to the products of row and column sums, are subtracted from the similarities to sharpen the "non-random" interactions.

One-by-one additive clustering An approach extending the one-by-one strategy of the principal component analysis to clustering so that an arbitrary eigenvector is changed for a binary cluster membership vector. This approach works well with the additive clustering model since the model extends the spectral decomposition of similarity matrices to the case of binary membership vectors.

Positive semidefinite matrix A square matrix A such that the quadratic form $f^T A f = \sum_{i,j} a_{ij} f_i f_j$ is non-negative for any vector f of the corresponding dimension. An equivalent property is that all the eigenvalues of A are not negative.

Pseudo-inverse Laplacian (Lapin) A similarity data standardization at which the pseudo-inverse matrix transformation is applied to the result of Laplacian matrix transformation. This transformation frequently sharpens the data cluster structure. It is said that Lapin effectively converts the similarly values to those of the currents in a corresponding electric circuit.

Rayleigh quotient Given a matrix A and vector x, Rayleigh quotient is $r(x) = x^T A x / (x^T x)$. The maximum (or minimum) value of $r(x)$ is the maximum (or minimum) eigenvalue of A and x, the corresponding eigenvector.

Similarity data scatter decomposition A decomposition of the similarity data scatter, the sum of squares of all the similarities, in two parts, that explained by a cluster structure obtained and that remaining unexplained. The latter is the sum of squared residuals being minimized in the data recovery approach. The decomposition allows taking into account the contributions of various elements of the cluster structure to the data scatter.

Similarity shift Value subtracted from all the similarity values in a data set. A value a_{ij} made negative with this effectively works against putting i and j together in the same cluster: any within-cluster similarity measure would decrease because of that. Therefore, the similarity shift can be considered a threshold of significance. Yet it works in a soft manner so that a negative a_{ij} impact can be overcome if simultaneously highly positive $a_{ij'}$ values are added.

Single linkage clustering A method in clustering in which the between-cluster distance is defined by the nearest entities. The method is related to the minimum spanning trees and connected components in the corresponding graphs.

Spectral approach to clustering Since many optimization clustering problems can be considered as those of combinatorial optimization of related Rayleigh quotients, the following heuristic can be applied. Let us relax the combinatorial constraints, find an eigenvector maximizing or minimizing the corresponding Rayleigh quotient and then apply the combinatorial constraints to that, to obtain an admissible cluster solution. This idea lies in the heart of the spectral approach.

Threshold graph A graph obtained by removal of all the links that are smaller than a pre-specified threshold. Here the threshold applies in a hard way, in contrast to the soft approach in the uniform clustering.

Uniform transformation Subtraction of a noise threshold, or similarity shift, from the similarities so that insignificant similarities become negative and, in this way, the significant similarities are sharpened.

Uniform clustering Application of the minimum cut, or equivalently, the maximum within-cluster similarity, criterion to similarity data uniformly transformed.

5.1 Summary Similarity Clustering

The very first clustering algorithms developed in Psychology just before the WWII were similarity clustering algorithms, looking for clusters of highly correlated variables in psychological tests. The sum of within-cluster similarities seems a good criterion for clustering—it is simple, intuitive and it satisfies the compactness principle: the greater the within-cluster similarities, the smaller are those between clusters. A drawback of the criterion is that at non-negative similarities its maximum is reached at a silly partition separating the weakest linked entity from the rest. This drawback is easily overcome by removing or subtracting background similarity patterns from the data. Two of the preprocessing options are

a. Subtracting a constant "noise" level [133,135]

b. Subtracting random interactions [164,165]

The summary criterion with the uniform noise subtracted is referred to as uniform clustering criterion, and that with the random interaction noise subtracted is referred to as modularity function.

In the remainder of this section, examples of summary similarity clustering are considered in application to the following three most popular similarity data formats:

i. genuine similarity,

ii. networks or graphs (flat similarity), and

iii. affinity data derived from distances according to an entity-to-feature table.

Each of these formats has its specifics: unpredictable quirks in similarities as raw data in format (i); many zeros and flat positive similarity values in format (ii); and geometric nature of affinities in format (iii).

5.1.1 Summary Similarity Clusters at Genuine Similarity Data

Example 5.1: Uniform Clustering Confusion Data Set

Consider a similarity data set such as Confusion between numerals in Table 1.10. A symmetric version of the Confusion data is presented in Table 5.1: the sum of $A + A^T$ without further dividing it by 2, for the sake of wholeness of the entries. In this table, care has been taken of the main diagonal. The diagonal entries are by far the largest and considerably differ among themselves, which may highly affect further computations. Since we are interested in patterns of confusion between different numerals, this would be an unwanted effect so that the diagonal entries are changed for zeros in Table 5.1.

Consider this matrix uniformly pre-processed by subtracting its average, $m = 66.91$, as presented in Table 5.2, with its rows and columns permuted according to the order of "natural" confusion clusters, $p = 1$–4–7–3–5–9–6–8–0–2.

TABLE 5.1

Confusion Made Symmetric

	2	3	4	5	6	7	8	9	0
1	21	36	171	18	40	329	11	29	22
2		76	26	62	61	18	57	36	22
3			11	61	7	61	57	263	22
4				18	22	51	25	87	11
5					176	14	25	208	21
6						4	225	22	61
7							11	25	32
8								149	243
9									64

Note: Confusion between the segmented numeral digits in Table 1.10 made symmetric by summing with its transpose. Only the upper triangle is presented because of the symmetry and removal of the diagonal.

TABLE 5.2

Confusion Uniform

	4	7	3	5	9	6	8	0	2
1	104.09	262.09	−30.91	−48.91	−37.91	−26.91	−55.91	−44.91	−45.91
4		−15.91	−55.91	−48.91	20.09	−44.91	−41.91	−55.91	−40.91
7			−5.91	−52.91	−41.91	−62.91	−55.91	−34.91	−48.91
3				−5.91	196.09	−59.91	−9.91	−44.91	9.09
5					141.09	109.09	−41.91	−45.91	−4.91
9						−44.91	82.09	−2.91	−30.91
6							158.09	−5.91	−5.91
8								176.09	−9.91
0									−44.91

Note: Symmetric Confusion data after the average $m = 66.91$ is subtracted from all the non-diagonal entries. The rows/columns are reordered according to permutation $p = $ 1–4–7–3–5–9–6–8–0–2.

One can clearly see two clusters with positive within cluster summary similarity, $S_1 = \{1, 4, 7\}$ with the within-cluster similarity sum $a(S_1, S_1) = $ 700.5 and $S_2 = \{3, 5, 6, 8, 9, 0\}$ with $a(S_2, S_2) = 1200.7$. The entity 2 has negative similarities to almost all other entities and makes a cluster of its own, with a 0 within-cluster similarity.

By further increasing the value of the noise threshold, say to $\pi = 100$, one gets the cluster S_2 fragmented in two subclusters, $S2_1 = \{3, 5, 9\}$ and $S2_2 = \{6, 8, 0\}$. The summary similarities on the set of obtained four clusters are presented in Table 5.3. ∎

Example 5.2: Modularity Clustering Confusion Data Set

Consider the Confusion matrix pre-processed by subtracting the random interactions $r_{ij} = a_{i+}a_{+j}/a_{++}$ from elements a_{ij}, where a_{i+} and a_{+j} are the summary similarities of entities i and j with the rest, and a_{++} is the sum of similarities over all non-diagonal elements a_{ij}. This is presented in Table 5.4, with its rows and columns permuted according to the order of "natural" confusion clusters, $p = 1$–4–7–3–5–9–6–8–0–2.

TABLE 5.3

Summary Similarity

	1–4–7	3–5–9	6–8–0	2
1–4–7	502	−601	−722	−235
3–5–9		464	−357	−126
6–8–0			458	−160

Note: Summary similarities between and within the four found Confusion clusters at the similarity shift $a = 100$.

TABLE 5.4

Confusion Modularity

	4	7	3	5	9	6	8	0	2
1	123.56	267.73	−30.78	−49.79	−70.27	−29.48	−79.27	−33.99	−21.61
4		12.81	−30.63	−24.26	25.12	−21.31	−31.27	−23.90	−0.56
7			7.24	−40.57	−54.91	−51.93	−61.67	−13.07	−16.30
3				1.52	175.90	−53.96	−22.21	−27.12	38.62
5					119.58	114.12	−55.41	−28.87	24.05
9						−68.62	31.26	−9.02	−19.57
6							142.59	9.89	22.11
8								176.59	6.46
0									−9.34

Note: Symmetric Confusion data after the random interactions are subtracted from all the non-diagonal entries. The rows/columns are reordered according to permutation $p = 1$–4–7–3–5–9–6–8–0–2.

The modularity matrix manifests two clusters too, $S_1 = \{1, 4, 7\}$ and $\{3, 5, 6, 8, 9, 0, 2\}$, with within-cluster similarities 808.2 and 1137.2, respectively. This differs from the uniform clusters by the fact that entity 2 here is part of the larger cluster. This is because the similarities of 2 with the rest are relatively small so that its random interactions are smaller as well. In contrast to the uniform subtraction of the same threshold value, here the subtracted values reflect the within-row values level and are relatively small, thus leading to positive similarity differences.

Is it possible to get the larger cluster broken in smaller parts under the modularity transformation? Well, yes, if one changes the intensity of random interactions making them a noise, μr_{ij}, from $\mu = 1$ to say $\mu = 1.5$. Then indeed, in matrix $a_{ij} - \mu r_{ij}$ there are three clusters, $\{1, 4, 7\}$, $\{3, 5, 9, 2\}$, and $\{6, 8, 0\}$, all with positive within-cluster similarities while the between cluster similarities are all negative. ∎

5.1.2 Summary Similarity Criterion at Flat Network Data

Example 5.3: Clustering a Flat Network

Consider graph in Figure 2.6b, p. 80, and its similarity matrix

$$A = \begin{pmatrix} 0 & 1 & 1 & 1 & 0 & 0 \\ 1 & 0 & 1 & 0 & 0 & 0 \\ 1 & 1 & 0 & 1 & 0 & 0 \\ 1 & 0 & 1 & 0 & 1 & 0 \\ 0 & 0 & 0 & 0 & 0 & 1 \\ 0 & 0 & 0 & 0 & 1 & 0 \end{pmatrix}$$

Let us pre-process this matrix by subtracting a threshold equal to the average non-diagonal entry, $\bar{a} = 0.467$:

$$A - \bar{a} = \begin{pmatrix} & 2 & 3 & 4 & 5 & 6 \\ 1 & 0.533 & 0.533 & 0.533 & -0.467 & -0.467 \\ 2 & & 0.533 & -0.467 & -0.467 & -0.467 \\ 3 & & & 0.533 & -0.467 & -0.467 \\ 4 & & & & 0.533 & -0.467 \\ 5 & & & & & 0.533 \end{pmatrix}$$

There are two islands of positive entries in the data reflecting two clusters, 1–2–3 and 5–6. The position of node 4 may seem somewhat indefinite because it brings a positive total similarity to either of them. Yet the positive value brought to 1–2–3, $2(0.533 - 0.467 + 0.533)$ is much greater than the sum $2(0.533 - 0.467)$ added to the within-cluster similarity of 5–6. Therefore, the optimal partition is $\{\{1, 2, 3, 4\}, \{5, 6\}\}$. The decision would be easier if the threshold is taken to be $a = 1/2$, the midrange rather than the average. Then the matrix would be:

$$A - a = \begin{pmatrix} & 2 & 3 & 4 & 5 & 6 \\ 1 & 0.5 & 0.5 & 0.5 & -0.5 & -0.5 \\ 2 & & 0.5 & -0.5 & -0.5 & -0.5 \\ 3 & & & 0.5 & -0.5 & -0.5 \\ 4 & & & & 0.5 & -0.5 \\ 5 & & & & & 0.5 \end{pmatrix}$$

The contribution of 4 to 5–6 according to this matrix is zero, thus making 4 to join 1–2–3 more obvious.

Let us consider now the modularity pre-processing for matrix A. The matrix of its random interactions, $r_{ij} = a_{i+} * a_{+j}/a_{++}$ where a_{i+} and a_{+j} are sums of A elements in row i and column j, and a_{++} the total sum, is as follows:

$$R = \begin{pmatrix} 0.64 & 0.43 & 0.64 & 0.64 & 0.43 & 0.21 \\ 0.43 & 0.29 & 0.43 & 0.43 & 0.29 & 0.14 \\ 0.64 & 0.43 & 0.64 & 0.64 & 0.43 & 0.21 \\ 0.64 & 0.43 & 0.64 & 0.64 & 0.43 & 0.21 \\ 0.43 & 0.29 & 0.43 & 0.43 & 0.29 & 0.14 \\ 0.21 & 0.14 & 0.21 & 0.21 & 0.14 & 0.07 \end{pmatrix}$$

One can note once again that the values differ depending on the numbers of interactions in A. By subtracting this from A, one obtains the modularity pre-processed data (with the diagonal entries zeroed):

$$Am = \begin{pmatrix} 0.00 & 0.57 & 0.36 & 0.36 & -0.43 & -0.21 \\ 0.57 & 0.00 & 0.57 & -0.43 & -0.29 & -0.14 \\ 0.36 & 0.57 & 0.00 & 0.36 & -0.43 & -0.21 \\ 0.36 & -0.43 & 0.36 & 0.00 & 0.57 & -0.21 \\ -0.43 & -0.29 & -0.43 & 0.57 & 0.00 & 0.86 \\ -0.21 & -0.14 & -0.21 & -0.21 & 0.86 & 0.00 \end{pmatrix}$$

In this matrix, one can clearly see the positive similarities within clusters 1–2–3 and 5–6, with node 4 positively contributing to either of them. Yet the total similarity of 4 to 5–6, $2(0.57 - 0.21) = 0.72$, is greater here than that to 1–2–3, $2(0.36 - 0.43 + 0.36) = 0.58$, thus determining the optimal clustering to be $\{\{1, 2, 3\}, \{4, 5, 6\}\}$. This seems contradicting the intuition of the structure of graph in Figure 2.6b. The cause is clear: the lesser values of the random interactions between the nodes 4, 5, and 6 subtracted from A. ∎

Example 5.4: Spectral Clusters at a Flat Network

Let us take a look at the eigenvectors corresponding to the maximum eigenvalues of the modularity and uniformly pre-processed similarity matrices. First row in Table 5.5 presents the first eigenvector for the uniform matrix $A - \bar{a}$, and the second and third rows correspond to first eigenvectors of the modularity matrix as is, $A - R$, and with the diagonal zeroed, Am. The vectors support the analysis above: the clusters of the same sign components at $A - \bar{a}$ are 1–2–3–4 and 5–6, whereas the same sign clusters at the modularity data are 1–2–3 and 4–5–6. As one can see the zeroing of the diagonal leads to little effect over the first eigenvector components, which is typical for any similarity data. ∎

The analysis reported in this section shows that the two criteria—or, better to say, the same summary similarity criterion at the two different data pre-processing formulas—should be applied in different contexts: the uniform criterion is better when the meaning of similarity is uniform across the table, whereas the modularity criterion works better when the similarities should be scaled depending on the individual entities.

Example 5.5: Uniform Clustering of Cockroach Graph

Let us take on the Cockroach graph in Figure 2.7; its structure seems a bit difficult for clustering, although one should consider sets 1–2–3, 7–8–9 and the rest as good candidates for clusters. The corresponding similarity

TABLE 5.5

First Eigenvectors

Matrix	1	2	3	4	5	6
$A - \bar{a}$	−0.47	−0.41	−0.47	−0.16	0.39	0.45
$A - R$	0.36	0.42	0.36	−0.07	−0.56	−0.50
Am	0.42	0.44	0.42	−0.06	−0.53	−0.41

Note: First eigenvectors of the uniform and modularity data for the network in Figure 2.6b.

matrix is as follows:

$A =$

	1	2	3	4	5	6	7	8	9	10	11	12
1	0.00	1.00	0.00	0.00	0.00	0.00	0.00	0.00	0.00	0.00	0.00	0.00
2	1.00	0.00	1.00	0.00	0.00	0.00	0.00	0.00	0.00	0.00	0.00	0.00
3	0.00	1.00	0.00	1.00	0.00	0.00	0.00	0.00	0.00	0.00	0.00	0.00
4	0.00	0.00	1.00	0.00	1.00	0.00	0.00	0.00	0.00	1.00	0.00	0.00
5	0.00	0.00	0.00	1.00	0.00	1.00	0.00	0.00	0.00	0.00	1.00	0.00
6	0.00	0.00	0.00	0.00	1.00	0.00	0.00	0.00	0.00	0.00	0.00	1.00
7	0.00	0.00	0.00	0.00	0.00	0.00	0.00	1.00	0.00	0.00	0.00	0.00
8	0.00	0.00	0.00	0.00	0.00	0.00	1.00	0.00	1.00	0.00	0.00	0.00
9	0.00	0.00	0.00	0.00	0.00	0.00	0.00	1.00	0.00	1.00	0.00	0.00
10	0.00	0.00	0.00	1.00	0.00	0.00	0.00	0.00	1.00	0.00	1.00	0.00
11	0.00	0.00	0.00	0.00	1.00	0.00	0.00	0.00	0.00	1.00	0.00	1.00
12	0.00	0.00	0.00	0.00	0.00	1.00	0.00	0.00	0.00	0.00	1.00	0.00

The average of the non-diagonal elements is $\bar{a} = 0.197$. After subtracting that, the unity entry changes for 0.803 and the zero entry, for -0.197. Let us apply the agglomerative clustering to this data. Since all positive entries are the same, one can propose to cluster first neighbors in the order from 1 to 12. That produces six pairs, 1–2, 3–4, 5–6, 7–8, 9–10, and 11–12, as intermediate clusters. The summary similarity matrix between them is

$A1 =$

	1–2	3–4	5–6	7–8	9–10	11–12
1	1.61	0.21	−0.79	−0.79	−0.79	−0.79
2	0.21	1.61	0.21	−0.79	0.21	−0.79
3	−0.79	0.21	1.61	−0.79	−0.79	1.21
4	−0.79	−0.79	−0.79	1.61	0.21	−0.79
5	−0.79	0.21	−0.79	0.21	1.61	0.21
6	−0.79	−0.79	1.21	−0.79	0.21	1.61

There are positive similarities off diagonal in $A1$, which makes further mergers desirable. The greatest similarity, 1.21, is between clusters 5–6 and 11–12. There are positive similarities between 1–2 and 3–4, as well as between 7–8 and 9–10. By combining these, one gets just three clusters, 1–2–3–4, 5–6–11–12, and 7–8–9–10. The summary similarity matrix between them is

$A2 =$

	1–2–3–4	5–6–11–12	7–8–9–10
S_1	3.64	−2.15	−2.15
S_2	−2.15	5.64	−2.15
S_3	−2.15	−2.15	3.64

Since all the off-diagonal entries are negative, further mergers can only decrease the summary criterion and, thus, undesirable.

Yet the current within-cluster summary similarity, $a = 3.64 + 5.64 + 3.64 = 12.92$, can be further increased with hill-climbing moves of individual entities. Specifically, moving 4 and 10 to the cluster in the middle makes the summary matrix equal to

$$A3 = \begin{pmatrix} & 1\text{–}2\text{–}3 & 4\text{–}5\text{–}6\text{–}10\text{–}11\text{–}12 & 7\text{–}8\text{–}9 \\ \hline S_1 & 2.82 & -2.55 & -1.78 \\ S_2 & -2.55 & 8.09 & -2.55 \\ S_3 & -1.78 & -2.55 & 2.82 \end{pmatrix}$$

with the total within-cluster similarity equal to $a = 2.82 + 8.09 + 2.82 = 13.73$, which is the global maximum of the summary similarity criterion. The solution is in accord with the intuition expressed above.

The very same result can be obtained straightforwardly by applying the one-by-one clustering algorithm AddRemAdd(j) from Section 7.4.2. This algorithm starts a cluster from singleton $\{j\}$ and then adds entities one-by-one, each time selecting the maximum increase of the criterion. Starting from $j = 1$, this would lead to cluster 1–2–3. Then, applying it at $j = 4$, cluster 4–5–6–10–11–12 is found. Cluster 7–8–9 is obtained at the next step. ∎

Example 5.6: Modularity Criterion on Cockroach Graph

Consider Cockroach similarity matrix A transformed by subtracting background random interactions and the diagonal zeroed:

$Am =$

	1	2	3	4	5	6	7	8	9	10	11	12
1	0.00	0.92	-0.08	-0.12	-0.12	-0.08	-0.04	-0.08	-0.08	-0.12	-0.12	-0.08
2	0.92	0.00	0.85	-0.23	-0.23	-0.15	-0.08	-0.15	-0.15	-0.23	-0.23	-0.15
3	-0.08	0.85	0.00	0.77	-0.23	-0.15	-0.08	-0.15	-0.15	-0.23	-0.23	-0.15
4	-0.12	-0.23	0.77	0.00	0.65	-0.23	-0.12	-0.23	-0.23	0.65	-0.35	-0.23
5	-0.12	-0.23	-0.23	0.65	0.00	0.77	-0.12	-0.23	-0.23	-0.35	0.65	-0.23
6	-0.08	-0.15	-0.15	-0.23	0.77	0.00	-0.08	-0.15	-0.15	-0.23	-0.23	0.85
7	-0.04	-0.08	-0.08	-0.12	-0.12	-0.08	0.00	0.92	-0.08	-0.12	-0.12	-0.08
8	-0.08	-0.15	-0.15	-0.23	-0.23	-0.15	0.92	0.00	0.85	-0.23	-0.23	-0.15
9	-0.08	-0.15	-0.15	-0.23	-0.23	-0.15	-0.08	0.85	0.00	0.77	-0.23	-0.15
10	-0.12	-0.23	-0.23	0.65	-0.35	-0.23	-0.12	-0.23	0.77	0.00	0.65	-0.23
11	-0.12	-0.23	-0.23	-0.35	0.65	-0.23	-0.12	-0.23	-0.23	0.65	0.00	0.77
12	-0.08	-0.15	-0.15	-0.23	-0.23	0.85	-0.08	-0.15	-0.15	-0.23	0.77	0.00

In contrast to the uniform transformation, this matrix leads to an unequivocal choice of agglomerations: the maximum similarity of 0.92 is reached at pairs 1–2 and 7–8, after which 6 and 12 merge with the second best 0.86 similarity. Then, at the 9-cluster partition, an ambiguity emerges. The maximum summary similarity 0.77 links entity 3 with both cluster 1–2 and singleton 4, and entity 9 with both cluster 7–8 and entity 10. Following the rule that the winner of a tie is the minimum label, we merge the entities with the clusters to form clusters 1–2–3 and 7–8–9. The summary similarity matrix for the 7-cluster partition is

$$Am7 = \begin{pmatrix} & 1\text{–}2\text{–}3 & 7\text{–}8\text{–}9 & 6\text{–}12 & 4 & 5 & 10 & 11 \\ \hline 1\text{–}2\text{–}3 & 3.38 & -0.96 & -0.77 & 0.42 & -0.58 & -0.58 & -0.58 \\ 7\text{–}8\text{–}9 & -0.96 & 3.38 & -0.77 & -0.58 & -0.58 & 0.42 & -0.58 \\ 6\text{–}12 & -0.77 & -0.77 & 1.69 & -0.46 & 0.54 & -0.46 & 0.54 \\ 4 & 0.42 & -0.58 & -0.46 & 0.00 & 0.65 & 0.65 & -0.35 \\ 5 & -0.58 & -0.58 & 0.54 & 0.65 & 0.00 & -0.35 & 0.65 \\ 10 & -0.58 & 0.42 & -0.46 & 0.65 & -0.35 & 0.00 & 0.65 \\ 11 & -0.58 & -0.58 & 0.54 & -0.35 & 0.65 & 0.65 & 0.00 \end{pmatrix}$$

The maximum between-cluster similarity here is 0.65 so that one can merge together 4 and 5, as well as 10 and 11. The 5-cluster partition will show that the summary similarities between clusters 6–12, 4–5, and 10–11 are all positive, thus leading to the same three-cluster partition, as with the uniform criterion, with $S1 = \{1, 2, 3\}$, $S2 = \{7, 8, 9\}$, and $S3 = \{4, 5, 6, 10, 11, 12\}$. The summary similarity matrix at this partition is

$$Am3 = \begin{pmatrix} & S_1 & S_2 & S_3 \\ \hline S_1 & 3.38 & -0.96 & -2.08 \\ S_2 & -0.96 & 3.38 & -2.08 \\ S_3 & -2.08 & -2.08 & 5.85 \end{pmatrix}$$

with the summary within cluster similarity 12.62.

Yet the partition found with the agglomeration process is not optimal. A better partition comprises four clusters, $S_1 = \{5, 6, 11, 12\}$, $S_2 = \{1, 2, 3\}$, $S_3 = \{7, 8, 9\}$, and $S_4 = \{4, 10\}$, as found by using the one-by-one clustering algorithm AddRemAdd from p. 312. Its summary similarity matrix is

$$Am4 = \begin{pmatrix} & S_1 & S_2 & S_3 & S_4 \\ \hline S_1 & 5.15 & -1.92 & -1.92 & -0.31 \\ S_2 & -1.92 & 3.38 & -0.96 & -0.15 \\ S_3 & -1.92 & -0.96 & 3.38 & -0.15 \\ S_4 & -0.30 & -0.15 & -0.15 & 1.31 \end{pmatrix}$$

with a somewhat larger value of the summary within cluster similarity, 13.23.

This counterintuitive separation of the intermediate pair 4–10 in the Cockroach graph as an individual cluster under the modularity criterion once again conforms to the view that the uniform normalization is more suitable for structuring flat networks. ∎

5.1.3 Summary Similarity Clustering at Affinity Data

Affinity data are similarities between entities in an entity-to-feature table. They are usually defined by a kernel function depending on entity-to-entity distances such as a Gaussian kernel function $G(x, y) = \exp\{-qd(x, y)\}$ where $d(x, y)$ is the squared Euclidean distance between x and y if $x \neq y$. For $x = y$, $G(x, y)$ is set to be zero. The coefficient q characterizes so-called width of the Gaussian. It may greatly affect results and is subject to the user's choice. According to the advice in Section 2.5.5, let us follow the following rule of thumb: take $q = 18/M$ where M is the number of variables.

Example 5.7: Summary Similarity Clusters at Affinity Data for Colleges

The affinity data for eight entities in Colleges data table (range normalized with the last three columns further divided by $\sqrt{3}$, see Section 2.4) form the following matrix A:

$$A = \begin{pmatrix}
0.00 & 0.27 & 0.10 & 0.05 & 0.00 & 0.00 & 0.00 & 0.00 \\
0.27 & 0.00 & 0.14 & 0.02 & 0.01 & 0.00 & 0.01 & 0.00 \\
0.10 & 0.14 & 0.00 & 0.01 & 0.05 & 0.01 & 0.01 & 0.00 \\
0.05 & 0.02 & 0.01 & 0.00 & 0.08 & 0.11 & 0.04 & 0.00 \\
0.00 & 0.01 & 0.05 & 0.08 & 0.00 & 0.15 & 0.12 & 0.01 \\
0.00 & 0.00 & 0.01 & 0.11 & 0.15 & 0.00 & 0.01 & 0.00 \\
0.00 & 0.01 & 0.01 & 0.04 & 0.12 & 0.01 & 0.00 & 0.21 \\
0.00 & 0.00 & 0.00 & 0.00 & 0.01 & 0.00 & 0.21 & 0.00
\end{pmatrix}$$

The diagonal entries are made 0 according to the definition of affinity data.

The average off-diagonal entry value is 0.0509. The corresponding uniform matrix is

$$A - \bar{a} = \begin{pmatrix}
0.00 & 0.22 & 0.05 & 0.00 & -0.05 & -0.05 & -0.05 & -0.05 \\
0.22 & 0.00 & 0.09 & -0.03 & -0.04 & -0.05 & -0.04 & -0.05 \\
0.05 & 0.09 & 0.00 & -0.04 & 0.00 & -0.04 & -0.04 & -0.05 \\
0.00 & -0.03 & -0.04 & 0.00 & 0.03 & 0.06 & -0.01 & -0.05 \\
-0.05 & -0.04 & 0.00 & 0.03 & 0.00 & 0.10 & 0.07 & -0.04 \\
-0.05 & -0.05 & -0.04 & 0.06 & 0.10 & 0.00 & -0.04 & -0.05 \\
-0.05 & -0.04 & -0.04 & -0.01 & 0.07 & -0.04 & 0.00 & 0.16 \\
-0.05 & -0.05 & -0.05 & -0.05 & -0.04 & -0.05 & 0.16 & 0.00
\end{pmatrix}$$

That clearly shows the structure of the three clusters corresponding to the subjects: Science, Engineering, and Arts—all the positive entries are within these clusters and all the negative entries are between them.

A similar structure can be as easily seen at the modularity transformed matrix (the similarity matrix after subtraction of the random interactions background):

$$Am = \begin{pmatrix}
0.00 & 0.20 & 0.06 & 0.00 & -0.06 & -0.04 & -0.06 & -0.03 \\
0.20 & 0.00 & 0.09 & -0.03 & -0.06 & -0.04 & -0.06 & -0.03 \\
0.06 & 0.09 & 0.00 & -0.03 & 0.00 & -0.02 & -0.04 & -0.02 \\
0.00 & -0.03 & -0.03 & 0.00 & 0.04 & 0.08 & -0.00 & -0.02 \\
-0.06 & -0.06 & 0.00 & 0.04 & 0.00 & 0.11 & 0.06 & -0.03 \\
-0.04 & -0.04 & -0.02 & 0.08 & 0.11 & 0.00 & -0.03 & -0.02 \\
-0.06 & -0.06 & -0.04 & -0.00 & 0.06 & -0.03 & 0.00 & 0.18 \\
-0.03 & -0.03 & -0.02 & -0.02 & -0.03 & -0.02 & 0.18 & 0.00
\end{pmatrix}$$

Here, however, there is a positive entry, $(5,7)$, between the clusters; a few others are too small at this level of resolution. Yet applying the agglomerative approach, the initial mergers will go according to maxima, 0.20 to merge 1 and 2, 0.18 to merge 7 and 8, and 0.11 to merge 5 and 6. After this, cluster 5–6 cannot merge with cluster 7–8 because the positive link 0.06 between 5 and 7 cannot overcome the other three negative links, −0.03, −0.03, and −0.02 totaling to −0.08. Therefore the same three clusters will make the optimal three-cluster partition. ∎

5.2 Normalized Cut and Spectral Clustering

In this section, another popular approach is presented—the normalized cut, a heuristic criterion that can be considered within the data recovery perspective after the original similarity data is preprocessed with the Lapin transformation. The normalized cut approach works in a divisive manner by splitting sets in two parts to minimize the similarity between them as measured by the normalized cut criterion in Equation 7.64 on p. 305. This criterion can be equivalently represented by the normalized tightness criterion. The latter sums individual cluster normalized tightness scores. The normalized tightness score is the ratio of the summary within-cluster similarities to the cluster's volume, that is the summary similarity of the cluster with all the entities.

> **Example 5.8: Normalized Cut and Normalized Tightness**
>
> Consider the affinity similarity matrix A for Colleges on p. 173 and compute the normalized cut and tightnesses for a two-cluster partition, the first comprising entities 1, 2, and 3, $S_1 = \{1, 2, 3\}$, and the second the rest, $S_2 = \{4, 5, 6, 7, 8\}$. The within cluster sums are $a1 = 1.03$ and $a2 = 1.47$, respectively. The cut, that is, the summary similarity between S_1 and S_2, is $a12 = 0.18$. The volumes, that is, the summary similarities of clusters with all, are $v1 = 1.21$ and $v2 = 1.64$. Therefore, the normalized within-cluster tightness values are $d1 = a1/v1 = 0.85$ and $d2 = a2/v2 = 0.89$. The normalized cut is $a12/v1 + a12/v2 = 0.15 + 0.11 = 0.26$. The three together sum to 2, $0.85 + 0.89 + 0.26 = 2$, as proven in Section 7.4.1. Similar computations for two other subject clusters generating splits: (i) $R_1 = \{4, 5, 6\}$ and R_2 comprising the rest, and (ii) $T_1 = \{7, 8\}$ and T_2 comprising the rest, lead to normalized cuts equal to $0.34 + 0.19 = 0.52$ and $0.33 + 0.10 = 0.43$, respectively. Therefore, of the three clusters, the Science-based cluster makes the best split versus the rest, according to the normalized cut criterion. ∎

The Laplace transformation provides for shifting the focus from a complex combinatorial optimization problem to the better developed spectral decomposition problem. To define the Laplace transformation for a similarity matrix $A = (a_{ij})$, one has to compute the row and column totals, $a_{i+} = \sum_j a_{ij}$ and $a_{+j} = \sum_i a_{ij}$. These are the same if A is symmetric, that is, $a_{i+} = a_{+i}$, as is the case of the affinity matrix A between the Colleges on p. 173. Then each entry in A is to be related to the square root of the product, $r_{ij} = \sqrt{a_{i+} a_{+j}}$. Thus obtained matrix $M = (m_{ij})$ where $m_{ij} = (a_{ij})/(\sqrt{a_{i+} a_{+j}})$, then is subtracted from the diagonal identity matrix $E = (e_{ij})$ such that $e_{ij} = 1$ if $i = j$ and $e_{ij} = 0$, otherwise. The result is the (normalized) Laplacian matrix L, which is semi-positive definite so that all its eigenvalues are non-negative reals.

Finding the optimal normalized cut is akin to finding the eigenvector of L corresponding to the minimum non-zero eigenvalue of L because the optimal cut minimizes the Rayleigh quotient of L, $z^T L z / z^T z$ at specific binary vectors z, as shown in Section 7.4.1. That means that by minimizing the quotient over arbitrary vectors z, one may approximate the optimal normalized cut.

To change the framework to a more convenient setting—finding the maximum, rather than minimum, eigenvalue—the Laplacian Transformation can be further modified. Authors of [151] propose using for this a pseudo-inverse L^+ of L which has the same eigenvectors but the corresponding eigenvalues are reciprocals of non-zero eigenvalues of L.

Example 5.9: Laplace, Lapin, and Spectral Clustering for the Affinity College Data

The upper triangle of matrix in Table 5.6 represents matrix $M = L + E$ complementary to Laplacian matrix L, for the affinity similarity matrix A between Colleges on p. 173. The lower triangle in the table represents the pseudo-inverse Laplacian matrix, that is shortened to Lapin.

The Lapin transformation is performed by finding all the eigenvalues and eigenvectors of matrix L (see Table 5.7), then removing those of zero value and taking reciprocals to the remaining eigenvalues. These are combined according to formula $L^+ = Z_- \Lambda_-^{-1} Z_-^T$ explained in Section 7.4.1.

Although the pattern of three subject clusters can be clearly seen in both of the matrices, M and L^+ in Table 5.6, it seems sharper in the Lapin matrix L^+. A nice feature of this matrix is that no background noise needs to be subtracted—somehow it has been done already within the Lapin transformation itself. Obviously, any similarity clustering criterion described so far will arrive at this partition as the best output, although even the Lapin matrix is not spotless in this regard: entity 5 has positive Lapin-similarities to both 7 and 8 forming a different cluster. But this is not that important here. According to the spectral approach, it is the eigenvector corresponding to the minimum non-zero eigenvalue highlighted with the bold font in Table 5.7 that determines the best split. And this is blameless to separate the first three positive components from the negative others, corresponding to the Science-versus-the-rest split. After the first split is

TABLE 5.6

Laplace and Lapin

	1	2	3	4	5	6	7	8
1	0.00	0.61	0.28	0.14	0.01	0.01	0.01	0.00
2	0.81	0.00	0.36	0.05	0.02	0.00	0.02	0.01
3	0.42	0.49	0.00	0.02	0.14	0.03	0.03	0.01
4	−0.31	−0.38	−0.30	0.00	0.23	0.36	0.12	0.01
5	−0.61	−0.63	−0.37	0.12	0.00	0.43	0.29	0.03
6	−0.50	−0.54	−0.36	0.27	0.35	0.00	0.04	0.00
7	−0.84	−0.85	−0.60	−0.08	0.17	−0.04	0.00	0.70
8	−0.70	−0.71	−0.51	−0.14	0.03	−0.11	0.94	0.00

Note: Upper triangle represents matrix $M = E - L$, the lower triangle is L^+, pseudo-inverse Laplacian.

TABLE 5.7

Deriving Lapin

	1	2	3	4	5	6	7	8
	0.39	0.46	−0.12	−0.31	0.04	0.67	0.26	−0.08
	0.40	0.48	−0.16	−0.09	−0.05	−0.71	0.26	0.06
	0.34	0.31	−0.07	0.60	0.05	0.11	−0.63	0.13
	0.33	−0.11	0.38	−0.57	−0.11	−0.12	−0.55	−0.27
	0.38	−0.24	0.34	0.45	−0.32	0.03	0.35	−0.49
	0.31	−0.19	0.53	0.00	0.19	0.04	0.19	0.72
	0.38	−0.45	−0.39	0.00	0.68	−0.05	0.02	−0.19
	0.28	−0.40	−0.51	−0.10	−0.61	0.06	−0.06	0.33
λ	0.00	**0.22**	0.50	1.12	1.76	1.63	1.34	1.44
1-λ	1.00	**0.78**	0.50	−0.12	−0.76	−0.63	−0.34	−0.44
1/λ		**4.58**	2.02	0.90	0.57	0.61	0.75	0.69

Note: The eight eigenvectors of matrices L, M, and L^+, and corresponding eigenvalues λ of L, along with their complements to 1 (eigenvalues of M) and
reciprocals (eigenvalues of L^+); those of interest are highlighted in bold.

identified, one should try to see whether the split parts can be clustered in
the same way. Specifically, the Laplacian transformation of the similarity
matrix over entities 4, 5, 6, 7, and 8 can be found. The matrices M and
L^+ are as presented below (M taking the upper triangle and L^+ the lower
triangle):

$$M/L^+ = \begin{pmatrix} & 4 & 5 & 6 & 7 & 8 \\ 4 & 0 & 0.28 & 0.43 & 0.15 & 0.01 \\ 5 & -0.03 & 0 & 0.48 & 0.32 & 0.03 \\ 6 & 0.15 & 0.08 & 0 & 0.04 & 0.00 \\ 7 & -0.41 & -0.35 & -0.53 & 0 & 0.72 \\ 8 & -0.44 & -0.42 & -0.54 & 0.37 & 0 \end{pmatrix}$$

Although the Lapin matrix does contain a negative entry between elements of the same cluster, (4,5), the eigenvector corresponding to the
minimum non-zero entry makes a correct split between clusters 4–5–6
and 7–8 as can be seen in Table 5.8. It is the second column in the
table corresponding to the minimum non-zero eigenvalue 0.44 of L and
maximum eigenvalue 2.28 of L^+.

An interesting feature of this example is that some eigenvalues of M
are negative, that is, some eigenvalues of L are greater than unity—and
the corresponding eigenvalues of L^+ are less than unity. If one continues
the splitting process and tries this approach further, say at the set $\{1, 2, 3\}$,
they will find out that this is quite possible, the corresponding eigenvector
separates 3 from the rest, but the corresponding eigenvalue of L is 1.3,
greater than the unity, which may be accepted as a rule of thumb for
halting the splitting process. ∎

TABLE 5.8

Deriving Lapin

	1	2	3	4	5
	−0.40	−0.36	−0.75	−0.36	−0.12
	−0.49	−0.27	0.64	−0.40	−0.35
	−0.43	−0.47	0.06	0.74	0.23
	−0.51	0.48	0.04	−0.21	0.68
	−0.39	0.59	−0.15	0.35	−0.59
λ	−0.00	**0.44**	1.28	1.48	1.81
$1-\lambda$	1.00	**0.56**	−0.28	−0.48	−0.81
$1/\lambda$		**2.28**	0.78	0.68	0.55

Note: The five eigenvectors of matrices L, M, and L^+ on set $\{4,5,6,7,8\}$, and corresponding eigenvalues λ of L, along with their complements to 1 (eigenvalues of M) and reciprocals (eigenvalues of L^+); those of interest are highlighted in bold.

Example 5.10: Spectral Clusters at Odds with the Structure of Lapin Transform of Cockroach Network

Take the Cockroach graph in Figure 2.7 with its similarity matrix on p. 169. Its Lapin matrix is presented in Table 5.9.

The matrix manifests a rather clear cut cluster structure of three clusters, 1–2–3–4, 7–8–9–10, and 5–6–11–12. Indeed, the positive entries are those within the clusters, except for two positive—yet rather small—entries, at (4,5) and (10,11).

Yet the first eigenvector reflects none of that; it cuts through by separating six nodes 1,2,3,4,5,6 (negative components) from the rest (positive components). This is an example of a situation in which the spectral

TABLE 5.9

Lapin Similarity Matrix for Cockroach Network

	2	3	4	5	6	7	8	9	10	11	12
1	2.43	1.18	0.05	−0.52	−0.58	−0.69	−0.92	−0.75	−0.59	−0.69	−0.63
2		1.75	0.16	−0.64	−0.74	−0.92	−1.22	−0.99	−0.74	−0.88	−0.81
3			0.44	−0.36	−0.51	−0.75	−0.99	−0.76	−0.46	−0.60	−0.58
4				0.14	−0.15	−0.59	−0.74	−0.46	0.02	−0.16	−0.24
5					0.68	−0.69	−0.88	−0.60	−0.16	0.46	0.44
6						−0.63	−0.81	−0.58	−0.24	0.44	0.94
7							2.43	1.18	0.05	−0.52	−0.58
8								1.75	0.16	−0.64	−0.74
9									0.44	−0.36	−0.51
10										0.14	−0.15
11											0.68

approach fails: the normalized cut criterion at the partition separating the first six nodes from the other six nodes is equal to 0.46, whereas its value at cluster 5–6–11–12 cut from the rest is 0.32. The same value of the criterion, 0.32, is attained at cluster 4–5–6–10–11–12 cut from the rest. These two cuts are optimal according to the criterion, and the spectral cut is not.

However these very splits are produced by the spectral approach applied to the uniform and modularity transformed Cockroach data directly based on the model in Section 7.4.1, with no Lapin transformation at all: the first eigenvector splits the uniform transform, with the mean 0. 197 subtracted from the Cockroach similarity data, by separating cluster 4–5–6–10–11–12, and the first eigenvector for the modularity transformed data separates cluster 5–6–11–12. This shows that the Laplacian transformation does not necessarily sharpen the data structure. However, there has been no experimental or theoretical investigation of this issue published as yet.   ∎

Indeed, the spectral approach can be applied to a similarity matrix directly because the approximate partitioning problem can be interpreted as a version of the model of additive clustering in Section 7.4.3.

Example 5.11: Spectral Clustering of Affinity Data with No Laplacian Transformation

Let us try the spectral approach to clustering by following the min cut criterion for splitting the entity set in two clusters—according to material in Section 7.4.1, this should follow the positive/negative split of components in the first eigenvector of each, uniformly transformed and modularity transformed data. Indeed, at the affinity Colleges data, both of the eigenvectors show cluster 1–2–3 split from the rest, as presented in the upper two lines of Table 5.10.

The third line contains the second eigenvector of the affinity matrix A. The idea is that the first eigenvector, consisting of all positive components, reflects the general similarity between entities and should be removed to make the cluster structure visible. Such a removal would leave the second eigenvector of A to be the first eigenvector of the matrix of residual similarities $a_{ij} - \lambda z_i z_j$. In fact, this transformation much resembles the modularity transformation. Indeed, the vector of summary similarities a_{i+}

TABLE 5.10

Relevant Eigenvectors for the Min Cut Split at Background Similarities Variously Subtracted

Standardization	1	2	3	4	5	6	7	8
Uniform	0.54	0.56	0.33	−0.10	−0.23	−0.19	−0.33	−0.28
Modularity	−0.52	−0.54	−0.29	0.12	0.30	0.21	0.37	0.27
Second	0.31	0.32	0.16	−0.25	−0.43	−0.33	−0.51	−0.39

can be considered a rough approximation of the first eigenvector, according to the so-called power method for finding the first eigenvalue and corresponding eigenvector [61]. All the eigenvectors in Table 5.10 point to the same split: cluster 1–2–3 against the rest. Applied to the remaining set $\{4, 5, 6, 7, 8\}$, all the three rules will split it in clusters 4–5–6 and 7–8. Therefore, the spectral approach here leads to good results. ∎

5.3 Additive Clustering

5.3.1 Additive Cluster Model

The additive clustering model has been proposed by R. Shepard and P. Arabie [185] in the context of cognitive modeling. Independently, a similar model was developed by B. Mirkin as an extension of the metric factor analysis onto non-metric structures such as partitions, subsets and orders (see [133] for references to publications made in 1970s in Russian).

To give intuition to the model, let us take an example of the situation at which the additive clustering model (7.77) is applicable [141]. Consider a taxonomy of the domain such as the ACM Classification of Computer Subjects (ACM-CCS) [1] and a set of topics from the taxonomy that are involved in the research conducted in a Computer Science Department:

- D.4.2 Operating systems: storage management
- D.4.7 Operating systems: organization and design
- I.2.4 Knowledge representation
- I.2.10 Vision and scene understanding
- I.3.5 Computational geometry and object modeling
- I.4.1 Digitization and image capture

The topics in the list are coded by the labels from the ACM-CCS.

Assume that a group undertakes research in Operating Systems, involving subjects constituting cluster OS = {D.4.2, D.4.7}, another group is doing Image Analysis, related to cluster IA = {I.2.10, I.3.5, I.4.1}, and the third group is engaged in Hierarchical Structures and their applications, corresponding to HS = {D.4.2, D.4.7, I.2.4, I.2.10, I.3.5}. Assume that the group research intensities differ; say, they are equal to 4 for OS, 3 for IA, and 2 for HS. Assume as well a background similarity between the topics, due to the fact that all belong to the Computer Science domain, as equal to 1. Then it is natural to define the similarity between topics D.4.2 and D.4.7 as the sum of intensities of the clusters containing both of them: 4 according to OS cluster, 2 according to HS and 1 as the background intensity, thus leading to the value of similarity equal to $4 + 2 + 1 = 7$. Analogously, the similarity between topics D.4.2 and I.3.5 will be $2 + 1 = 3$, and between topics D.4.2 and I.4.1, just the background similarity 1. A similarity matrix can be clearly derived in this way from the clusters and their intensity values. Yet the clustering problem is reverse: given a similarity matrix, find additive clusters and their intensities so that the derived matrix is as close as possible to the given one.

According to the additive clustering model in Equation 7.77, p. 313, similarities a_{ij} are defined by the intensity weights of clusters containing both i and j. In matrix terms, the model is $A = Z\Lambda Z^T + E$ where Z is an $N \times (K+1)$ matrix of cluster membership vectors, of which the $(K+1)$th vector corresponds to λ_0 and has unity in all of its components, Λ is a $K \times K$ diagonal matrix with $\lambda_1, \ldots, \lambda_K$ on its main diagonal, all other entries being zero, and $E = (e_{ij})$ the residual matrix. Each cluster membership vector is binary with unities corresponding to elements of the cluster, so that $(K+1)$th cluster is universal consisting of all the entities. This equation is a straightforward analogue to the spectral decomposition of a symmetric matrix A in Equation 7.10.

Unfortunately, fitting the model in a situation when clusters, hidden in the data, do overlap is a tricky business. A reasonable idea would be to use the alternating minimization approach over two groups of variables, clusters and their intensity weights. Given clusters, intensity weights are relatively easy to get determined, with the operation of orthogonal projection. Given a set of intensity weights, finding clusters is much more difficult. A further improvement of a set of clusters admits different locally optimal strategies. One of them would be to check, for every entity, at which of the clusters its state should be changed, and then to perform the best change (see more on this, in a slightly different situation, in [32]).

5.3.2 One-by-One Additive Clustering Strategy

In a situation when the clusters are not expected to overlap or, if expected, their contributions ought to much differ, the one-by-one PCA strategy may be applied as described in Section 7.4.2. Indeed, the strategy suggests that, provided that A has been preliminarily centered, minimizing $L(E) = \sum_{i,j} e_{ij}^2$ over unknown S_k and $\lambda_k, k = 1, \ldots, K$, can be done by finding one cluster S at a time. Given a similarity matrix A, an optimal cluster S should maximize the product $g(S) = a(S)|S|$ where $a(S)$ is the average within-S similarity, see Section 7.4.2. Then the next cluster is to be sought with the same criterion applied to a residual similarity matrix. After each step, the average similarity $a(S)$ is to be subtracted from entries a_{ij} at $i, j \in S$ in the matrix A of the previous step.

Here is an example of this approach.

Example 5.12: Additive Clustering of Algebraic Functions

Consider similarity data between algebraic functions from Table 1.6 reproduced here as Table 5.11.

The average of non-diagonal entries is $\bar{a} = 2.56$. After subtracting this from the data, in the manner of the uniform transformation, the following

TABLE 5.11

Algebraic Functions

#	Function	e^x	lnx	$1/x$	$1/x^2$	x^2	x^3	$\sqrt{x}$	$\sqrt[3]{x}$
1	lnx	6							
2	$1/x$	1	1						
3	$1/x^2$	1	1	6					
4	x^2	2	2	1	1				
5	x^3	3	2	1	1	6			
6	$\sqrt{x}$	2	4	1	1	5	4		
7	$\sqrt[3]{x}$	2	4	1	1	5	3	5	
8	$|x|$	2	3	1	1	5	2	3	2

Note: Similarities between nine elementary functions from Table 1.6.

similarity matrix emerges:

$$A - \bar{a} =$$

$$
\begin{pmatrix}
 & 2 & 3 & 4 & 5 & 6 & 7 & 8 & 9 \\
\hline
1 & 3.44 & -1.56 & -1.56 & -0.56 & 0.44 & -0.56 & -0.56 & -0.56 \\
2 & & -1.56 & -1.56 & -0.56 & -0.56 & 1.44 & 1.44 & 0.44 \\
3 & & & 3.44 & -1.56 & -1.56 & -1.56 & -1.56 & -1.56 \\
4 & & & & -1.56 & -1.56 & -1.56 & -1.56 & -1.56 \\
5 & & & & & 3.44 & 2.44 & 2.44 & 2.44 \\
6 & & & & & & 1.44 & 0.44 & -0.56 \\
7 & & & & & & & 2.44 & 0.44 \\
8 & & & & & & & & -0.56
\end{pmatrix}
$$

In contrast to the case of the summary similarity criterion, though, this should be considered not a uniform data standardization but rather extraction of the universal cluster I consisting of all the nine functions and supplied with the intensity $\bar{a} = 2.56$. The contribution of the universal cluster to the similarity data scatter is $207.8/676 = 30.4\%$. The denominator and numerator are the sums of the squares of elements in A, Table 5.11, and in $A - \bar{a}$ above, respectively.

Let us give an example of working of the algorithm AddRemAdd(j) (p. 312) for finding an additive cluster S containing entity j. The algorithm adds entities one by one according to their average similarity to the current cluster, with removal of weakly attracted entities if needed. For convenience, take $j = 5$. This constitutes the initial contents of S, $S = \{5\}$. Now we find an entity that has the largest similarity with S, which is entity 6 with $a_{56} = 3.44$ to make $S = \{5, 6\}$. Now we find an entity that has the largest positive summary similarity with the current S. Obviously, entities 1, 2, 3, and 4 cannot do, because they have negative summary similarities with S. Entities 7, 8, and 9 have summary similarities with S equal to $2.44 + 1.44 = 3.88$, $2.44 + 0.44 = 2.88$, and $2.44 - 0.56 = 1.88$, respectively. The maximum is 3.88 at entity 7. To see whether 7 can be added to S, we must find out whether its average similarity to S, $3.88/2 = 1.94$ is greater than half

TABLE 5.12

Additive Clusters Found at Algebraic Functions

No.	Cluster	Intensity	Weight (%)	Interpretation
1	1–9	2.56	30.44	Functions
2	5,6,7,8	2.11	10.55	Growing power
3	1,2	3.44	7.02	Natural
4	3,4	3.44	7.02	Decreasing power
5	5,9	2.44	3.54	Growing even
6	2,7,8	1.07	1.53	Slow growing

TABLE 5.13

Clusters and Their Intensities, Both as Found and Optimally Adjusted

Cluster	All	5–6–7–8	1–2	3–4	5–9	2–7–8	Contr (%)
Former	2.56	2.11	3.44	3.44	2.44	1.07	48.16
Adjusted	1.61	2.76	4.39	4.39	3.39	1.81	94.16

the within-S similarity, that is, $3.44/2 = 1.72$. The answer "yes" warrants joining 7 to S. Now $S = \{5, 6, 7\}$ and the positive summary similarities with S are $2.44 + 0.44 + 2.44 = 5.32$, at 8, and $2.44 - 0.56 + 0.44 = 2.32$, at 9. The maximum, at 8, leads to the average similarity $a(8, S) = 5.32/3 = 1.77$. The average within-S similarity is $(3.44 + 2.44 + 1.44)/3 = 7.32/3 = 2.44$, so that its half, $2.44/2 = 1.22$, is <1.77. Therefore, 8 joins in S. Now entity 9 is the only one to have a positive summary similarity with $S = \{5, 6, 7, 8\}$, that is, $2.44 - 0.56 + 0.44 - 0.56 = 1.76$. However, the average similarity $a(9, S) = 1.76/4 = 0.44$ is less than the half of the within-S average similarity, $a(S) = 2.11$. Moreover, no entity should be removed from S, because each its element has a positive attraction to S. This leads to halting the process and outputting $S = \{5, 6, 7, 8\}$, together with its intensity $a(S) = 2.11$ and the contribution $w(S) = a(S)^2 * 16/676 = 10.55\%$. It is not difficult to see that this S is the best among all the clusters found at different j.

Table 5.12 presents seven additive clusters found with the one-by-one process of additive clustering.

Unfortunately, the contribution weights do not lead to any sound advice for halting the process. Here, the process stops at six clusters. It is nice that all the found clusters have more or less reasonable, and in fact exclusive, interpretations.

The same process, under the constraint that the next clusters are not to overlap those previously found, would lead to just first three clusters in Table 5.13 leaving entity 9 a singleton. ∎

Example 5.13: Readjusting the Cluster Intensity Weights

One-by-one clustering, as it stands, does not allow adjustment or readjustment of the intensity weights of clusters after they have been found.

Yet the intensities can be optimally readjusted either after each step or in the end, with a price—the price being the loss of the individual cluster contributions. These are additive, even when clusters are overlapping, because of the sequential fitting procedure.

Let us take the clustering results presented in Table 5.12 and try to see whether the intensities can be readjusted based on the set of all clusters found.

To do so, let us reshape the similarity data in a one-dimensional vector format. The matrix in Table 5.11 can be considered as a 36-dimensional vector y by putting first eight entries of the first row, then seven entries of the second row, and so on. Then the six clusters in Table 5.12 can be set as columns of a 36×6 binary matrix X. The first column of X corresponds to the universal cluster thus embracing all the similarity links so that all its components are unities. The second column of X corresponds to cluster 5–6–7–8, thus leaving the $26 = 8 + 7 + 6 + 5$ components corresponding to first four lines being zero; all the other components are also zero, except for six entries 27, 28, 29, 31, 32, and 34 that correspond to associations in cluster 5–6–7–8 and are set to be equal to 1. Columns 3, 4 and 5 correspond to two-element clusters and, thus, have all the components zero except for components 1 in column 3, 16 in column 4 and 30 in column 5 corresponding to clusters 1–2, 3–4 and 5–9, respectively. Cluster 2–7–8 sets the column 6 as that with three non-zero components, those 13, 14, and 34.

Now the problem is to find a vector c of such six coefficients that the weighted combination of the six columns of X is as close as possible to the vector y representing the similarity matrix. This problem is known to have a simple solution involving the orthogonal projection matrix $P_X = X(X^TX)^{-1}X^T$ so that the nearest linear combination is $yy = P_Xy$ and $c = (X^TX)^{-1}X^Ty$. With intensity weights from c the individual contribution of clusters cannot be estimated. However, the total contribution can be computed as the complement to unity of the residual variance. This is the total square of the difference between yy and y related to the sum of squared components of y.

Using these rules, Table 5.13 has been computed.

The adjustment leads to an impressive increase in the coverage of the similarities by the clusters, more than 94% of the quadratic similarity scatter, about 50% greater than the original contribution. Regarding the intensity weights, this, basically, takes a unity off the universal cluster's intensity and makes other clusters' intensities somewhat larger. ∎

Obviously, a brute force one-by-one deriving a cluster after a cluster along with the follow-up subtraction of the average may grossly obscure the existing subtle overlaps among the hidden clusters. Therefore, the method of one-by-one clustering should be applied only when clusters are not expected to much overlap. Because of its sequential character, the method would be expected to find correct clusters if their contributions to the data scatter much differ. However, the experimental evidence does not support this expectation. Even with the different contributions, the clusters extracted at later stages can be much blurred, as the following example demonstrates.

Example 5.14: One-by-One Clustering Does Not Work at Strongly Overlapping Clusters

Consider a set of seven entities and the similarity matrix generated by a cluster structure of three clusters whose membership vectors are columns of matrix

$$Z = \begin{pmatrix} 1 & 0 & 0 \\ 1 & 0 & 0 \\ 1 & 1 & 1 \\ 1 & 1 & 1 \\ 0 & 1 & 0 \\ 0 & 1 & 0 \\ 0 & 0 & 1 \end{pmatrix}$$

and the intensity weights are the respective diagonal elements of matrix

$$\Lambda = \begin{pmatrix} 1 & 0 & 0 \\ 0 & 1 & 0 \\ 0 & 0 & 1 \end{pmatrix}$$

These lead to similarity matrix $A = Z\Lambda Z^T$:

	1	2	3	4	5	6	7
1	1	1	1	1	0	0	0
2	1	1	1	1	0	0	0
3	1	1	3	3	1	1	1
$A =$ 4	1	1	3	3	1	1	1
5	0	0	1	1	1	1	0
6	0	0	1	1	1	1	0
7	0	0	1	1	0	0	1

Indeed $a_{34} = 3$ because the pair of entities 3 and 4 belongs to all the three clusters, with unity intensity each, whereas $a_{31} = 1$ because the pair 1 and 3 belong only to one cluster S_1.

To proceed with application of the one-by-one clustering, let us utilize the conventional option at which the diagonal entries are not involved. Indeed, in many applications, the diagonal entries are either undefined or irrelevant or have rather arbitrary values. Another thing recommended within the framework of one-by-one extraction clustering is to start with defining λ_0 as the average similarity, which is 0.6122, as corresponding to the universal cluster of all the seven entities. Therefore, the next cluster is found at $A - 0.6122$.

Using AddRemAdd(j) at all $j = 1, \ldots, 7$ and choosing the best result, after five steps one obtains the following (see Table 5.14).

The intensities are rather small because the average similarity has been subtracted beforehand. The result, with its most contributing cluster {3,4},

TABLE 5.14

Clusters Found Using the One-by-One Clustering
Approach after Centering Matrix A

Step	Contribution (%)	Intensity	List
1	51.4	2.30	3, 4
2	3.6	0.25	3, 4, 5, 6
3	2.5	0.21	1, 2, 3, 4
4	0.9	0.30	3, 7
5	0.9	0.30	4, 7

TABLE 5.15

Clusters Found at the Different Weights

Step	Contribution (%)	Intensity	List
1	78.1	5.53	3, 4, 7
2	3.1	0.78	1, 2, 3, 4
3	1.3	1.22	3, 4

is not quite compatible with the cluster structure behind A, though the
second and third clusters do coincide with two of the clusters making A.
On the other hand, cluster {3,4} may be distinguished by the fact that it is
part of all the three clusters constituting matrix Z.

Changing the weights, say, to

$$\Lambda = \begin{pmatrix} 2 & 0 & 0 \\ 0 & 1 & 0 \\ 0 & 0 & 6 \end{pmatrix}$$

leads to an output of just three clusters (Table 5.15).

Two clusters, 1–2–3–4 and 3–4–7, do coincide with those original. Yet
the cluster with the minimum intensity weight is lost in the output because
of the subtracted cluster intensities. ∎

**Example 5.15: One-by-One Extraction May Work at Not So Much
Overlapping Clusters**

Consider a set of six entities and the similarity matrix in Table 5.16
generated by a cluster structure of three clusters whose membership
vectors are columns of matrix

$$Z = \begin{pmatrix} 1 & 0 & 0 \\ 1 & 0 & 0 \\ 1 & 1 & 1 \\ 0 & 1 & 0 \\ 0 & 1 & 0 \\ 0 & 0 & 1 \end{pmatrix}$$

TABLE 5.16

Similarity Matrix with Noise Added

		1	2	3	4	5	6
$A =$	1	1.02	1.23	1.00	0.12	−0.18	0.09
	2	1.05	1.15	1.11	−0.39	0.01	0.25
	3	0.99	1.04	1.73	1.12	1.13	−0.00
	4	0.05	0.08	1.24	2.15	1.36	1.26
	5	0.26	−0.29	1.03	0.85	1.23	0.23
	6	0.05	−0.28	0.82	1.07	−0.79	1.40

TABLE 5.17

Clusters Found at Pre-Centered Matrix A

Step	Contribution (%)	Intensity	List
1	29.3	0.61	1, 2, 3
2	29.3	0.61	3, 4, 5
3	9.8	0.61	4, 6

and the intensity weight matrix

$$\Lambda = \begin{pmatrix} 1 & 0 & 0 \\ 0 & 1 & 0 \\ 0 & 0 & 1 \end{pmatrix}$$

These clusters do overlap, but not that much and not on the same core. The similarity matrix $A = Z\Lambda Z^{\mathsf{T}}$ is

		1	2	3	4	5	6
	1	1	1	1	0	0	0
	2	1	1	1	0	0	0
$A =$	3	1	1	2	1	1	0
	4	0	0	1	2	1	1
	5	0	0	1	1	1	0
	6	0	0	0	1	0	1

The one-by-one cluster extracting method involving the best result by AddRemAdd(j) produces exactly the clusters that form the similarity matrix A, see Table 5.17.

When matrix A is slightly distorted by addition of the random Gaussian noise (with 0 mean and 0.3 standard deviation), it may be like this:

The noise does not much affect the results (see Table 5.18). The contribution to the data scatter of the fourth cluster, $\{1, 2\}$, is so minor that the cluster should be ignored. ■

TABLE 5.18

Clusters Found at the Similarity Matrix with
Added Noise

Step	Contribution (%)	Intensity	List
1	25.1	0.69	3, 4, 5
2	21.4	0.64	1, 2, 3
3	9.45	0.74	4, 6
4	0.08	0.07	1, 2

5.4 Consensus Clustering

5.4.1 Ensemble and Combined Consensus Concepts

Consensus, or ensemble, clustering is an activity of summarizing a set of clusterings into a single clustering. Back in 1970s, the current author considered the problem in the context of deriving clusters of different life styles by combining partitions expressing different aspects of the life style (for a review, see [135]). Nowadays the popularity of the subject is based on the current situation in real world data analysis. There are many clustering programmes available, which may lead to different clustering results found on the same data set. After applying different clustering algorithms, or even the same algorithm at different settings, to the same data set, one gets a number of different solutions, a clustering ensemble. The question that arises then: is there a cluster structure behind the wealth of solutions found? And if yes, as one would hope, what is that structure? A recent review on this can be found in [58].

According to the author's views, mathematically expressed in Section 7.4.3 in the framework of the data recovery approach, there are two ways to define a consensus partition: (a) ensemble consensus corresponding to the view that all found clusterings must be some "noisy" versions of the hidden partition we are looking for and (b) combined consensus corresponding to the view that the partition we are looking for should combine some "majority" patterns derived from the partitions that are available.

These concepts lead to using the available partitions as the data to derive consensus partitions. The least-squares criteria for such derivations can be expressed both in terms of pair-wise partition-to-feature distributions, which are contingency or confusion tables, and in terms of pair-wise similarities (see Section 7.4.3). Only this latter framework of entity-to-entity similarity data will be considered in this section.

It appears, to find an ensemble consensus partition, one should build the so-called consensus similarity matrix A at which a_{ij} is the number of those of the given partitions on I in which both i and j are in the same cluster. After this, the ensemble consensus partition $S = \{S_1, \ldots, S_K\}$ is to be found over the consensus matrix A by maximizing criterion (7.85) which is

$$g(S) = \sum_{k=1}^{K} \sum_{i,j \in S_k} a_{ij}/N_k \tag{5.1}$$

This criterion also appeared in this book in two somewhat differing frameworks. First, this criterion is equivalent to the square-error criterion of K-Means clustering if: (a) the classes of given partitions are represented by 1/0 dummy categories for an entity-to feature table and (b) the inner product measures the similarity between entities (see Section 7.4.2). Second, the criterion emerges when using an arbitrary similarity matrix A and the additive clustering model with non-overlapping clusters (see Section 7.4.3).

On the other hand, to find a combined consensus partition, a projection consensus matrix $P = (p_{ij})$ is to be defined in a less straightforward manner so that p_{ij} is the sum of reciprocals of the cardinalities of those given clusters that contain both i and j. Then the average $\bar{p} = T/N$ is subtracted from all the p_{ij} where T is the number of given partitions and the summary similarity criterion (7.90)

$$f(S, \bar{p}) = \sum_{k=1}^{K} \sum_{i,j \in S_k} (p_{ij} - \bar{p}) \tag{5.2}$$

is to be maximized over a partition S.

The following examples illustrate the procedures and show some specifics in them.

Example 5.16: Ensemble Clustering of Results of Multiple Runs of K-Means

Consider the Colleges data set and cluster them with Batch K-Means at $K = 3$ initializing at random triplets of entities as the initial centroids. Consider 12 random triplets forming the following series:

$$\begin{pmatrix}
1 & 2 & 5 & 8 \\
2 & 8 & 3 & 5 \\
3 & 5 & 1 & 7 \\
4 & 7 & 5 & 6 \\
5 & 4 & 7 & 1 \\
6 & 5 & 3 & 1 \\
7 & 2 & 1 & 6 \\
8 & 6 & 3 & 5 \\
9 & 3 & 8 & 7 \\
10 & 2 & 7 & 3 \\
11 & 8 & 4 & 6 \\
12 & 3 & 5 & 6
\end{pmatrix}$$

Each row in the array above consists of the indexes of three entities, in the order of generation, to be taken as the initial seeds for a run of K-Means.

TABLE 5.19

Partitions of the Colleges Set Found with K-Means at the 12 Random Initializations

No.	1	2	3	4	5	6	7	8
1	1	1	1	2	2	2	3	3
2	2	2	2	3	3	3	1	1
3	2	2	2	1	1	1	3	3
4	2	2	2	3	2	3	1	1
5	3	3	3	1	2	1	2	2
6	3	3	2	2	1	2	1	1
7	2	1	1	3	3	3	3	1
8	2	2	2	1	3	1	3	3
9	1	1	1	3	3	3	3	2
10	1	1	3	3	3	3	2	2
11	2	2	3	2	3	3	1	1
12	1	1	1	3	2	3	2	2

Note: Each row presents labels of clusters at the entities in columns.

Each line in the Table 5.19 represents one of the 12 partitions found starting from the triplets above.

Ensemble consensus matrix for the ensemble of partitions in Table 5.19 is presented in Table 5.20.

This matrix demonstrates the three Subject clusters—Science, Engineering and Arts—in the Colleges data set quite clearly. Even the concept of concordant partition will be successful with this consensus matrix. Let us remind ourselves that the concordant partition is a minimizer of the summary distance to the given partitions. As shown in Section 7.5.3, this maximizes the summary similarity criterion at the uniform data transformation, in this case, by subtracting the threshold $a = T/2 = 6$. When looking at the similarity data in Table 5.20, one cannot help but notice that

TABLE 5.20

Ensemble Consensus Matrix for Partitions in Table 5.19

	1	2	3	4	5	6	7	8
1	12	11	8	1	1	0	0	0
2	11	12	9	1	1	0	0	1
3	8	9	12	2	3	3	0	1
4	1	1	2	12	6	11	2	0
5	1	1	3	6	12	7	6	4
6	0	0	3	11	7	12	2	0
7	0	0	0	2	6	2	12	10
8	0	1	1	0	4	0	10	12

TABLE 5.21

Projection Consensus Matrix for Partitions in Table 5.19

	1	2	3	4	5	6	7	8
1	3.42	2.42	1.08	−1.17	−1.25	−1.50	−1.50	−1.50
2	2.42	2.75	1.42	−1.17	−1.25	−1.50	−1.50	−1.17
3	1.08	1.42	2.33	−0.92	−0.67	−0.58	−1.50	−1.17
4	−1.17	−1.17	−0.92	2.92	0.25	2.58	−1.00	−1.50
5	−1.25	−1.25	−0.67	0.25	2.17	0.58	0.33	−0.17
6	−1.50	−1.50	−0.58	2.58	0.58	2.92	−1.00	−1.50
7	−1.50	−1.50	−1.50	−1.00	0.33	−1.00	3.33	2.83
8	−1.50	−1.17	−1.17	−1.50	−0.17	−1.50	2.83	4.17

none of the within cluster similarities will become negative after subtraction of this value from them, and none of the between cluster similarities will be positive after the subtraction. This ensures that the three-cluster partition is a maximizer, in fact, the maximizer of the summary similarity criterion, that is, makes the concordant partition indeed. One should recall that this concept has not passed the Muchnik's test described in Section 7.5.4, thus cannot be considered universal. ∎

Example 5.17: Combined Consensus Clustering

Consider the same set of partitions of colleges and apply the concept of combined consensus. The projection-based similarity matrix between colleges is in Table 5.21.

This matrix sums votes of the K-Means clusters weighted by the reciprocals of the cluster cardinalities. That is, the larger the cluster, the less its contribution to the similarity. After all votes have been summed, the average value, which is equal to the number of partitions related to the number of entities—in this case, $12/8 = 1.5$—is subtracted from all the entries (see Section 7.5.2). Then the combined consensus is determined by maximizing the summary similarity criterion. This will be the subject-based cluster partition of Colleges again, even in spite of the positive between-cluster-similarity $p_{57} = 0.33$. This is the only oddity among the similarity values, and it is small enough to make no difference. ∎

Example 5.18: Ensemble Consensus Clustering at Iris Data Set

Consider the 150×4 Iris data set and range standardize it. To generate an ensemble of partitions, let us randomly generate 51 triplets of indexes to serve as initial centroids to run K-Means at $K = 3$ (see Table 5.22).

With these 51 initializations, K-Means applied to the range-standardized Iris data set from Table 1.7 produces 51 partitions leading to such an ensemble consensus matrix that the one-by-one additive non-overlapping clustering method produces just three clusters that keep the

TABLE 5.22

Random Triplets of Indices Generated for Initialization of K-Means Clustering over Range-Standardized Iris Data Set

Triplet			Triplet			Triplet		
115	34	46	67	104	114	146	5	3
63	18	117	112	79	73	13	29	18
110	150	69	23	45	64	124	88	9
13	85	40	12	50	15	85	51	125
71	134	13	71	34	120	41	85	68
69	54	147	133	89	2	129	131	17
50	15	80	125	98	139	102	122	85
138	122	32	130	94	102	123	6	70
65	117	52	49	63	57	134	2	97
17	97	6	17	5	109	18	127	26
77	57	27	127	48	37	76	144	57
57	48	6	25	128	90	123	101	88
25	38	140	121	128	122	140	43	98
67	117	81	112	106	41	99	118	4
32	108	90	39	104	101	36	57	122
80	17	142	111	77	65	132	139	42
57	35	116	1	137	51	133	83	82

three-taxon structure of the data set, yet up to a number of errors. Specifically, consensus cluster 1 contains 61 specimens coinciding with taxon 2 up to the following errors: 14 entities (113, 115, 117, 118, 119, 121, 125, 126, 127, 133, 136, 138, 139, 148) are taken from taxon 3, whereas three entities from taxon 2 (56, 63, 70) are missed. Consensus cluster 2 coincides with taxon 1 exactly. Accordingly, cluster 3 consists of just 39 entities of which 36 belong to taxon 3 and 3 entities are taken from taxon 2.

The same clustering appears to be optimal for the combined consensus matrix in this case.

There are 17 errors altogether, a rather mediocre result which is typical at a crisp or fuzzy K-Means run. Minkowski metric iK-Means with weighted features leads to clustering results much better matching the Iris taxa, with five or six errors only, as described on p. 118. ■

Example 5.19: Combined Consensus Clustering on Digits Data

Consider the seven binary variables at Digits data in Table 1.9 as categorical features of the numerical digits. Let us form a combined consensus partition expressing them in an optimal way.

According to the theory described in Section 7.5.2, the combined consensus partition maximizes the summary similarity clustering criterion over the projection consensus matrix. For any two digits i and j, the value

TABLE 5.23

Digits

Digit	e1	e2	e3	e4	e5	e6	e7
1	0	0	1	0	0	1	0
2	1	0	1	1	1	0	1
3	1	0	1	1	0	1	1
4	0	1	1	1	0	1	0
5	1	1	0	1	0	1	1
6	1	1	0	1	1	1	1
7	1	0	1	0	0	1	0
8	1	1	1	1	1	1	1
9	1	1	1	1	0	1	1
0	1	1	1	0	1	1	1

Note: Seven binary features for 10 digits from Table 1.9.

TABLE 5.24

Combined Consensus Matrix for Digits Partitions in Table 5.23

	1	2	3	4	5	6	7	8	9	0
1	1.12	−0.32	−0.05	0.54	−0.42	−0.59	0.62	−0.46	−0.30	−0.13
2	−0.32	1.34	0.09	−0.43	−0.29	−0.04	−0.20	0.09	−0.16	−0.06
3	−0.05	0.09	0.36	−0.15	−0.01	−0.18	0.08	−0.05	0.11	−0.20
4	0.54	−0.43	−0.15	0.85	−0.11	−0.28	0.04	−0.15	0.01	−0.30
5	−0.42	−0.29	−0.01	−0.11	0.66	0.49	−0.30	−0.01	0.16	−0.15
6	−0.59	−0.04	−0.18	−0.28	0.49	0.74	−0.46	0.24	−0.01	0.10
7	0.62	−0.20	0.08	0.04	−0.30	−0.46	0.74	−0.34	−0.17	−0.01
8	−0.46	0.09	−0.05	−0.15	−0.01	0.24	−0.34	0.36	0.11	0.22
9	−0.30	−0.16	0.11	0.01	0.16	−0.01	−0.17	0.11	0.28	−0.03
0	−0.13	−0.06	−0.20	−0.30	−0.15	0.10	−0.01	0.22	−0.03	0.55

p_{ij} in this matrix sums the reciprocals of all the categories containing both i and j short of the value 0.7, the ratio of the number of features, 7, and the number of entities, 10.

This matrix is presented in Table 5.24. For example, digits 1 and 2 fall in 0-category of e2 and in 1-category of e3. The frequency of the former is 4 and that of the latter is 8, so that $p_{12} = 1/4 + 1/8 − 0.7 = −0.325$.

Let us apply the agglomerative clustering procedure to maximize the summary similarity criterion. The maximum off-diagonal value in matrix p is $p_{17} = 0.62$. Therefore, let us add row 7 to row 1, after which column 7 is added to column 1, and both row 7 and column 7 are removed from the matrix. The result is the summary similarity matrix after 1 and 7 are merged together. This matrix' maximum is $p_{1-7,4} = 0.57$ leading to merging 1, 4, and 7 together. The summary similarity matrix is

TABLE 5.25

Summary Similarity Matrix after Two Mergers for Table 5.24

	1–4–7	2	3	5	6	8	9	0
1–4–7	5.09	−0.96	−0.12	−0.83	−1.33	−0.96	−0.46	−0.43
2	−0.96	1.34	0.09	−0.29	−0.04	0.09	−0.16	−0.06
3	−0.12	0.09	0.36	−0.01	−0.18	−0.05	0.11	−0.20
5	−0.83	−0.29	−0.01	0.66	0.49	−0.01	0.16	−0.15
6	−1.33	−0.04	−0.18	0.49	0.74	0.24	−0.01	0.10
8	−0.96	0.09	−0.05	−0.01	0.24	0.36	0.11	0.22
9	−0.46	−0.16	0.11	0.16	−0.01	0.11	0.28	−0.03
8	−0.43	−0.06	−0.20	−0.15	0.10	0.22	−0.03	0.55

TABLE 5.26

Summary Similarity Matrix between Four Clusters

	1–4–7	2	3	5–6–8–9–0
1–4–7	5.09	−0.96	−0.12	−4.01
2	−0.96	1.34	0.09	−0.46
3	−0.12	0.09	0.36	−0.33
5–6–8–9–0	−4.01	−0.46	−0.33	4.80

obtained by adding row 4 to row 1, then column 4 to column 1 and removing item 4 after that. This summary similarity matrix is presented in Table 5.25.

Now the off-diagonal maximum is $p_{56} = 0.49$, so that 5 and 6 are to be merged. Next maximum is 0.23 to add entity 8 to cluster 5–6. Then maximum, 0.26, is at the summary similarity of the merged cluster 5–6–8 and digit 9. The next maximum, 0.13, moves 0 into the merged cluster, thus leading to the summary similarity matrix in Table 5.26.

This matrix has only one positive value, $p_{23} = 0.09$, thus leading to merging 2 and 3 together. After this, all the between-cluster similarities are negative, thus preventing them from merging together. Indeed, let us recall that the goal is to maximize the summary similarity, which would be undermined by further mergers.

Therefore, the partition $S = \{1–4–7, 2–3, 5–6–8–9–0\}$ is the combined consensus found with the agglomerative algorithm at the summary within-cluster similarity equal to $a(S) = 5.09 + 1.87 + 4.80 = 11.76$. The value of the criterion at the four-cluster ensemble consensus partition $R = \{1–4–7, 3–5–9, 6–8–0, 2\}$ is somewhat smaller, $a(R) = 11.01$.

This result is at odds with other clustering results on this data set. This perhaps is because the combined consensus criterion is at odds with the nature of data. ∎

5.4.2 Experimental Verification of Least-Squares Consensus Methods

Verification of a consensus clustering method should address the following two issues:

1. Are consensus clustering results any better than those found by a clustering algorithm over the data table?
2. Is the proposed method competitive over its competition?

To address these, the following computational experiment has been designed [153]. First, a data set is generated of K^* M-dimensional Gaussian clusters. Second, K-Means clustering method, with K not necessarily equal to K^*, is applied to the data set many times starting from random initial centroids. The obtained partitions constitute the input to a consensus clustering method. Then a consensus partition is found with each of the consensus methods at hand. This partition is evaluated, first, with respect to its similarity to the Gaussian cluster partition generated at the data set, and, second, with respect to the ensemble of partitions used to build the consensus. This is repeated a number of times to see how stable the results are regarding the randomness of the data generator. The averaged results are used to address both of the issues above.

Let us present the set of consensus clustering methods under consideration:

1. LSEC—The least-squares ensemble consensus clustering method utilizing one-by-one clustering with AddRemAdd(j) over the consensus matrix.
2. LSCC—The least-squares combined clustering method utilizing the agglomerative process with the summary similarity criterion over the projection consensus matrix.
3. Vote—The Voting algorithm from [214] that intends to minimize the total number of misclassification errors in the combined partition by updating one of the input partitions with the data from other partitions.
4. FT—The Fusion-Transfer algorithm that approximates the concordant partition by using a mix of cluster agglomerative clustering and intermediate moves of entities to the most attracting clusters [67].
5. Borda—The Borda-Condorcet ranking rule that assigns each entity with the Borda summary ranks of clusters, depending on a preliminary alignment of the input clusterings [184].
6. MCLA—A heuristic from [201] that first builds "meta-clusters" using Jaccard similarity index between clusters and then assigns entities to those "nearest" to them.

The Table 5.27 presents results of an experiment in which the data have been generated 10 times as 300×8 matrices of five Gaussian clusters with

TABLE 5.27

Comparing Consensus Partitions

Algorithms	LSEC	LSCC	Vote	FT	Borda	MCLA
ARIG	0.90	0.85	0.68	0.64	0.74	0.64
Std	0.15	0.17	0.10	0.14	0.14	0.13
ARII	0.73	0.72	0.73	0.73	0.73	0.71
Std	0.06	0.05	0.05	0.06	0.04	0.07

Note: The average values of Adjusted Rand Index to the data generated partition (ARIG) and to all the input partitions (ARII), along with their standard deviations.

diagonal covariance matrices and random cardinalities. Results of 50 runs of Batch K-Means at $K = 5$ have been used as input partitions for the consensus methods. The accuracy is evaluated by using the Adjusted Rand Index (ARI) from p. 245. Other generating options lead to similar results [153].

As one can see, all the methods lead to results that are stable: the standard deviations are comparatively low. The least-squares methods outperform the others with a good margin regarding the recovery of the generated partition. Surprisingly, regarding the average similarity to the input partitions, all the methods show similar results.

Moreover, in this experiment, the consensus clustering seems superior to the random K-Means runs: it gets as close as at ARI $= 0.9$ to the generated partition whereas the K-Means results themselves are lagging much behind on the level of ARI at 0.6–0.7.

5.5 Single Linkage, Minimum Spanning Tree, and Connected Components

The single linkage, or nearest neighbor, method is based on entity-to-entity dissimilarities like those presented in the Primates data (Table 1.2) or in the distance matrix calculated in Section 3.2.2 when computing initial seeds for the Colleges data (Table 3.9). It can be applied to the case of similarity data as well.

It should be pointed out that the dissimilarities can be computed from the original data table at each step so that there is no need to maintain the dissimilarity matrix as a separate data file. This may save quite a lot of memory, especially when the number of entities is large. For instance, if the size of the original data table is 1000×10, it takes only 10,000 numbers to store, whereas the entity-to-entity distances may take up to half a million numbers. There is always a trade-off between memory and computation in distance-based approaches that may require some efforts to balance.

The single linkage approach is based on the principle that the dissimilarity between two clusters is defined as the minimum dissimilarity (or, maximum similarity) between entities of one and the other cluster [65]. This can be implemented into the agglomerative distance-based algorithm described in Section 4.5.1 by leaving it without any change except for the formula (4.6) that calculates distances between the merged cluster and the others and must be substituted by the following:

$$d_{w1 \cup w2, w} = \min(d_{w1, w}, d_{w2, w}) \tag{5.3}$$

Formula (5.3) follows the principle of minimum distance, which explains the method's name.

In general, agglomerative processes are rather computationally intensive because the minimum of inter-cluster distances must be found at each merging step. However, for single linkage clustering, there exists a much more effective implementation involving the concept of the minimum spanning tree (MST) of a weighted graph.

A weighted graph, in our context, is a representation of a dissimilarity or distance matrix $D = (d_{ij})$ with nodes corresponding to entities $i \in I$ and edges connecting any i and j from I with weight d_{ij}. Such a graph for the Primates data is in Figure 5.1a: only edges whose weights are smaller than 3 are shown.

A spanning tree of a weighted graph is a subgraph T without cycles such that it covers all its nodes (as in Figure 5.1b). The length of T is defined as the sum of all weights d_{ij} over edges $\{i, j\}$ belonging to T. An MST T must have minimum length. The concept of MST is of prime importance in many applications in the Computer Sciences. Let us take a look at how it works in clustering.

First of all, let us consider the so-called Prim algorithm for finding an MST T. The algorithm processes nodes (entities), one at a time, starting with $T = \emptyset$ and updating T at each step by adding to T an element i and edge $\{i, j\}$ minimizing d_{ij} over all $i \in I - T$ and $j \in T$. An exact formulation is this.

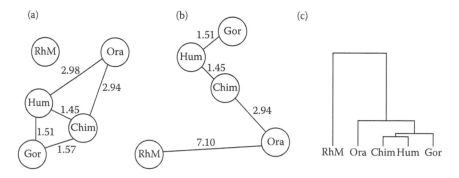

FIGURE 5.1
Weighted graph (a), minimum spanning tree (b), and single-linkage hierarchy for Primates data (c).

PRIM ALGORITHM

1. *Initialization.* Start with set $T \subset I$ consisting of an arbitary $i \in I$ with no edges.
2. *Tree update.* Find $j \in I - T$ minimizing $d(i,j)$ over all $i \in T$ and $j \in I - T$. Add j and (i,j) with the minimal $d(i,j)$ to T.
3. *Stop-condition.* If $I - T = \emptyset$, halt and output tree T. Otherwise go to 2.

To build a computationally effective procedure for the algorithm may be a cumbersome issue, depending on how $d(i,j)$ and their minima are handled, to which a lot of published work has been devoted. A simple pre-processing step can be useful: in the beginning, find a nearest neighbor for each of the entities; only they may go to MST.

Example 5.20: Building an MST for Primate Data

Let us apply Prim algorithm to the Primates distance matrix in Table 1.2, p. 5.

Let us start, for instance, with $T=\{$Human$\}$. Among remaining entities, Chimpanzee is the closest to Human (distance 1.45), which adds Chimpanzee corresponding edge to T as shown in Figure 5.1b. Among the three other entities, Gorilla is the closest to one of the elements of T (Human, distance 1.57). This adds Gorilla to T and the corresponding edge in MST in Figure 5.1b. Then comes Orangutan as the closest to Chimpanzee in T. The only remaining entity, Monkey, is nearest to Orangutan, as shown in the drawing. ∎

Curiously, in spite of its quantitative definition, MST depends only on the order of dissimilarities, not their quantitative values. Moreover, the algorithm can be applied to similarity data for building an MST—the only difference would be that, at each step, the maximum, rather than minimum, link is taken to the tree.

To see the relation between an MST and single linkage clustering, let us build an MST-based distance between nodes (entities). This distance, $T(i,j)$, is defined as the maximum weight of an edge on the only path joining i and j in the tree T. (If another path existed, the two paths would create a cycle through these two nodes, which is impossible because T is a tree.) The definition implies that $d_{ij} \geq T(i,j)$ for all $i,j \in I$ because, otherwise, the tree T can be made shorter by adding the shorter (i,j) and removing the edge at which $T(i,j)$ is reached. Moreover, $T(i,j)$ can be proven to satisfy the so-called ultra-metric inequality $T(i,j) \leq \max(T(i,k), T(j,k))$ for all $i,j,k \in I$. The metric $T(i,j)$ coincides with that implied by the single linkage method. An important property of the ultra-metric $T(i,j)$ is that it approximates the distance d_{ij} in the best way among all the ultra-metrics U satisfying inequality $U(i,j) \le d_{ij}$ for all $i,j \in I$ [72].

An MST T allows finding the single linkage hierarchy by a divisive method: sequentially cutting edges in MST beginning from the largest and continuing in descending order. The result of each cut creates two connected components that are children of the cluster in which the cut occurred (see, for instance, Figure 5.1c).

A feature of the single linkage method is that it involves just $N - 1$ dissimilarity entries occurring in an MST rather than all $N(N - 1)/2$ of them. This results in a threefold effect: (1) a nice mathematical theory, (2) fast computations, and (3) poor application capability.

Example 5.21: MST and Single Linkage Clusters in the Colleges Data

To illustrate point (3) above, let us consider an MST built on the distances between Colleges in Table 3.9 (see Figure 5.2a). Cutting the tree at the longest two edges, we obtain clusters presented in part (b) of the figure. Obviously these clusters do not reflect the data structure properly, in spite of the fact that the structure of (a) does correspond to the fields-of-study division.   ∎

One more graph-theoretical interpretation of single-linkage clusters is in terms of a prime cluster concept in graph theory, the connected component. Let us consider a dissimilarity matrix $D = (d_{ij})$, $i,j \in I$. Then, given any real u, the so-called threshold graph G_u is defined on the set of vertices I as follows: $i \in I$ and $j \in I$ are connected with edge (i,j) if and only if $d_{ij} < u$. Obviously, the edges in this graph are not directed since $d_{ij} = d_{ji}$. A subset S is referred to as connected in a graph if for any different $i,j \in S$ there exists a path between i and j within S. A connected component is a maximal connected $S \subseteq I$: it is connected and either coincides with I or loses this property if any vertex $i \in I - S$ is added to S.

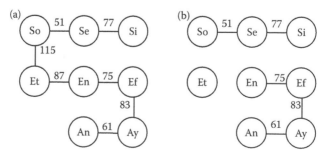

FIGURE 5.2
Minimum spanning tree for Colleges (a) and a result of cutting it (b).

When a minimum spanning tree T is split over its greatest link leading to a partition of I in two clusters, S_1 and S_2, each of them is obviously a connected component of the threshold graph G_u with u equal to the weight of the link removed from T. For the sake of simplicity, we consider that no other edge weight in T is equal to u. Then any i and j in S_1 can be joined by a path belonging to the relevant part of T, thus G_u, because all the weights along T are smaller than u by definition. On the other hand, for any $k \in S_2$, the dissimilarity $d_{ik} \geq u$ so that edge (i, k) is absent from G_u, thus making S_1 a connected component. The fact that S_2 is a connected component of G_u can be proven similarly. When the maximum weight u on the tree T is not unique, T must be split along all maximum weight edges to make the split parts connected components. It can be proven that the connected components of threshold graphs G_u (for u being the weight of an edge from a minimum spanning tree) are single linkage clusters.

5.6 Overall Assessment

Clustering of similarity data attracts an increasing attention because of three different, although intermingling, processes. One is of expanding the role of networks and other similarity-based data in the world. Another process is of further progress in developing of hardware to allow processing massive data previously unthinkable. Yet one more process relates to developing mathematical approaches to clustering relating it to concepts of the classical mathematics.

Two criteria straightforwardly implement the two-fold goal of clustering: make clusters as cohesive as possible while keeping them as far away from each other as possible—the minimum cut and the minimum normalized cut that are described in the first two sections. Conventionally, the summary similarity, that is, minimum cut, clustering criterion is rejected because it leads to silly results at usual, positive, similarity values. Yet the criterion appears quite good when the similarity data is pre-processed by subtracting background similarities. Two concepts of background similarities are considered, a constant noise value and the random interactions, leading to the uniform and modularity transformations, respectively. The former is better suitable for flat, unweighted, network data, whereas the latter is better at diverse link weighting. Heuristics to optimize the summary similarity criterion are considered in terms of local changes, including hierarchical clustering, as well as the spectral approach of finding the eigenvector corresponding to a maximum or minimum eigenvalue. Although the eigenvector does not necessarily leads to an optimal partition, in most situations spectral clusters are reasonable. The minimum normalized cut is conventionally handled with the Laplacian and pseudo-inverse Laplacian transformations of the similarities to make the spectral approach applicable.

A related clustering criterion, the summary average similarity, emerges both at the least-squares clustering of entity-to-feature data and additive clustering. It does not require similarity pre-processing and allows to find clusters one-by-one, which is useful when a complete clustering is not adequate or when clusters may overlap.

It appears that this framework covers the problem of consensus clustering as well. The consensus clustering problem can emerge in two different contexts: (a) ensemble clustering, when different algorithms have been applied to find a hidden partition, resulting in different partitions, and (b) combined clustering, when there are different partitions characterizing different aspects of the phenomenon under consideration. These perspectives lead to somewhat different consensus similarity matrices and criteria that appear to be specifications of the criteria analyzed in the previous sections. The presented approaches are consistent with the other data recovery clustering models; moreover, they show good performance in computational experiments.

The last section is devoted to description of another, graph-theoretic, approach, single link clustering, which appears to be related to another popular concept, the MST.

6

Validation and Interpretation

After reading through this chapter, the reader will know about

1. A controversy between the statistics and knowledge discovery approaches regarding what is validation
2. Internal and external validity
3. Cross-validation of clustering results and algorithms
4. Conventional interpretation aids
5. Least-squares interpretation aids based on the contribution of cluster structure elements to the data scatter
6. Relation between the contributions and statistical associations
7. Conceptual description of clusters
8. Mapping clusters to external knowledge including
   a. Comparing clusters and categories
   b. Comparing partitions
   c. Describing a partition with decision trees
   d. Parsimoniously lifting a cluster within a hierarchical tree

Key Concepts

Association rule A production rule $A \rightarrow B$ for a set of, for example, warehouse transactions in which A and B are nonoverlapping sets of goods, or any categories in general. It is characterized by its support (proportion of transactions containing A in the set of all transactions) and precision (proportion of transactions containing B among all those containing A): the greater the support and precision, the better the quality of the association rule.

Chi-squared An index of relationship between two sets of categories or clusters on the same entity set. Originally was proposed to measure the extent of deviation of the observed bivariate distribution from the statistical independence. The index has also a different meaning in the context of data analysis as an averaged value of the change of the frequency of categories when a crossing category becomes known (Quetelet indexes).

Cluster representative An entity representing its cluster. Conventionally, such an entity is drawn as the nearest to the cluster centroid in the Euclidean space. The data recovery approach suggests that the criterion should be the inner product rather than the distance—this would extend the cluster tendencies over the grand mean.

Comprehensive description A conjunctive predicate P describing a subset $S \subset I$ in such a way that P should hold for an entity if and only if it belongs to S. The description may have errors of two types: false positives, the entities satisfying P but not belonging to S, and false negatives, the entities from S that do not satisfy P.

Conceptual description Description of clusters using feature-based logical predicates.

Decision tree A conceptual description of a subset $S \subset I$ or a partition on I in a tree-like manner. The root of the tree corresponds to all entities in I, and each node is divided according to values of a feature so that a tree leaf is described by a conjunction of the decision criteria along the path from the root leading to the leaf. The tree is built in such a way that the leafs are homogeneous, to an extent.

Contributions to the data scatter Additive items representing parts of the data scatter that are explained by certain elements of a cluster structure such as feature–cluster pairs. The greater the contribution, the more important the element. Summary contributions coincide with statistical measures of correlation and association, which lends a theoretical support to the recommended data standardization rules.

Cross validation A procedure for testing consistency of a clustering algorithm or its results by the comparison of cluster results found on subsamples obtained by a random partition of the sample in a prespecified number of parts. Each of the parts is used for testing the results found at the rest.

External validation Evaluation of the extent of correspondence between clusters and data or knowledge that has not been used for obtaining the clusters.

Generalization Deriving or inducing a general conception from particulars.

Internal validation Evaluating the extent of correspondence between clusters found in a data set and the data set.

Interpretation Explaining the meaning of an expression or a cluster in understandable terms.

Interpretation aids Computational tools for helping the user to interpret clusters in terms of features, external or used in the process of clustering. Conventional interpretation aids include cluster centroids and bivariate distributions of cluster partitions and features. Contribution-based interpretation aids such as ScaD and QScaD tables are derived from the decomposition of the data scatter into parts explained and unexplained by the clustering.

Parsimoniously lifting Representation of a thematic set by higher rank nodes in a conceptual taxonomy of the domain in such a way that the number of emerging inconsistencies such as "head subjects," "gaps," and "offshoots" is minimized.

Production rule A predicate $A \rightarrow B$ which is true for those and only those entities that should satisfy B whenever they satisfy A. A conceptual description of a cluster S can be a production rule $A \rightarrow B$ in which B expresses belongingness of an entity to S.

ScaD and QScaD tables Interpretation aids helping to capture cluster-specific features that are relevant to K-Means clustering results. ScaD is a cluster-to-feature table whose entries are cluster-to-feature contributions to the data scatter. QScaD is a table of the relative Quitelet coefficients of the ScaD entries to express how much they differ from the average.

Visualization Mapping clusters to a ground image such as Cartesian plane or geographic map in such a way that important properties of clusters translate into visual properties over the image.

6.1 General: Internal and External Validity

After applying a clustering algorithm, the user would like to know whether the results indeed reflect an innate property of the data or they are just an artifact generated by the algorithm. In other words, the user wants to know whether the cluster structure found is valid or not. It is not a difficult question when there is a probabilistic mechanism for data generation, which is known to the user. In this case, the issue can be solved by testing a corresponding statistical hypothesis. Unfortunately, this is not typical in cluster analysis. Typically, the data come from a source that cannot be modeled this way if it can be at all. There is one more cause of uncertainty here: the user's wish to have an aggregate representation of the data, at a level of resolution, which cannot be straightforwardly formalized and, moreover, may depend on the task at hand.

There are two types of considerations that may help in validation of the clustering results. They are referred to as internal and external ones. The internal validity view looks at how well the cluster structure reflects the data from which it has been generated. The external validity view looks at how well the cluster structure fits into the domain knowledge available to the user. Therefore, the external approach much relates to the issues of interpretation of clustering results, that is, mapping them to the domain knowledge.

One may say that the internal view relates to the statistics approach to data analysis, whereas the external view is more compatible with the knowledge discovery paradigm.

6.2 Testing Internal Validity

This author distinguishes between three ways for testing internal validity of clustering results:

1. Directly scoring the correspondence between the cluster structure and data
2. Estimating stability of cluster solutions under random data perturbations
3. Estimating stability of cluster solutions with respect to changes in parameter values of a cluster-producing algorithm

Let us consider them in turn.

6.2.1 Scoring Correspondence between Clusters and Data

An internal validity index in clustering is a measure of correspondence between a cluster structure and the data from which it has been generated. The better the index value, the more reliable the cluster structure. The following is a selection from various formulations that have experimental or model-based substantiation.

6.2.1.1 Measures of Cluster Cohesion versus Isolation

6.2.1.1.1 Silhouette Width

The silhouette width of an entity $j \in I$ [91] is a popular measure defined as

$$sil(j) = (b(j) - a(j)) / \max(a(j), b(j))$$

where $a(j)$ is the average dissimilarity of j with its cluster and $b(j)$ the smallest average dissimilarity of j from the other clusters. Values $a(j)$ and $b(j)$ measure cohesion and isolation, respectively. Entities with large silhouette width are well clustered, while those with small width can be considered intermediate.

The greater the average silhouette width, the better the clustering. The measure has shown good results in some experiments [170].

6.2.1.1.2 Point-Biserial Correlation

The point-biserial correlation between a partition and distances is a global measure that has shown very good results in experiments described in [148]. As any other correlation coefficient, it can be introduced in the context of the data recovery approach similar to that of the linear regression in Section 7.1.3. We follow the presentation in [135].

Let us denote the matrix of between-entity distances by $D = (d_{ij})$. This can be just an input dissimilarity matrix. In our context, D is a matrix of squared Euclidean distances. For a partition $S = \{S_1, \ldots, S_K\}$ on I, let us consider the corresponding "ideal" dissimilarity matrix $s = (s_{ij})$ where $s_{ij} = 0$ if i and j belong to the same cluster S_k for some $k = 1, \ldots, K$, and $s_{ij} = 1$ if i and j belong to different clusters. Both matrices are considered here at unordered pairs of different i and j; the number of these pairs is obviously $N_* = N(N-1)/2$. Consider the coefficient of correlation (2.6) between these matrices as N_*-dimensional vectors. With elementary transformations, it can be proven that the coefficient can be expressed as follows:

$$r(D, s) = \frac{(d_b - d_w)\sqrt{N_b N_w}}{N_* \sigma_D} \tag{6.1}$$

where d_b is the average between cluster dissimilarity, d_w the average within cluster dissimilarity, N_b the number of unordered pairs of entities from different clusters and N_w the total number of unordered pairs of entities taken from the same cluster; σ_D is the standard deviation of all the distances from their grand mean.

This coefficient perhaps should be referred to as the uniform correlation coefficient because it is the coefficient of correlation in the problem of uniform partitioning [135], that is, finding a partition in which all within cluster distances are equal to the same number α and all between cluster distances are equal to the same number β, so that its distance matrix has the format $\lambda s + \mu$ where $\mu = \alpha$, $\lambda = \beta - \alpha$ and $s = (s_{ij})$ is the dissimilarity matrix of partition S defined above. In the framework of the data recovery approach, the problem of uniform partitioning is the problem of approximating matrix D with an unknown matrix $\lambda s + \mu$. This is the problem of regression analysis of D over s with the added stance that s is also unknown. It can be proven that the least squares solution to the uniform partitioning problem is that one that maximizes the correlation coefficient $r(D, s)$ (6.1).

It appears that the problem indirectly involves the requirement that the cluster sizes should be balanced, which works well when the underlying clusters are of more or less similar sizes [135]. The criterion may fail, however, when the underlying clusters drastically differ in sizes.

6.2.1.2 Indexes Derived Using the Data Recovery Approach

6.2.1.2.1 Attraction

This is a data-recovery-based analogue to the concept of silhouette width above. The attraction of entity $j \in I$ to its cluster is defined as $\beta(j) = a(j) - a/2$ where $a(j)$ is the average similarity of j to entities in its cluster and a the average within cluster similarity. If the data are given in feature-based format, the similarity is measured by the inner product of corresponding row vectors. Otherwise, it is taken from data as described in Section 7.4.2.3.

Attraction indexes are positive in almost all clusterings obtained with the K-Means and Ward-like algorithms; still, the greater they are, the better the clustering. The average attraction coefficient can be used as a measure of cluster tightness. However, in our preliminary experiments, the silhouette width outperforms the attraction index.

6.2.1.2.2 Contribution of the Cluster Structure to the Data Scatter

A foremost index of validity of a cluster structure in the data recovery approach is the measure of similarity between the observed data and that arising from a clustering model. Criterion $B(S, c)$, the cluster structure's contribution to the data scatter according to decomposition (7.20) in Section 7.2, measures exactly this type of similarity: the greater it is, the better the fit between the partition and the data.

It should be pointed out that the data recovery-based criteria are formulated in such a way that they seem to score cluster cohesion only. However, they implicitly do take into account cluster isolation as well, as shown in Chapter 7.

6.2.1.3 Indexes Derived from Probabilistic Clustering Models

A number of indexes have been suggested based on probabilistic models of cluster structure; for reviews, see [17,62,141]. Results of using these indexes highly depend on the accepted model which, typically, the user has no possibility of verifying.

6.2.2 Resampling Data for Validation

Resampling is a procedure oriented at testing and improving reliability of data-based estimators and rules with respect to changes in data. Basically, it produces an empirical distribution of clustering results and related indexes scoring the extent of dispersion with regard to random changes in the data. This type of analysis has become feasible with the development of computer hardware and is in its infancy as a discipline. After having read through a number of publications involving resampling and done some computational experiments, the author can think of the following system.

Resampling involves the following steps:

A: Generation of a number of data sets, *copies*, related to the data under consideration.

B: Running an accepted algorithm (such as regression, classification tree, or clustering) independently at each of the copies.

C: Evaluating the results.

D: Combining the results and evaluations.

Let us describe these steps in a greater detail for the standard entity-to-feature data format:

A. *Copy generation* can be done in various ways including *subsampling* or *splitting* or *bootstrapping* or *perturbing* the data set:

A1 *Subsampling*: A proportion α, $0 < \alpha < 1$, is specified and αN entities are selected randomly without replacement in a subsample; the copy data is constituted by the rows corresponding to selected entities.

A2 *Splitting*: Set I is randomly split in two parts of pre-specified sizes: the training and testing parts. This can be done as is, by generating a copy with a *train/testing split*, or via the so-called *cross validation* at which the entity set is randomly split in a pre-specified number of parts Q of approximately equal size, and Q copies are created simultaneously. A copy is created by considering one of the Q parts as the testing set, and its complement as the training set. Usually, Q is taken as either 2, 5 or 10. However, the case when Q is equal to the number of entities is also popular; it is called *leave-one-out* cross validation [76].

A3 *Bootstrapping*: In a bootstrapping process, exactly N row indices are selected but some rows may be left out of that since some rows may occur there two or more times [42]. The probability of a row not being selected can be estimated as follows. In each trial, the probability of a row not being selected is $1 - 1/N$, so that that the approximate proportion of rows never selected in the sampling process is $(1 - 1/N)^N \approx 1/e = 36.8\%$ of the total number of entities. That means that any bootstrap copy on average includes only about 63.2% of the entities, some of them several times. A similar process can be applied to columns of the data matrix [48].

A4 *Perturbing*: The size of a copy here is the same as in the original data set but the entries are perturbed, usually, by adding noise. In this process, a random real number is added to each entry. The added number is typically generated from the normal distribution with zero mean and a relatively small variance. In [93], a combination of bootstrapping and adding noise is described. For a K-Means found cluster structure (S, c), let us calculate the residuals e_{ik} in the data recovery model (7.18). An individual trial is defined then by randomly sampling with replacement among the residuals e_{iv} and putting the sampled values in Equation 7.18 to calculate the bootstrapped data.

B. *Running an algorithm* on a copy is done the same way as on the original data set except that if the copy is obtained by splitting, the algorithm is run over its training part only.

C. *Evaluating results* depends on the copy type. If copies are obtained by subsampling or bootstrapping or perturbing, the results on copies are compared with the results found at the original data set. Usually, this involves analysis of the distribution of the results found at the copies. If copies are obtained by splitting, the training part-based results are evaluated on the test part. Let us describe this in more detail for each of the copy types.

C1 *Subsample case.* The cluster structure found from a subsample is compared with the cluster structure on the same subsample resulting from the application of the algorithm to the original data set. The comparison can be done with any index such as *Rand* (discussed in Section 6.2.1) or chi-squared contingency coefficients (discussed in Section 2.2.3). For instance, in [112] the evaluation is done according to the averaged overlap index (6.12); in [162] relative distance between "sample" and "real" centroids is computed. This concerns situations in which subsample results do not depend on the subsample size, which is not so sometimes. For instance, if the fitted model involves a feature interval, the interval's length on an αN subsample will be, on average, α times the interval's length on the original data. Then the interval must be stretched out in the proportion $1/\alpha$ [112].

C2 *Split case.* The evaluation of a model fitted to the training data is rather easy when the model is that of prediction or description as in the case of the regression or classification tree. In these, the training part fitted model is applied as is to the testing part and the error score of the prediction/description in the testing part is calculated. When the fitted model is a cluster structure such as produced by K-Means, its evaluation in the testing part is not straightforward and can be done differently. For instance, in [39] the following method, Clest, is proposed. First, apply K-Means to the testing parts anew. Then build a prediction model, such as a classification tree, on the training part and apply it to the testing part to predict, for each testing part entity, what cluster it should belong to. This also can be done by using the minimum distance rule applied to the testing parts at the training part-based centroids [121]. In this way, two partitions are found on the testing part: that predicted from the training part and that found anew. The evaluation then is done by comparing these two partitions: the closer the better. An original use of the leave-one-out model for measuring similarity between entities is described in [78]; the authors use the minimum rather than average correlation over pair-wise correlations with one column removed; this also cleans the outlier effects.

C3 *Bootstrap case.* In the case of bootstrap, evaluation can be done by comparing those parts of the two cluster structures, which relate to that aspect of data that was not changed by the sampling. In particular, Felsenstein [48] bootstraps columns (features) rather than entities in building hierarchical clusters and more generally phylogenetic trees. To evaluate the results, entity clusters found on the bootstrap trial data are compared with those found on the original data: the greater the number of common clusters, the better.

C4 *Perturbation case.* The evaluation is especially simple when the data have been perturbed by adding noise, because the entities remain the same. Thus, the evaluation can be done by comparing the cluster structure found with the original data and cluster structure found at the perturbed data. In the case of partitions, any of the indexes of Section 6.2.1 would fit. In [93], evaluation is done by testing the perturbed data partition against the original centroids; the error is proportional to the number of cases in which the minimum distance rule assigns an entity to a wrong cluster.

D. *Combining results.* This can be done in various ways such as the following:

D1 *Combining evaluations.* Typically, this is just averaging the evaluations for each individual copy. The averaged score can be considered as a test result for the algorithm. In this way, one can select the best performing algorithm among those tested. This can also be applied to selection of parameters, such as the number of clusters, with the same algorithm [39,112]. The results on perturbed copies can be used to score confidence in various elements of the cluster structure found at the original data set. For instance, in [48], each of the hierarchic clusters is annotated by the proportion of copies on which the same algorithm produced the same cluster: the greater the proportion, the greater the confidence.

D2 *Averaging models.* Models found with different copies can be averaged if they are of the same format. The averaging is not necessarily done by merely averaging numerical values. For instance, a set of hierarchical clustering structures can be averaged into a structure that holds only those clusters that are found in a majority of the set structures [119]. In [36], centroids found at subsamples are taken as a data set which is clustered on its own to produce the averaged centroids.

D3 *Combining models.* When models have different formats, as in the case of decision trees that may have different splits over different features at different copies, the models can be combined to form a "committee" in such a way that it is their predictions are averaged rather than the copies. Such is the procedure referred to as bagging in [76].

6.2.3 Cross Validation of iK-Means Results

To illustrate the points above, let us consider a popular validation method, m-fold cross validation. According to this method, the entity set is randomly partitioned into m equal parts and m pairs of training and testing sets are formed by taking each one of the m parts as the testing set, with the rest used as the training set.

This scheme is easy to use regarding the problems of learning of decision rules: a decision rule is formed using a training set and then tested on the corresponding testing set. Then testing results are averaged over all m train-test experiments. How can this line of thought be applied to clustering?

In the literature, several methods for extension of the cross-validation techniques to clustering have been described (see references in Section 6.2.2). Some of them fall in the machine-learning framework and some in the data-mining framework. In the machine-learning framework, one tests the consistency of a clustering algorithm. To do this, results of the algorithm run over each of the m training sets are compared. But how can two clusterings be compared if they partition different sets? One way to do this is by extending each clustering from the training set to the full entity set by assigning appropriate cluster labels to the test set elements. Another way would be to compare partitions pairwise over the overlap of their training sets. The overlap is not necessarily small. If, for instance, $m = 10$, then each of the training sets covers 90% of entities and the pairwise overlap is 80% of the data set.

In data mining, the clustering results are tested rather than algorithms. In this framework, the selected clustering method is applied to the entire data set before the set is split into m equal-sized parts. Then m training sets are formed as usual, by removing one of the parts and combining the other parts. These training sets are used to verify the clustering results found on the entire data set. To do this, the clustering algorithm can be applied to each of the m training sets and the found clustering is compared with that obtained on the entire data set.

Here are examples of how these strategies can be implemented.

Example 6.1: Cross-Validation of iK-Means Clusters of the Market Towns Data

Let us address the issue of consistency of clustering results, a data-mining approach. We already have found a set of clusters in the Market towns data, see Example 3.12 on p. 112. This will be referred to as the base clustering. To explore how stable base clusters are, let us do 10-fold cross-validation. First, randomly partition the set of 45 towns in 10 parts of approximately same size, five parts of four towns each and five parts of five towns each. Taking out each of the parts, we obtain ten 90% subsamples of the original data as the training sets and run iK-Means on each of them. To see how much these clusterings differ from the base clustering found using the entire set, one can use three scoring functions, as follows.

1. *Average distance between centroids (adc).* Let c_k $(k = 1, \ldots, 7)$ be base centroids and c'_l $(l = 1, \ldots, L)$ centroids of the clustering found on a 90% sample. For each c_k find the nearest c'_l over $l = 1, \ldots, L$, calculate $d(c_k, c'_l)$ and average the distance over all $k = 1, \ldots, 7$. (The correspondence between c_k and c'_l can also be established with the so-called best matching techniques [6].) This average distance scores the difference between base clusters and sample clusters. The smaller it is the more consistent is the base clustering.

2. *Relative distance between partitions of samples (M).* Given a 90% training sample, let us compare two partitions of it: (a) the partition found on it with the clustering algorithm and (b) the base partition constrained to the sample. Cross classifying these two partitions, one can obtain a contingency table $P = (p_{tu})$ of frequencies p_{tu} of sample entities belonging to the t-th class of one partition and the u-th class of the other. The distance (or mismatch coefficient or Mirkin distance [126]) is

$$M = \left(\sum_t p_{t+}^2 + \sum_u p_{+u}^2 - 2 \sum_{l,u} p_{tu}^2 \right) / 2$$

where p_{t+} and p_{+u} are summary frequencies over rows and columns of P, as is introduced later in formula (6.13).

3. *Relative chi-squared contingency coefficient (T).* This is computed over the cross-classification too. The chi-squared coefficient (2.12), (2.13) formula is

$$\chi^2 = \sum_{t,u} p_{tu}^2 / (p_{t+} p_{+u}) - 1$$

The recommended version is a normalized index $T = \chi^2 / \sqrt{(K-1)(L-1)}$, the Tchouproff coefficient which cannot be greater than 1.

Averaged results of 15 independent 10-fold cross validation tests are presented in the left column of Table 6.1 together with the standard deviations (in the parentheses).

One can see that the distances *adc* and *M* are low and contingency coefficient *T* is high. But how low and how high are they? Can any cornerstones or benchmarks be found?

One may wish to compare *adc* with the average distance between uniformly random vectors. This is not difficult, because the average squared difference $(x - y)^2$ between numbers x and y that are uniformly random in a unity interval is $1/6$. This implies that the average distance in a 12-dimensional space is 2 which is by far greater than the observed 0.064.

TABLE 6.1

Averaged Results of 15 Cross Validations of Market
Towns Clusters with Real and Random Data

Method	Real Data	Random Data
adc	0.064 (0.038)	0.180 (0.061)
M	0.018 (0.018)	0.091 (0.036)
T	0.865 (0.084)	0.658 (0.096)

This difference, however, should not impress anybody, because the distance 2 refers to an unclustered set. Let us generate thus a uniformly random 45×12 data table and simulate the same computations as with the real data. Results of these computations are in the column on the right in Table 6.1. We can see that distances *adc* and M_s over random data are small too; however, they are three to five times greater than those on the real data. If one accepts the hypothesis that the average distances at random and real data may be considered as sampling averages of normal or chi-squared distributions, one may consider a statistical test of difference such as that by Fisher [76,92] to be appropriate and lead to a statistically sound conclusion that the hypothesis that the clustering of real data differs from that of random data can be accepted with a great confidence level. ∎

Example 6.2: Cross Validation of iK-Means Algorithm on the Market Towns Data

In this example, the cross-validation techniques are applied in the machine-learning framework, that is to say, we are going to address the issue of consistency of the clustering algorithm itself rather than just its results.

Thus, the partitions found on the training samples will be compared not with the base clustering but with each other. A 10-fold cross validation is applied here as in the previous example. Ten 90% cross-validation subsamples of the original data are produced and iK-Means is applied to each of them. Two types of comparison between the 10 subsample partitions are used, as follows:

1. *Comparing partitions on common parts.* Two 90% training samples' overlap comprises 80% of the original entities, which suffices to build their contingency table over common entities. Then both the distance M and chi-squared T coefficients can be used.

2. *Comparing partitions by extending them to the entire entity set.* Given a 90% training sample, let us first extend it to the entire entity set. To do so, each entity from the 10% testing set is assigned to the cluster whose centroid is the nearest to the entity. Having all ten 90% partitions extended in this way to the entire data set, their pair-wise contingency tables are built and scoring functions, the distance M and chi-squared T coefficients, are calculated.

TABLE 6.2

Averaged Comparison Scores between iK-Means
Results at 80% Real Market Towns and Random Data

Method	Real Data	Random Data
M_s	0.027 (0.025)	0.111 (0.052)
T	0.848 (0.098)	0.604 (0.172)

TABLE 6.3

Averaged Comparison Scores between iK-Means
Results Extended to All Real Market Towns and
Random Data

Method	Real Data	Random Data
M_s	0.032 (0.028)	0.128 (0.053)
T	0.832 (0.098)	0.544 (0.179)

Tables 6.2 and 6.3 present results of the pair-wise comparison between partitions found by iK-Means applied to the Market towns data in both ways, on 80% overlaps and on the entire data set after extension, averaged over fifteen 10-fold cross-validation experiments. The cluster discarding threshold in iK-Means has been set to 1 as in the previous examples. We can see that these are similar to figures observed in the previous example though the overall consistency of clustering results decreases here, especially when comparisons are conducted over extended partitions.

It should be noted that the issue of consistency of the algorithm is treated somewhat simplistically in this example, with respect to the Market towns data only, not to a pool of data structures. ∎

6.3 Interpretation Aids in the Data Recovery Perspective

Interpretation is an important part of clustering, especially from the knowledge discovery perspective in which it is a validation tool as well. The data recovery view of clustering allows us to fill in some gaps here as described in this section.

6.3.1 Conventional Interpretation Aids

Two conventional tools for interpreting K-Means clustering results, partition S and cluster centroids $c = \{c_1, \ldots, c_K\}$ are

1. Analysis of cluster centroids c_k
2. Analysis of bivariate distributions between cluster partition $S = \{S_k\}$ and various features

In fact, under the zero–one coding system for categories, cross-classification frequencies are but cluster centroids, which allows us to safely suggest that the cluster centroid at various feature spaces is the only conventional interpretation aid.

> **Example 6.3: Conventional Interpretation Aids Applied to Market Towns Clusters**
>
> Consider Table 6.4 displaying centroids of the seven clusters of Market towns data both in real and range standardized scales. These show some tendencies rather clearly. For instance, the first cluster is a set of larger towns that score 30–50% higher than average on almost all of the 12 features. Similarly, cluster 3 obviously relates to smaller than average towns. However, in other cases, it is not always clear what features cause the separation of some clusters. For instance, both clusters 6 and 7 seem too close to the average to have any real differences at all. The issue of computationally finding out the differences is addressed in this section later on.   ■

6.3.2 Contribution and Relative Contribution Tables

Given a partition S of data set I over data matrix Y, the data recovery approach leads to further helpful interpretation aids:

1. Decomposition of the explained part of the data scatter in contributions of cluster–feature pairs (table ScaD)
2. Quetelet coefficients for the decomposition (table QScaD)

According to Equations 3.4 and 3.5, clustering decomposes the data scatter $T(Y)$ in the explained and unexplained parts, $B(S, c)$ and $W(S, c)$, respectively. The explained part can be further presented as the sum of additive items $B_{kv} = N_k c_{kv}^2$, which account for the contribution of every pair S_k ($k = 1, \ldots, K$) and $v \in V$, a cluster and a feature. The unexplained part can be further additively decomposed in contributions $W_v = \sum_{k=1}^{K} \sum_{i \in S_k} (y_{iv} - c_{kv})^2$, which can be differently expressed as $W_v = T_v - B_{+v}$ where T_v and B_{+v} are parts of $T(Y)$ and $B(S, c)$ related to feature $v \in V$, $T_v = \sum_{i \in I} y_{iv}^2$ and $B_{+v} = \sum_{k=1}^{K} B_{kv}$.

This can be displayed as a decomposition of $T(Y)$ in a table ScaD whose rows correspond to clusters, columns to variables and entries to the contributions (see Table 6.5).

Summary rows, Expl and Total, and column, Total, are added to the table; they can be expressed as percentages of the data scatter $T(Y)$. The notation follows the notation of the contingency and redistribution data. The row Unex accounts for the "unexplained" differences $W_v = T_v - B_{+v}$. The contributions highlight relative roles of features both at individual clusters and in total.

These can be extended to within cluster levels (see Table 6.10 further on as an example).

TABLE 6.4

Patterns of Market Towns in the Cluster Structure Found with iK-Means

k/#	Centr	P	PS	Do	Ho	Ba	Su	Pe	DIY	SP	PO	CAB	FM
1	Real	18,484	7.63	3.63	1.13	11.63	4.63	4.13	1.00	1.38	6.38	1.25	0.38
8	Stand	0.51	0.38	0.56	0.36	0.38	0.38	0.30	0.26	0.44	0.47	0.30	0.17
2	Real	5268	2.17	0.83	0.50	4.67	1.83	1.67	0.00	0.50	1.67	0.67	1.00
6	Stand	−0.10	−0.07	−0.14	0.05	0.02	−0.01	−0.05	−0.07	0.01	−0.12	0.01	0.80
3	Real	2597	1.17	0.50	0.00	1.22	0.61	0.89	0.00	0.06	1.44	0.11	0.00
18	Stand	−0.22	−0.15	−0.22	−0.20	−0.16	−0.19	−0.17	−0.07	−0.22	−0.15	−0.27	−0.20
4	Real	11245	3.67	2.00	1.33	5.33	2.33	3.67	0.67	1.00	2.33	1.33	0.00
3	Stand	0.18	0.05	0.16	0.47	0.05	0.06	0.23	0.15	0.26	−0.04	0.34	−0.20
5	Real	5347	2.50	0.00	1.00	2.00	1.50	2.00	0.00	0.50	1.50	1.00	0.00
2	Stand	−0.09	−0.04	−0.34	0.30	−0.12	−0.06	−0.01	−0.07	0.01	−0.14	0.18	−0.20
6	Real	8675	3.80	2.00	0.00	3.20	2.00	2.40	0.00	0.00	2.80	0.80	0.00
5	Stand	0.06	0.06	0.16	−0.20	−0.06	0.01	0.05	−0.07	−0.24	0.02	0.08	−0.20
7	Real	5593	2.00	1.00	0.00	5.00	2.67	2.00	0.00	1.00	2.33	1.00	0.00
3	Stand	−0.08	−0.09	−0.09	−0.20	0.04	0.10	−0.01	−0.07	0.26	−0.04	0.18	−0.20

Note: First column displays cluster labels (top) and cardinalities (bottom).

TABLE 6.5

ScaD: Decomposition of the Data Scatter over a K-Means
Cluster Structure

Feature Cluster	$f1$	$f2$	fM	Total
S_1	B_{11}	B_{12}	B_{1M}	B_{1+}
S_2	B_{21}	B_{22}	B_{2M}	B_{2+}
S_K	B_{K1}	B_{K2}	B_{KM}	B_{K+}
Expl	B_{+1}	B_{+2}	B_{+M}	$B(S,c)$
Unex	W_1	W_2	W_M	$W(S,c)$
Total	T_1	T_2	T_M	$T(Y)$

Example 6.4: Contribution Table ScaD for Market Towns Clusters

Table 6.6 presents the Market towns data scatter decomposed over both
clusters and features, as an example of the general Table 6.5.

The table shows that, among the variables, the maximum contribution
to the data scatter is reached at FM. The least contributing is DIY. The
value of the ratio of the explained part of DIY to the total contribution,
$0.79/1.75 = 0.451$, amounts to the correlation ratio between the partition
and DIY, as explained in Sections 6.3.4 and 7.2.3.

The entries in the table actually combine together the cardinalities of
clusters with the squared differences between the grand mean vector and
within-cluster centroids. Some show an exceptional value such as contribution 3.84 of FM to cluster 2, which covers more than 50% of the
total contribution of FM and more than 90% of the total contribution

TABLE 6.6

Table ScaD at Market Towns

Cl-r	P	PS	Do	Ho	Ba	Su	Pe	DIY	SP	PO	CAB	FM	Total	Tot. (%)
1	2.09	1.18	2.53	1.05	1.19	1.18	0.71	0.54	1.57	1.76	0.73	0.24	14.77	35.13
2	0.06	0.03	0.11	0.01	0.00	0.00	0.02	0.03	0.00	0.09	0.00	3.84	4.19	9.97
3	0.86	0.43	0.87	0.72	0.48	0.64	0.49	0.10	0.85	0.39	1.28	0.72	7.82	18.60
4	0.10	0.01	0.07	0.65	0.01	0.01	0.16	0.07	0.20	0.00	0.36	0.12	1.75	4.17
5	0.02	0.00	0.24	0.18	0.03	0.01	0.00	0.01	0.00	0.04	0.06	0.08	0.67	1.59
6	0.02	0.02	0.12	0.20	0.02	0.00	0.01	0.03	0.30	0.00	0.03	0.20	0.95	2.26
7	0.02	0.02	0.03	0.12	0.00	0.03	0.00	0.02	0.20	0.00	0.09	0.12	0.66	1.56
Expl	3.16	1.69	3.96	2.94	1.72	1.88	1.39	0.79	3.11	2.29	2.56	5.33	30.81	73.28
Unex	0.40	0.59	0.70	0.76	0.62	0.79	1.02	0.96	1.20	0.79	1.52	1.88	11.23	26.72
Total	3.56	2.28	4.66	3.70	2.34	2.67	2.41	1.75	4.31	3.07	4.08	7.20	42.04	100.00

Note: Decomposition of the data scatter over clusters and features.

of the cluster. Still, overall they do not give much guidance in judging whose variables' contributions are most important in a cluster because of differences between relative contributions of individual rows and columns. ∎

To measure the relative value of contributions B_{kv}, let us utilize the property that they sum to the total data scatter and, thus, can be considered an instance of the redistribution data. The table of contributions can be analyzed in the same way as a contingency table (see Section 2.2.3). Let us define, in particular, the relative contribution of feature v to cluster S_k, $B(k/v) = B_{kv}/T_v$, to show what part of the variable contribution goes to the cluster. The total explained part of T_v, $B_v = B_{+v}/T_v = \sum_{k=1}^{K} B(k/v)$ is equal to the correlation ratio $\eta^2(S, v)$ introduced in Section 2.2.3.

More sensitive measures can be introduced to compare the relative contributions $B(k/v)$ with the contribution of cluster S_k, $B_{k+} = \sum_{v \in V} B_{kv} = N_k d(0, c_k)$, related to the total data scatter $T(Y)$. These are similar to Quetelet coefficients for redistribution data: the difference $g(k/v) = B(k/v) - B_{k+}/T(Y)$ and the relative difference $q(k/v) = g(k/v)/(B_{k+}/T(Y)) = \frac{T(Y)B_{kv}}{T_v B_{k+}} - 1$. The former compares the contribution of v with the average contribution of variables to S_k. The latter relates this to the cluster's contribution. Index $q(k/v)$ can also be expressed as the ratio of the relative contributions of v: within S_k, B_{kv}/B_{k+}, and in the whole data, $T_v/T(Y)$. We refer to $q(k/v)$ as the Relative contribution index, $\text{RCI}(k, v)$.

For each cluster k, features v with the largest $\text{RCI}(k, v)$ should be presented to the user for interpretation as those most relevant.

Example 6.5: Table QScaD of the Relative and Quetelet Indexes

All three indexes of association above, $B(k/v)$, $g(k/v)$, and RCI $q(k/v)$, applied to the Market towns data in Table 6.6, are presented in Table 6.7 under the cluster centroids.

Now the differences have become visible indeed. One can see, for instance, that variable Do highly contributes to cluster 5: RCI is 219.9. Why? As the upper number in the cell, 0, shows, this is a remarkable case indeed: no Doctor surgeries in the cluster at all.

The difference between clusters 6 and 7, that was virtually impossible to spot with other interpretation aids, now can be explained by the high RCI values of SP, in excess of 100%, reached at these clusters. A closer look at the data shows that there is a swimming pool in each town in cluster 7 and none in cluster 6. If the variable SP is removed, then clusters 6 and 7 will not differ anymore and merge together.

Overall, the seven non-trivial clusters can be considered as reflecting the following four tiers in the settlement system: the largest towns (Cluster 1), small towns (Cluster 3), large towns (Clusters 4 and 6) and small-to-average towns (Clusters 2, 5, and 7). In particular, the largest town Cluster 1 consists of towns whose population is two to three times larger than

TABLE 6.7

Tendencies of the Cluster Structure of Market Towns

k	C-d	P	PS	Do	Ho	Ba	Su	Pe	DIY	SP	PO	CAB	FM
1	Real	18,484.00	7.63	3.63	1.13	11.63	4.63	4.13	1.00	1.38	6.38	1.25	0.38
	Stand	0.51	0.38	0.56	0.36	0.38	0.38	0.30	0.26	0.44	0.47	0.30	0.17
	Rcnt	58.75	51.52	54.17	28.41	50.61	44.31	29.37	30.67	36.43	57.31	17.99	3.40
	Dcnt	23.62	16.39	19.04	−6.72	15.48	9.18	−5.76	−4.46	1.30	22.18	−17.14	−31.73
	RCI	67.23	46.65	54.21	−19.12	44.05	26.14	−16.40	−12.69	3.69	63.15	−48.80	−90.31
2	Real	5267.67	2.17	0.83	0.50	4.67	1.83	1.67	0.00	0.50	1.67	0.67	1.00
	Stand	−0.10	−0.07	−0.14	0.05	0.02	−0.01	−0.05	−0.07	0.01	−0.12	0.01	0.80
	Rcnt	1.54	1.33	2.38	0.41	0.09	0.05	0.73	1.88	0.00	2.79	0.02	53.33
	Dcnt	−8.43	−8.64	−7.59	−9.57	−9.88	−9.93	−9.24	−8.09	−9.97	−7.18	−9.95	43.36
	RCI	−84.52	−86.61	−76.08	−95.93	−99.10	−99.54	−92.72	−81.17	−99.96	−72.05	−99.82	434.89
3	Real	2597.28	1.17	0.50	0.00	1.22	0.61	0.89	0.00	0.06	1.44	0.11	0.00
	Stand	−0.22	−0.15	−0.22	−0.20	−0.16	−0.19	−0.17	−0.07	−0.22	−0.15	−0.27	−0.20
	Rcnt	24.11	18.84	18.60	19.46	20.31	24.06	20.38	5.63	19.60	12.70	31.39	10.00
	Dcnt	5.51	0.24	−0.00	0.86	1.71	5.46	1.79	−12.96	1.00	−5.90	12.79	−8.60
	RCI	29.62	1.30	−0.01	4.63	9.20	29.36	9.61	−69.71	5.39	−31.70	68.78	−46.23
4	Real	11,245.33	3.67	2.00	1.33	5.33	2.33	3.67	0.67	1.00	2.33	1.33	0.00
	Stand	0.18	0.05	0.16	0.47	0.05	0.06	0.23	0.15	0.26	−0.04	0.34	−0.20
	Rcnt	2.70	0.38	1.56	17.66	0.37	0.37	6.70	3.76	4.54	0.13	8.73	1.67
	Dcnt	−1.47	−3.79	−2.61	13.49	−3.80	−3.80	2.53	−0.41	0.38	−4.04	4.56	−2.50
	RCI	−35.32	−90.91	−62.62	323.75	−91.10	−91.19	60.68	−9.87	9.06	−96.94	109.47	−60.00

5	Real	5347.00	2.50	0.00	1.00	2.00	1.50	2.00	0.00	0.50	1.50	1.00	0.00
	Stand	−0.09	−0.04	−0.34	0.30	−0.12	−0.06	−0.01	−0.07	0.01	−0.14	0.18	−0.20
	Rcnt	0.48	0.17	5.09	4.86	1.26	0.29	0.00	0.63	0.00	1.28	1.55	1.11
	Dcnt	−1.12	−1.43	3.50	3.27	−0.33	−1.30	−1.59	−0.97	−1.59	−0.31	−0.04	−0.48
	RCI	−70.08	−89.58	219.92	205.73	−20.61	−81.96	−99.79	−60.66	−99.91	−19.48	−2.58	−30.17
6	Real	8674.60	3.80	2.00	0.00	3.20	2.00	2.40	0.00	0.00	2.80	0.80	0.00
	Stand	0.06	0.06	0.16	−0.20	−0.06	0.01	0.05	−0.07	−0.24	0.02	0.08	−0.20
	Rcnt	0.52	0.92	2.60	5.41	0.73	0.02	0.54	1.56	6.93	0.08	0.74	2.78
	Dcnt	−1.74	−1.34	0.34	3.15	−1.53	−2.24	−1.72	−0.69	4.67	−2.18	−1.52	0.52
	RCI	−77.04	−59.31	14.89	139.25	−67.69	−99.25	−76.27	−30.73	206.73	−96.44	−67.17	22.95
7	Real	5593.00	2.00	1.00	0.00	5.00	2.67	2.00	0.00	1.00	2.33	1.00	0.00
	Stand	−0.08	−0.09	−0.09	−0.20	0.04	0.10	−0.01	−0.07	0.26	−0.04	0.18	−0.20
	Rcnt	0.55	0.95	0.57	3.24	0.17	1.23	0.01	0.94	4.54	0.13	2.33	1.67
	Dcnt	−1.01	−0.61	−0.99	1.68	−1.39	−0.33	−1.56	−0.62	2.98	−1.43	0.76	0.11
	RCI	−64.79	−38.95	−63.22	107.78	−89.21	−20.98	−99.68	−39.84	191.16	−91.84	48.96	6.78
Ex.	Rcnt	88.64	74.11	84.97	79.45	73.54	70.32	57.72	45.07	72.05	74.42	62.74	73.96
	Dcnt	15.36	0.83	11.69	6.17	0.26	−2.95	−15.56	−28.21	−1.22	1.14	−10.54	0.68
	RCI	20.96	1.14	15.96	8.42	0.35	−4.03	−21.23	−38.49	−1.67	1.56	−14.38	0.93

Note: At each cluster, the first and second lines show the cluster's centroid in the raw and standardized scales; the other lines display the relative contribution $B(k/v)$ (Rcnt), difference $g(k/v)$ (Dcnt) and RCI $q(k,v)$, respectively, expressed as per cent. The last three lines show these three indexes applied to the explained parts of feature contributions.

the average, and they have respectively larger numbers of all the facilities, of which slightly over-represented are Post Offices, Doctors, Primary Schools and Banks. The small town Cluster 3 consists of the smallest towns with 2000–3000 residents. Respectively, the other facility values are also smaller, and some are absent altogether (such as DIY shops and Farmers' markets). Two large town clusters, Cluster 4 and Cluster 6, are formed by towns of 9000–12,000 residents. Although lack of such facilities as Farmers' market is common to them, Cluster 4 is by far richer with service facilities, that are absent in Cluster 6, which probably is the cause of the separation of the latter within the tier. Three small-to-average town clusters of about 5000 residents differ from each other by the presence of a few fancy objects that are absent from the small town cluster, as well as from the other two clusters of this tier. These objects are: a Farmers' market in Cluster 2, a Hospital in Cluster 5 and a Swimming pool in Cluster 7.　　　　　　　　　　　　　　　　　　　　　　　　　　　■

Example 6.6: ScaD and QScaD for Colleges

Tables 6.8 and 6.9 present similar decompositions with respect to subject-based clustering of the Colleges data in Table 2.13 on page 63. This

TABLE 6.8

ScaD for Colleges Data in Table 3.2

Subject	Stud	Acad	NS	EL	MSc	BSc	Cert.	Total	Total (%)
Science	0.03	0.05	0.04	1.17	0.09	0.00	0.06	1.43	24.08
Engin.	0.14	0.25	0.15	0.42	0.00	0.09	0.06	1.10	18.56
Arts	0.06	0.12	0.50	0.28	0.09	0.09	0.38	1.53	25.66
Expl	0.23	0.41	0.69	1.88	0.18	0.18	0.50	4.06	68.30
Unex	0.51	0.28	0.20	0.00	0.44	0.44	0.00	1.88	31.70
Total	0.74	0.69	0.89	1.88	0.63	0.63	0.50	5.95	100.00

TABLE 6.9

Relative Centroids

Subject	Stud	Acad	NS	EL	MSc	BSc	Cert.
Science	0.10	0.12	−0.11	−0.63	−0.02	0.17	−0.14
	−83.3	−70.5	−81.5	**158.1**	−100.0	−40.1	−50.5
Engin.	−0.21	−0.29	−0.22	0.38	0.17	−0.02	−0.14
	1.2	**91.1**	−9.8	20.2	−22.3	−100.0	−35.8
Arts	0.18	0.24	0.50	0.38	−0.22	−0.22	0.43
	−68.3	−33.0	**119.5**	−41.5	−43.3	−43.3	**197.0**

Note:　Cluster centroids standardized and Relative contribution indexes of variables, in cluster first and second lines, respectively; those highest are highlighted in bold.

time, only Quetelet indexes of variables, $RCI(k, v)$, are presented (in Table 6.9).

Table 6.9 shows feature EL as the most contributing to the Science cluster, feature Acad to the Engineering cluster and features NS and Certificate to the Arts cluster. Indeed, these clusters can be distinctively described by the predicates "EL = 0," "Acad < 280," and "NS > 3" (or "Course type is Certificate"), respectively. Note that the decisive role of Acad for the Engineering cluster cannot be recognized from the absolute contributions in Table 6.8: EL prevails over the Engineering cluster in that table. ∎

6.3.3 Cluster Representatives

The user can be interested in a conceptual description of a cluster, but they also can be interested in looking at the cluster via its representative, a "prototype." This is especially appealing when the representative is a well-known object. Such an object can give much better meaning to a cluster than a logical description in situations where entities are complex and the concepts used in description are superficial. This is the case, for instance, in mineralogy where a class of minerals can be represented by its stratotype, or in literary studies where a general concept can be represented by a popular literary character.

Conventionally, that entity is selected as a representative of its cluster, that is the nearest to its cluster's centroid. This strategy can be referred to as "the nearest in distance." It can be justified in terms of the square error criterion $W(S, c) = \sum_{k=1}^{K} \sum_{h \in S_k} d(y_h, c_k)$ (3.2). Indeed, the entity $h \in S_k$ which is the nearest to c_k contributes the least to $W(S, c)$, that is, to the unexplained part of the data scatter.

The contribution-based approach supplements this conventional distance-based approach. Decomposition of the data scatter (3.4) suggests a different strategy by relating to the explained rather than unexplained part of the data scatter. This strategy suggests that the cluster's representative must be the entity that maximally contributes to the explained part, $B(S, c) = \sum_{k=1}^{K} \sum_{v} c_{kv}^2 N_k$.

How can one compute the contribution of an entity to that? To manifest contributions of individual entities, let us take c_{kv}^2 in $B(S, c)$ as the product of c_{kv} with itself, and change one of the factors according to the definition, $c_{kv} = \sum_{i \in S_k} y_{iv} / N_k$. In this way, we obtain equation $c_{kv}^2 N_k = \sum_{i \in S_k} y_{iv} c_{kv}$. This leads to a formula for $B(S, c)$ as the summary inner product:

$$B(S, c) = \sum_{k=1}^{K} \sum_{i \in S_k} \sum_{v \in V} y_{iv} c_{kv} = \sum_{k=1}^{K} \sum_{i \in S_k} (y_i, c_k) \tag{6.2}$$

which shows that the contribution of entity $i \in S_k$ is (y_i, c_k).

The most contributing entity is "the nearest in inner product" to the cluster centroid, which may lead sometimes to different choices. Intuitively, the choice according to the inner product follows tendencies represented in c_k towards the whole of the data rather than c_k itself, which is manifested in the choice according to distance.

Example 6.7: Different Concepts of Cluster Representatives

The entity-based elements of the data scatter decomposition for the Science cluster from Table 6.8 are displayed in Table 6.10. Now some contributions are negative, which shows that a feature at an entity may be at odds with the cluster centroid. According to this table, the maximum contribution to the data scatter, 8.82%, is delivered by the Soliver college. Yet the minimum distance to the cluster's centroid is reached at a different college, Sembey.

To see why this may happen, let us take a closer look at the two colleges versus in their relation to within-cluster and grand means (Table 6.11).

Table 6.11 clearly shows that the cluster's centroid is greater than the grand mean on the first two components and smaller on the third one. These tendencies are better expressed in Sembey college over the first component and in Soliver college over the other two, which accords with the contributions in Table 6.10. Thus, Soliver wins over Sembey as better representing the differences between the cluster centroid and the overall gravity center, expressed in the grand mean. With the distance measure, no overall tendency can be taken into account. ■

TABLE 6.10

Decomposition of Feature Contributions to the Science Cluster in Table 6.8 (in Thousandth)

Name	Stud	Acad	NS	EL	MSc	BSc	Cert.	Cntr	Cntr (%)	Dist
Soliver	−19	29	37	391	61	5	21	524	8.82	222
Sembey	38	6	0	391	61	5	21	521	8.77	186
Sixpent	8	12	0	391	−36	−9	21	386	6.49	310
Science	27	46	37	1172	86	2	62	1431	24.08	0

Note: The right-hand column shows distances to the cluster's centroid.

TABLE 6.11

Two Science Colleges along with Features Contributing to Their Differences

Item	Stud	Acad	NS
Soliver	19.0	43.7	2
Sembey	29.4	36.0	3
Cluster mean	24.1	39.2	2.67
Grand mean	22.4	34.1	3.00

Example 6.8: Interpreting Bribery Clusters

Let us apply similar considerations to the five clusters of the Bribery data listed in Table 3.17. Since individual cases are not of interest here, no cluster representatives will be considered. However, it is highly advisable to consult the original data and their description on p. 19.

In cluster 1, the most contributing features are: Other branch (777%), Change of category (339%), and Level of client (142%). Here and further in this example the values in parentheses are relative contribution indexes RCI. By looking at the cluster's centroid, one can find specifics of these features in the cluster. In particular, all its cases appear to fall in Other branch, comprising such bodies as universities or hospitals. In each of the cases the client's issue was of a personal matter, and most times (six of the eight cases) the corrupt service provided was based on re-categorization of the client into a better category. The category Other branch (of feature Branch) appears to be distinctively describing the cluster: the eight cases in this category constitute the cluster.

Cluster 2 consists of 19 cases. Its most salient features are: Obstruction of justice (367%), Law enforcement (279%), and Occasional event (151%). By looking at the centroid values of these features, one can conclude: (1) all corruption cases in this cluster have occurred in the law enforcement system and (2) they are mostly done via obstruction of justice for occasional events. The fact (1) is not sufficient for distinctively describing the cluster since there are 34 cases, not just 19, that have occurred in the law enforcement branch. Two more conditions have been found by a cluster description algorithm, APPCOD (see in Section 6.4.3), to be conjunctively added to (1) to make the description distinctive: (3) the cases occurred at office levels higher than Organization and (4) no cover-up was involved.

Cluster 3 contains 10 cases for which the most salient categories are: Extortion in variable III Type of service (374%), Organization (189%), and Government (175%) in X Branch. Nine of the 10 cases occurred in the Government branch, overwhelmingly at the level of organization (feature I) and, also overwhelmingly, the office workers extorted money for rendering their supposedly free services (feature III). The client level here is always of an organization, though this feature is not that salient as the other three.

Cluster 4 contains seven cases, and its most contributing categories are: Favors in III (813%), Government in X (291%), and Federal level of Office (238%). Indeed, all its cases occurred in the government legislative and executive branches. The service provided was mostly Favors (six of seven cases). Federal level of corrupt office was not frequent, two cases only. Still, this frequency was much higher than the average, for the two cases are just half of the total number, four, of cases in which Federal level of office was involved.

Cluster 5 contains 11 cases and pertains to two salient features: Cover-up (707%) and Inspection (369%). All of the cases involve Cover-up as the service provided, mostly in inspection and monitoring activities (9 cases of 11). A distinctive description of this cluster can be defined to conjunct two statements: it is always a cover-up but not at the level of Organization.

Overall, the cluster structure leads to the following synopsis:

The most important is Branch which is the feature defining Russian corruption when looked at through the media glass. Different branches tend to supply different corrupt services. The government corruption involves either Extortion for rendering their free services to organizations (Cluster 3) or Favors (Cluster 4). The law enforcement corruption in higher offices is for either Obstruction of justice (Cluster 2) or Cover-up (Cluster 5). Actually, Cover-up does not exclusively belong in the law enforcement branch: it relates to all offices that are to inspect and monitor business activities (Cluster 5). Corruption cases in Other branch involve re-categorization of individual cases into more suitable categories (Cluster 1).   ∎

6.3.4 Measures of Association from ScaD Tables

Here, we are going to show that the summary contributions of clustering towards a feature in ScaD tables are compatible with traditional statistical measures of correlation considered in Section 2.2.

6.3.4.1 Quantitative Feature Case: Correlation Ratio

As proven in Section 7.2.3, the total contribution $B_{+v} = \sum_k B_{vk}$ of a quantitative feature v to the cluster-explained part of the scatter, presented in the ScaD tables, is proportional to the correlation ratio between v and cluster partition S, introduced in Section 2.2.2. In fact, the correlation ratios can be found by relating the row Expl to row Total in the general ScaD Table 6.5.

Example 6.9: Correlation Ratio from a ScaD Table

The correlation ratio of the variable P (Population resident) over the clustering in Table 6.6 can be found by relating the corresponding entries in rows Expl and Total; it is $3.16/3.56 = 0.89$. This relatively high value shows that the clustering closely follows this variable. In contrast, the clustering has rather little to do with variable DIY, the correlation ratio of which is equal to $0.79/1.75 = 0.45$.   ∎

6.3.4.2 Categorical Feature Case: Chi-Squared and Other Contingency Coefficients

The summary contribution of a nominal feature l to the clustering partition S has something to do with contingency coefficients introduced in Section 2.2.3. Denote the set of its categories by V_l. The contribution is equal to

$$B(S, l) = \frac{N}{|V_l|} \sum_{k=1}^{K} \sum_{v \in V_l} \frac{(p_{kv} - p_{k+}p_{+v})^2}{p_{k+}b_v^2} \tag{6.3}$$

where b_v stands for the scaling coefficient at the data standardization, as proven in Section 7.2.4. Divisor $|V_l|$, the number of categories, comes from the rescaling stage introduced in Section 2.4.

The coefficient $B(S, l)$ in Equation 6.3 can be further specified depending on the scaling coefficients b_v. The item under the summation in Equation 6.3 is

1. $(p_{kv} - p_k p_v)^2 / p_k$ if $b_v = 1$, the range
2. $(p_{kv} - p_k p_v)^2 / p_k p_v (1 - p_v)$ if $b_v = \sqrt{p_v(1 - p_v)}$, the Bernoullian standard deviation
3. $(p_{kv} - p_k p_v)^2 / p_k p_v$ if $b_u = \sqrt{p_u}$, the Poissonian standard deviation

Expressions 1 and 3 above lead to $B(S, l)$ being equal to the summary Quetelet coefficients introduced in Section 2.2.3. In this way, the Quetelet coefficients appear to relate to the data standardization. Specifically, G^2 corresponds to $b_v = 1$, and $Q^2 = X^2$ to $b_v = \sqrt{p_v}$. Case 2, the Bernoullian standardization, leads to an association coefficient which has not been considered in the literature as yet. The Bernoullian standardization coincides with that of z-scoring for the case of binary features.

Example 6.10: ScaD-Based Association between a Feature and Clustering

Let us consider the contingency table between the subject-based clustering of Colleges and the only nominal variable in the data, Course type (Table 6.12).

In this example, the dummy variables have been range normalized and then rescaled with $b'_v = \sqrt{3}$, which is consistent with Formula (6.3) at $b_v = 1$ and $|V_l| = 3$ for the summary contribution $B(S, l)$. Table 6.13 presents the values of $(p_{kv} - p_k p_v)^2 / (3 p_k / N)$ in each cell of the cross classification. In fact, these are entries of the full ScaD table in Table 6.8, p. 220, related to the categories of Course type (columns) and the subject-based clusters (rows), with row Total corresponding to row Expl in Table 6.8. In particular, the total contribution of the clustering and variable Course type is equal to $0.18 + 0.18 + 0.50 = 0.86$, or about 14.5% of the data scatter. ∎

TABLE 6.12

Cross Classification of the Subject-Based Partition and Course Type at the Eight Colleges (in Thousandth)

Class	MSc	BSc	Certificate	Total
Science	125	250	0	375
Engineering	250	125	0	375
Arts	0	0	250	250
Total	375	375	250	1000

TABLE 6.13

Elements of Calculation $B(S, l)$ According to
Formula (6.3) (in 10-thousandth)

Class	MSc	BSc	Certificate	Total
Science	17	851	625	1493
Engineering	851	17	625	1493
Arts	938	938	3750	5626
Total	1806	1806	5000	8606

6.3.5 Interpretation Aids for Cluster Up-Hierarchies

In the literature, the issue of interpretation of hierarchic clustering has not received much attention as yet. In this aspect, this section is unique. A number of interpretation aids derived within the data recovery approach [135] are described here.

Ward-like divisive clustering algorithms produce cluster trees rooted at the entire entity set whose leaves are not necessarily singletons, what was referred to as cluster up-hierarchies. Ward divisive clustering model leads to a bunch of interpretation aids specifically oriented at cluster up-hierarchies [135]. In contrast, Ward agglomerative tree is left without specific interpretation aids because agglomeration steps contribute to the unexplained part of clustering, as clearly seen in Chapter 7 (see decomposition (7.48)), and thus cannot be used in the manner in which contributions to the explained part are used. Divisive steps do contribute to the explained part. This allows us to propose useful split interpretation aids.

There can be three different aspects of the contribution-based interpretation aids with regard to a cluster up-hierarchy **S** because each split can be considered either in terms of

  i. Features, or

 ii. Covariances between features, or

iii. Individual entities.

Let us briefly describe the corresponding contributions.

(i) *Split-to-Feature.* At this level, one can take a look at contributions of cluster splits to the total contribution of a feature to the data scatter. As stated, data scatter $T(Y) = \sum_{v \in V} T_v$ where $T_v = (y_v, y_v) = \sum_{i \in I} y_{iv}^2$ is the contribution of feature $v \in V$. The denotation y_v refers to column v of the pre-processed data matrix. According to Equation 7.45 at $u = v$, one has

$$T_v = (y_v, y_v) = \sum_w \frac{N_{w1} N_{w2}}{N_w} (c_{w1,v} - c_{w2,v})^2 + (e_v, e_v) \tag{6.4}$$

where summation goes over all internal hierarchy clusters S_w split in parts S_{w1} and S_{w2}. Each split contributes, therefore, $\frac{N_{w1}N_{w2}}{N_w}(c_{w1,v} - c_{w2,v})^2$; the larger a contribution the greater the variable's effect to the split.

The overall decomposition of the data scatter in Equation 7.49,

$$T(Y) = \sum_w \frac{N_{w1}N_{w2}}{N_w} d(c_{w1}, c_{w2}) + W(S, c) \qquad (6.5)$$

where S is the set of leaf clusters of an upper cluster hierarchy **S**, shows contributions of both splits and leaf clusters.

Both parts of the decomposition can be used for interpretation:

- Ward distances $\frac{N_{w1}N_{w2}}{N_w} d(c_{w1}, c_{w2})$ betwen split parts, to express differences between them for taxonomic purposes, and
- decomposition of the square error criterion $B(S, c)$ over leaf clusters, which has been employed in the analysis of results of K-Means clustering in the previous section; although they serve to purposes of typology rather than taxonomy.

Example 6.11: Split-to-Feature Aids on a Box Chart and in Table ScaD

Split-to-feature interpretation aids are displayed at the box chart in Figure 4.4. The upper split contributes 34.2% to the data scatter. Between cluster differences $(c_{w1v} - c_{w2v})^2$ contributing most are at variables NS and Certificate. The next split contributes 34.1% and is almost totally due to EL (25.2 out of 34.1).

A more complete picture can be seen in Table 6.14, which extends Table 6.8 ScaD by adding one more aspect: split contributions. Maximum

TABLE 6.14

ScaD Extended

Aid	Item	Stud	Acad	NS	EL	BSc	MSc	Cert	Total	Total (%)
Typ	Science	0.03	0.05	0.04	**1.17**	0.09	0.00	0.06	1.43	24.08
	Engineering	0.14	0.25	0.15	**0.42**	0.00	0.09	0.06	1.10	18.56
	Arts	0.06	0.12	**0.50**	0.28	0.09	0.09	**0.38**	1.53	25.66
	Expl	0.23	0.41	0.69	1.88	0.18	0.18	0.50	4.06	68.30
Tax	Split1	0.09	0.16	**0.67**	0.38	0.12	0.12	**0.50**	2.03	34.22
	Split2	0.14	0.25	0.02	**1.50**	0.06	0.06	0	2.03	34.09
	Expl	0.23	0.41	0.69	1.88	0.18	0.18	0.50	4.06	68.30
	Unex	0.51	0.28	0.20	0.00	0.44	0.44	0.00	1.88	31.70
	Total	0.74	0.69	0.89	1.88	0.63	0.63	0.50	5.95	100.00

Note: Decomposition of the data scatter over the subject-based hierarchy for Colleges data in Table 3.2. Contributions are provided for both clusters, also known as typology (upper part), and splits, also known as taxonomy (bottom part).

contributions are highlighted in bold-face. The upper part of the table supplies aids for typological analysis, treating each cluster as is, and the middle part for taxonomical analysis, providing aids for interpretation of splits. Both parts take into account those contributions that relate to the explained part of the leaf cluster partition.   ■

(ii) *Split-to-Covariance.* Feature-to-feature covariances are decomposed over splits according to cluster contributions equal to $\frac{N_{w1}N_{w2}}{N_w}(c_{w1,v} - c_{w2,v})(c_{w1,u} - c_{w2,u})$ according to Equation 7.45, where $u, v \in V$ are co-variate features. At this level, not only the quantities remain important for the purposes of comparison, but also one more phenomenon may occur. A covariance coefficient entry may appear with a different sign at a cluster, which indicates that in this split the association between the variables concerned changes its direction from positive to negative or vice versa. Such an observation may lead to insights into the cluster's meaning.

Example 6.12: Decomposition of Covariances over Splits

Let us consider covariances between variables Acad, NS and EL. Their total values, in thousandth, are presented in the left-hand part of the matrix equation below, and corresponding items related to the first and second splits in Figure 4.3. Entries on the main diagonal relate to the data scatter as discussed above.

	Acad	NS	EL		S1				S2	
Acad	87	58	−47	20	40	30		32	9	−77
NS	58	111	42	= 40	83	63	+	9	2	−21	+ ⋯
EL	−47	42	234	30	63	47		−77	−21	188

The decomposition of an entry may tell us a story. For instance, the global positive correlation between NS and EL (+42) becomes more expressed at the first split (+63) and negative at the second split (−21). Indeed, these two are at their highest in the Arts cluster and are in discord between Science and Engineering.   ■

(iii) *Split-to-Entry.* Any individual row-vector y_i in the data matrix can be decomposed according to a cluster up-hierarchy S into the sum of items contributed by clusters $S_w \in S$ containing i, plus a residual, which is zero when i itself constitutes a singleton cluster belonging to the hierarchy. This is guaranteed by the model (7.44). Each cluster S_{w1} containing i contributes the difference between its centroid and the centroid of its parent, $c_{w1} - c_w$, as described on p. 289. The larger the cluster, the more aggregated its contribution is.

Example 6.13: Decomposition of an Individual Entity over a Hierarchy

The decomposition of an individual entity vector, such as Sixpent row (entity 3 in the Colleges data), into items corresponding to individual

TABLE 6.15

Single Entity Data Decomposed Over Clusters Containing It; the Last Line Is the Residuals

Item	Stud	Acad	NS	EL	BSc	MSc	Cert
Sixpent	23.9	38	3	0	0	1	0
Grand Mean	22.45	34.13	3.00	0.63	0.38	0.38	0.25
1 split	−1.03	−3.32	−0.50	−0.13	0.13	0.13	−0.25
2 split	2.68	8.43	0.17	−0.50	0.17	−0.17	0.00
Residual	−0.20	−1.23	0.33	0.00	−0.67	0.67	0.00

clusters from the hierarchy presented in Figure 4.3 on p. 145 is illustrated in Table 6.15. The entity data constitute the first line in this table. The other lines refer to contributions of the three largest clusters in Figure 4.3 containing entity 3: the root, the not-Arts cluster, and the Science cluster. These clusters are formed before the splitting, after the first split, and after the second split, respectively. The last line contains residuals, due to the aggregate nature of the Science cluster. One can see, for instance, that $EL = 0$ for the entity initially grows to the grand mean, 0.63, and then step by step declines—the most drastic being at the second split separating Science from the rest. ∎

This type of analysis, which emphasizes unusually high negative or positive contributions, can be applied to a wide variety of hierarchical clustering results.

6.4 Conceptual Description of Clusters

Individual clusters can be interpreted at the following three levels: (1) cluster representative, (2) statistical description, and (3) conceptual description. Of these, we considered (1) and (2) in the previous sections. Here, we concentrate on the conceptual description of clusters.

6.4.1 False Positives and Negatives

A conceptual description of a cluster is a logic predicate over features defined on entities that should be true on entities belonging to the cluster and false on entities out of it. For instance, the cluster of Science colleges, according to Colleges data in Table 1.11, can be described conceptually with predicate "$EL = 0$" or predicate "$3800 \leq Stud \leq 5880 \& 2 \leq NS \leq 3$." These descriptions can be tested for any entity from the table. Obviously, the former predicate distinctively describes the cluster with no errors at all, while the latter admits one false-positive error: the entity Efinen satisfies the description but belongs to a different cluster. False-positive errors are entities that do not belong to the cluster but satisfy its description; in contrast, an entity from the cluster that does not satisfy its description is referred to as a false negative. Thus, the problem of conceptual description of clusters can be formalized as that of finding as brief and clear descriptions of them as possible while keeping the false-positive and false-negative errors as low as possible.

Conventionally, the issue of interpretation of clusters is considered art rather than science. The problem of finding a cluster description is supposed to have little interest on its own and be of interest only as an intermediate tool in cluster prediction: a cluster description sets a decision rule which is then applied to predict, given an observation, which class it belongs to. This equally applies to both supervised and unsupervised learning, that is, when clusters are pre-specified or found from data. In data mining and machine learning, the prediction problem is frequently referred to as that of classification so that a decision rule, which is not necessarily based on a conceptual description, is referred to as a classifier (see, for instance, [40]). In clustering, the problem of cluster description is part of the interpretation problem. Some authors even suggest conceptual descriptions to be a form of cluster representation [84].

6.4.2 Describing a Cluster with Production Rules

Given a cluster S_k, one can be interested in finding such a predicate A that the rule "if A, then S_k" holds with a high accuracy. Usually, predicates A are formulated as conjunctions of simple individual-feature-based predicates such as "value of feature x is greater than a." The problem of producing such production rules for a single-cluster S_k, without any relevance to other clusters, has attracted considerable attention of researchers. The problem fits well into the context of describing single clusters.

A production rule technique produces a set of statements "if A_p then S" $(p \in P)$ at which A_p is a conceptual, typically conjunctive, description of a subset in the feature space, that is, a predicate. A more precise formulation of a production rule would be "for all x in the feature space, $A_p(x) \rightarrow x \in S$." Rectangles surrounding black circles in Figure 6.1 correspond to such production rules. A production rule is characterized by two numbers: (a) the support, that is the number of entities in the set I satisfying A_p, (b) the precision, the proportion of entities from S among those entities of I that satisfy A_p. In some methods, descriptions A_p may overlap [96,198], in others they do not [3,28]. Production rules may have rather small supports. Moreover, different production rules may cover different parts of S and leave some parts not covered as shown in Figure 6.1.

Production rule techniques conventionally have been developed for prediction rather than for description. Especially popular are the so-called association rules [40,76,174]—a cornerstone of data-mining activities.

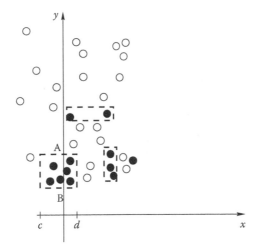

FIGURE 6.1
Rectangles in the feature space corresponding to production rules.

6.4.3 Comprehensive Conjunctive Description of a Cluster

Methods for deriving what can be called a comprehensive description is a less popular subject in the machine-learning literature. A comprehensive description of a subset $S \subset I$ is a statement "$x \in S$ if and only if $A(x)$" where $A(x)$ is a predicate defined for every entity in the feature space. Thus, a comprehensive description A is a unity of two production rules, "if $A(i)$ then $i \in S$" and "if $i \in S$ then $A(i)$."

Producing a comprehensive description is of interest in the situations in which S has been compiled in a process involving intuition and informal considerations, such as in the case of groups of protein folding or other microbiological taxonomy classes. This seems a good idea for K-Means clusters because they are, in general, compact and convex.

The extent of accuracy of a comprehensive description is characterized by the numbers of false positives and false negatives. A false positive for a description A of subset S is an element $i \in I - S$ satisfying $A(i)$, and a false negative is an element i from S, at which $A(i)$ is false. These errors correspond to the errors of the first and second kinds in the theory of statistical hypotheses if $A(x)$ is considered as a hypothesis about S.

A comprehensive description of S in the example of black and blank circles in Figure 6.2 is obtained by enclosing the subset S in a rectangle representing predicate $(a \leq x \leq b)$ and $(c \leq y \leq d)$.

Obviously, the problem of finding a unified comprehensive description of a group of entities can be solved by finding production rules predicting it and combining them disjunctively. However, when the number of disjunctive terms is constrained from above, in an extreme case just by 1, the problem becomes much more challenging especially if S is not "compact"

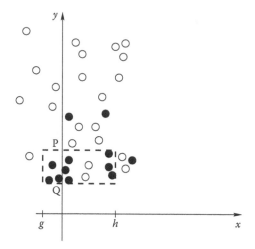

FIGURE 6.2
Rectangular description defining the numbers of false positives (blank circles within the box) and false negatives (black dots outside of the rectangle).

geometrically but spread over the data set I in the feature space. Methods for finding a comprehensive description of a cluster, based on forward and backward search strategies, have been considered in [137] and [168]. In [168], the difficulty of the problem was partly avoided by applying the techniques to cohesive clusters only and in [137] transformation of the space by arithmetically combining features was used as a device to reduce the spread of S over I.

An algorithm outlined in [137] has the following input: set I of entities described by continuously valued features $v \in V$ and a group $S \subset I$ to be described. The algorithm operates with interval feature predicates $P_f(a,b)$ defined for each feature f and real interval (a,b), of values $a \leq x \leq b$: $P_f(a,b)(x) = 1$ if the value of feature f at x falls between a and b, and $P_f(a,b)(x) = 0$ otherwise. The output is a conjunctive description of S with the interval feature predicates along with its false-positive and false-negative errors.

The algorithm involves two building steps:

1. *Finding a conjunctive description in a given feature space V.* To do this, all features are initially normalized by their ranges or other coefficients. Then all features are ordered according to their contribution weights B_{kv} which are proportional to the squared differences between their within-group $S \subset I$ averages and grand means, c_{kv}, as described in Section 6.3.2. A conjunctive description of S is then found by consecutively adding feature interval predicates $f_v(a,b)$ according to the sorted order, with (a,b) being the range of feature v within group S ($v \in V$). An interval predicate $f_v(a,b)$ is added to a current description

A only if it decreases the error.* This forward feature selection process stops after the last element of *V* is checked. Then a backward feature search strategy is applied to decrease the number of conjunctive items in the description, if needed.

2. *Expanding the feature space V.* This operation is applied if there are too many errors in the description *A* found on Step 1. It produces a new feature space by arithmetically combining original features $v \in V$ with those occurring in the found description *A*.

These two steps can be reiterated up to a pre-specified number of feature combining operations. Step 2 transforms and arithmetically combines the features to extend the set of potentially useful variables. Then Step 1 is applied to the feature space thus enhanced. The combined variables appearing in the derived decision rule are then used in the next iteration, for combining with the original variables again and again, until a pre-specified number of combining operations, *t*, is reached. For example, one iteration may produce a feature $v1 * v2$, the second may add to this another feature, leading to $v1 * v2 + v3$, and the third iteration may further divide this by $v4$, thus producing the combined feature $f = (v1 * v2 + v3)/v4$ with three combining operations.

This iterative combination process will be referred to as APPCOD (APProximate COmprehensive Description). An APPCOD iteration can be considered a specification of the recombination step in genetic algorithms [154].

In this way, given a feature set *V*, entity set *I* and class $S \subset I$, APPCOD produces a new feature set $F(V)$, a conjunction of the interval feature predicates based on $F(V)$, *A* and its errors, the numbers of false positives FP and false negatives FN. Since the algorithm uses within-*S* ranges of variables, it fits into the situation in which all features are quantitative. However, the algorithm can also be applied to categorical features presented with dummy zero–one variables; just intervals (a, b) here satisfy the condition $a = b$, that is, correspond to either of the values, 1 or 0.

It should be mentioned that a more conventional approach to finding a good description *A* with mixed scale variables involves building logical predicates over combinations of features [99].

Example 6.14: Combined Features to Describe Body Mass Groups

Let us apply APPCOD to the Body mass data in Table 1.8, with the restriction of no more than one feature combining operation. Depending on which of the groups, overweight or normal weight, is to be described, the algorithm first selects either Weight or Height. This gives a lousy description of the group with too many false positives. Then APPCOD combines the feature with the other one and produces the difference Height–Weight and the ratio Height/Weight for both the overweight and the normal-weight groups. The overweight group is described with these combined

---

* In this way, correlated features are eliminated without much fuss about it.

features with no errors, whereas the normal group cannot be described distinctively with them. This means that the former group is somewhat more compact in the combined feature space. The overweight group's comprehensive description is: $2.12 \leq H/W \leq 2.39$ and $89 \leq H - W \leq 98$, with no errors.

The second inequality can be reinterpreted as stating that the difference Height–Weight for the other, normal group, is about 100, which fits well into the known common sense rule: "Normally, the difference between Height (cm) and Weight (kg) ought to be about 100."

When the number of permitted feature combining operations increases to 2 and the number of conjunctive items is limited by 1, the method produces the only variable, $H * H/W$, which distinctively describes either group and, in fact, is the inverse BMI used to define the groups.   ∎

Example 6.15: Binary Data: Republicans and Democrats U.S. Congressional Voting

Let us describe results of applying the APPCOD method on the data set of 1984 United States Congressional Voting Records from [52] which contains records of 16 yes/no votings by the 267 democratic and 168 republican members of the House. It appears the set of republicans can be described by a rather simple predicate pff/esa $= 1$ that admits just 11 FP and 0 FN, 2.53% of total error. Here pff and esa are abbreviations of issues "physician-fee-freeze" and "export-administration-act-south-africa," and 1 codes "yes" and 0 "no."

This means that all republicans, and only 11 democrats, voted consistently on both issues, either both yes or both no, which cannot be described in such a short way without using the operation of arithmetic division. Since the method describes each individual class independently, we can check which of the classes, republican or democrat, is more coherent—that one admitting a simple description with minimum error. When applying APPCODE to various subsamples, the descriptions of democrats always had three to four times more errors than descriptions of republicans, which fits very well into the popular images of these parties [142].   ∎

6.4.4 Describing a Partition with Classification Trees

Having found a partition on the entity set, one can try to interpret it with the machinery of decision, or classification, trees that is used as a prediction tool in machine learning (see, e.g., [75,76,140]). However, the classification tree structure has been used, from the very moment of its inception back in 1960s, as a device for interpretation as well.

As illustrated in Figure 6.3, a classification tree recursively partitions the entity set into smaller clusters by splitting a parental cluster over a single feature. The root corresponds to the entire entity set and nodes, to subsets of entities, clusters, so that each cluster's children are the cluster's parts defined by values of a single interpreting feature x. Note that the trees in Figure 6.3

FIGURE 6.3
Decision trees for three classes of Colleges, Science, Engineering and Arts, made using categorical features, on the left, and quantitative features, on the right.

are binary: each interior node is split in two parts. This is a most convenient format, currently used in most popular programs.

Classification trees are very popular because they are simple to compute, understand, interpret and use. They are built from top to bottom in such a way that every split is made to maximize the homogeneity of the resulting subsets with respect to a desired target feature, in this case, the cluster-based partition. The splitting stops either when the homogeneity is enough for a reliable description of the clusters or when the subset of entities is too small to consider its splits reliable. A function scoring the extent of homogeneity to decide of the stopping is, basically, a measure of correlation between the partition of the entity set being built and the target partition.

When the process of building a tree is completed, each terminal node is assigned with a target class prevailing at that node. For example, both trees in Figure 6.3 are precise—each terminal class corresponds to one and only one product, which is a target class, so that each of the trees gives a precise conceptual description of all the teaching programs by conjunctions of the corresponding branch values. In particular, Science colleges can be described as those not in Certificate type, nor E-Learning utilized in the teaching process (left-side tree) or as that in which less than four courses are involved and the number of academics is >300. Both descriptions are fitting here since both give no errors at the set.

To build a binary classification tree, one needs the following information:

1. Set of input features X and a target partition T
2. A scoring function $F(S, T)$ that scores admissible partitions S against the target partition
3. A stopping criterion
4. A rule for the assignment of T-classes to terminal nodes

Let us comment on each of these items:

1. The input features are, typically, quantitative or nominal. Quantitative features are handled rather easily by testing all possible splits of their ranges in fragments "less than or equal to a" and "greater than a." More problematic are categorical features, especially those

with many categories because the number of possible binary splits of the set of all categories can be very large. However, this issue does not emerge at all if categorical features are preprocessed into the quantitative format of binary dummy variables corresponding to individual categories as advised in Section 2.4. Indeed, each of the dummy variables admits only one split—that separating the corresponding category from the rest, which reduces the number of possible splits to the number of categories—an approach becoming increasingly popular starting from [116].

2. Given a decision tree, its terminal nodes (leaves) form a partition S, which is considered then against the target partition T with a scoring function measuring the overall correlation $F(S, T)$. It appears that two most popular scoring functions, Gini index [20] and Pearson's chi-squared [66], fit perfectly in the framework of the data recovery approach. In fact, these two can be considered as implementations of the same approach of maximizing the contribution to the data scatter of the target categories—the only difference being the way the dummy variables representing the categories are normalized: (i) no normalization to make it Gini index or (ii) normalization by Poissonian standard deviations so that less frequent categories get more important, to make it Pearson's chi-squared. This sheds a fresh light on the criteria and suggests the user a way for choosing between the two. Indeed, if the user feels that the less fequent categories are more important than those more frequent, then $F(S, T)$ should be Pearson's chi-squared used in SPSS. If, in contrast, the user feels that the frequency has nothing to do with the category's importance, then they should use Gini or Gini-like criterion such as used in CART. Contemporary programs tend to split clusters in two parts only: (i) that corresponding to a category and the rest, for a categorical feature or (ii) given an a, those "less than or equal to a" and those "greater than a," for a quantitative feature. All possible splits are tested and that producing the largest value of the criterion is actually made, after which the process is reiterated.

3. Stopping rule typically assumes a degree of homogeneity of sets of entities, that is, clusters, corresponding to terminal nodes and, of course, their sizes: too small clusters are not stable and should be excluded.

4. Assigning a terminal node with a T category conventionally is done according to the maximum probability of a T class. The assignment can be of an issue in some situations: (i) if no obvious winning T class occurs in the cluster, (ii) if the category of interest is quite rare, that is, when T distribution is highly skewed. In this latter case, using Quetelet coefficients relating the node proportions with those in the entire set may help by revealing some great improvements in the proportions, thus leading to interesting tendencies discovered.

How one should define a function $F(S, T)$ to score correlation between the target partition T and partition S being built? Three possible approaches are:

1. *Reduction of uncertainty.* A popular idea is to use a measure of uncertainty, or impurity of a partition and score the goodness of split S by the reduction of uncertainty achieved when the split is made. If it is Gini index, or nominal variance, which is taken as the measure of uncertainty, the reduction of uncertainty is the popular impurity function utilized in a popular decision tree building program CART [20]. If it is entropy, which is taken as the measure of uncertainty, the reduction of uncertainty is the popular Information gain function utilized in another popular decision tree building program C4.5 [171].

2. *Association in contingency tables.* Let us use a correlation measure defined over the contingency table between partitions S and T such as Pearson's chi-squared. Indeed Pearson's chi-squared is used for building decision trees in a popular program, SPSS [66] as a statistical independence criterion, though, rather than a measure of association. In contrast, the latter view is advocated in this book (see pp. 225 and 239).

3. *Determinacy.* According to the analysis of variance approach in statistics, correlation can be measured by the proportion of the target feature variance taken into account by the partition S. This can be implemented if one considers a class of the target partition as represented by the corresponding dummy feature, which is equal to 1 at entities belonging to the class and 0 at the rest. It appears that both the impurity function and Pearson's chi-squared can be expressed as the summary explained proportion of the target variance, under different normalizations of the dummy variables course. To get the impurity function (Gini index), no normalization is needed at all, and Pearson's chi-squared emerges if each of the dummies is normalized by the square root of its frequency. That means that more frequent classes are less contributing according to Pearson's chi-squared. This might suggest the user to choose Pearson's chi-squared if they attend to this idea, or, in contrast, the impurity function if they think that the frequencies of target categories are irrelevant to their case.

Consider an entity set I with a pre-specified partition $T = \{T_l\}$ that is to be interpreted by producing a classification tree. At each step of the tree building process, a subset $J \subseteq I$ is to be split into a partition $S = \{S_k\}$ in such a way that S is as close as possible to $T(J)$ which is that part of T that holds on J. To measure the similarity between S and $T(J)$ take the confusion (contingency) table between S and $T(J)$, $P = (p_{kl})$ where p_{kl} is the proportion of J-entities in $S_k \cap T_l$, that expresses the confusion between S and $T(J)$.

Let us score the extent of reduction of uncertainty over $T(J)$ obtained when S becomes available, according to the first approach mentioned above. This works as follows: take a measure of uncertainty of partition $T(J)$, $v(T(J))$, and evaluate it at each of S-classes, $v(T(S_k))$, $(k = 1, \ldots, K)$. Then the average uncertainty on these classes will be $\sum_{k=1}^{K} p_k v(T(S_k))$, where p_k are proportions of in classes S_k in J, so that the reduction of uncertainty is equal to

$$v(T(J)/S) = v(T(J)) - \sum_{k=1}^{K} p_k v(T(S_k)) \qquad (6.6)$$

Two very popular measures defined according to Equation 6.6 are the so-called impurity function [20] and information gain [171].

The impurity function builds on Gini index as a measure of variance. The Gini index for partition T is defined as $G(T) = 1 - \sum_{l=1}^{L} p_l^2$ where p_l is the proportion of entities in T_l. If J is partitioned in clusters $S_k, k = 1, \ldots, K$, partitions T and S form a contingency table of relative frequencies $P = (p_{kl})$. Then the reduction (6.6) of the value of Gini coefficient due to partition S is equal to $\Delta(T(J), S) = G(T(J)) - \sum_{k=1}^{K} p_k G(T(S_k))$. This index $\Delta(T(J), S)$ is referred to as impurity of S over partition T. The greater the impurity, the better the split S.

It is not difficult to prove that $\Delta(T(J), S)$ relates to Quetelet indexes from Section 2.2.3. Indeed, $\Delta(T(J), S) = G^2(T, S)$ where $G^2(T, S)$ is the summary absolute Quetelet index defined by Equation 2.11 in Section 2.2.3. This can be proven with simple algebraic manipulations as described in [140].

The information gain function builds on entropy as a measure of uncertainty. Let us recall that entropy of partition T is $H(T) = -\sum_{l=1}^{L} p_l \log(p_l)$ where p_l is the proportion of entities in T_l. Then the reduction (6.6) of the value of entropy due to partition S is equal to $I(T(J), S) = H(T(J)) - \sum_k p_k H(T(S_k))$. This index is referred to as the information gain due to S. In fact, it is equal to a popular characteristic of the cross-classification of T and S, the mutual information defined as $I(T, S) = H(T) + H(S) - H(ST)$ where $H(ST)$ is entropy of the bivariate distribution represented by contingency table P. (The J argument is omitted here as irrelevant to the statement.)

The reduction of uncertainty measures are absolute differences that much depend on the measurement scale and, also, on the values of $v(T)$ and $v(S)$. Therefore, relative versions of the reduction of uncertainty measures normalized by $v(T)$ or $v(S)$ or both are frequently used. For example, popular program C4.5 [171] uses the information gain normalized by $H(S)$ and referred to as the information gain ratio.

To take a look at different approaches to scoring the similarity, let us assign each target class T_l with a binary variable x_l, a dummy, which is just a $1/0$ N-dimensional vector whose elements $x_{il} = 1$ if $i \in T_l$ and $x_{il} = 0$, otherwise. Consider, first, the average of x_l within cluster S_k: the number of unities among x_{il} such that $i \in S_k$ is obviously N_{kl}, the size of the intersection $S_k \cap T_l$ because $x_{il} = 1$ only if $i \in S_k$. That means that the within-S_k average of x_l is equal to $c_{kl} = N_{kl}/N_{k+}$ where N_{k+} stands for the size of S_k, or, p_{kl}/p_{k+} in terms of the relative contingency table P.

Let us now standardize each binary feature x_l by a scale shift a_l and rescaling factor $1/b_l$, according to the conventional standardization formula $y_l = (x_l - a_l)/b_l$. This will change the averages to $c_{kl} = (p_{kl}/p_{k+} - a_l)/b_l$. In data analysis, the scale shift is considered usually as positioning the data against a backdrop of the "norm," whereas the act of rescaling is to balance feature "weights." The feature means can be taken as the "norms" so that $a_l = p_{+l}$ because p_{+l} is the mean of the binary feature x_l. The choice of rescaling factors is somewhat less certain, though using all $b_l = 1$ should seem reasonable too, because all the dummies are just $1/0$ variables measured in the same scale. Incidentally, 1 is the range of x_l as well. Dispersion scores for x_l can be used as well. Therefore, the center's components are:

$$c_{kl} = \frac{p_{kl} - p_{k+}p_{+l}}{p_{k+}b_l} \tag{6.7}$$

According to formula (7.21) from Section 7.2.1, the pair (S_k, T_l) contributes to the data scatter the value $B_{kl} = N_k c_{kl}^2$ where N_k is the size of cluster S_k. From Equation 6.7, it follows that:

$$B_{kl} = N_k \frac{(p_{kl} - p_{k+}p_{+l})^2}{p_{k+}^2 b_l^2} = N \frac{(p_{kl} - p_{k+}p_{+l})^2}{p_{k+}b_l^2} \tag{6.8}$$

Therefore, the total contribution of the pair (S, T) to the total scatter of the set of standardized dummies representing T is equal to

$$B(T/S) = \sum_k \sum_l B_{kl} = N \sum_k \sum_l \frac{(p_{kl} - p_{k+}p_{+l})^2}{p_{k+}b_l^2} \tag{6.9}$$

This total contribution relates to both the averaged relative Quetelet coefficient (2.12) and the averaged absolute Quetelet coefficient (2.11). The latter, up to the constant N of course, emerges when all the rescaling factors $b_l = 1$. The former emerges when the rescaling factors are $b_l = \sqrt{p_l}$. This square root of the frequency has an appropriate meaning of an estimate of the standard deviation in Poisson's model of the variable. According to this model, N_{+l} unities are thrown randomly into the fragment of memory assigned for the storage of vector x_l. In fact, at this scaling system, $B(T/S) = \chi^2$, Pearson's chi-squared.

Let us summarize the proven facts about the impurity function and the Pearson chi-squared. Recall that the former is defined as the reduction (6.6) of Gini uncertainty index of T when S is taken into account. The latter is a measure of statistical independence between partitions T and S, a very different framework. Yet these have two other very similar meanings:

1. The impurity function is the averaged absolute Quetelet index G^2 and chi-squared is the averaged relative Quetelet index Q^2.

2. Both are equal to the total contribution of partition S to the summary data scatter of the set of dummy $1/0$ features corresponding

to T-classes and standardized by subtracting their means and (a) no rescaling for the impurity function and (b) further relating to $\sqrt{p_l}$ for the chi-squared.

Let us discuss, in brief, the problem of stopping the process of splitting when building a classification tree. A cluster is not to be split anymore if it is smaller than a user defined threshold or is homogeneous enough. Among homogeneity tests are: (a) large enough proportion of a target category in the cluster, say, above 80%; (b) small enough value of the scoring function which can be set to be 0.03 for Gini index, 0.08 for Pearson's chi-squared, and 0.15 for Information gain, that are chosen experimentally. These levels of magnitude reflect the functions' ranges: Gini index is very close to 0 hardly reaching 0.5 at all, Pearson's chi-squared, related to N, changes between 0 and 1 because it cannot be greater than the number of split parts minus 1, and Information gain can have larger values when the number of target categories is 3 or more.

Example 6.16: Classification Tree for Iris Taxa

At Iris dataset with its three taxa, *Iris setosa* and *Iris versicolor* and *Iris virginica*, taken as target classes, all the three scoring functions mentioned above, Impurity (Gini) function, Pearson's chi-squared and Information gain, lead to the same classification tree, presented in Figure 6.4.

The tree of 6.4 comprises three leaf clusters: One on the right consisting of all 50 *Iris setosa* specimens; that on the left containing 54 entities of which 49 are of *Iris versicolor* and five of *Iris virginica*; that in the middle containing 46 entities of which 45 are of *Iris virginica* and one of *Iris versicolor*. Altogether, this misplaces six entities leading to the accuracy of 96%. Of course, the accuracy would somewhat diminish if a cross-classification scheme is applied (see also [116] where a slightly different tree is found for Iris dataset).

Let us take a look at the action of each variable, w1 to w4, at each of the two splits in Table 6.16. Each time features w3 and w4 appear to be most contributing, so that at the first split, at which w3 and w4 give the same

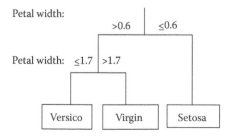

FIGURE 6.4
Classification tree for the three-taxon partition at Iris data set found by using each of Gini, Pearson's chi-squared and Information gain scoring functions.

TABLE 6.16

Values of Gini Index at the Best Split of Each
Feature on Iris data Set Clusters in Figure 6.4

Feature	First Split		Second Split	
	Value	Gini	Value	Gini
w1	5.4	0.228	6.1	0.107
w2	3.3	0.127	2.4	0.036
w3	1.9	0.333	4.7	0.374
w4	0.6	0.333	1.7	0.390

TABLE 6.17

Confusion Tables between Splits and the Target Taxa on Iris data set

Target Taxon	Iris Setosa	Iris Versicolor	Iris Virginica	Total
Full set				
$w4 \leq 1.7$	50	49	5	104
$w4 > 1.7$	0	1	45	46
Total	50	50	50	150
$w4 \leq 1.7$	50	49	5	104
$w4 > 1.7$	0	1	45	46
Total	50	50	50	150

impurity value, w4 made it through just because it is the last maximum
which is remembered by the program.

The tree involves just one feature, w4: Petal width, split over twice, first
at w4 = 0.6 and then at w4 = 1.7. The Pearson chi-squared value (related
to N of course) is 1 at the first split and 0.78 at the second. The Impurity
function grows by 0.33 at the first split and 0.39 at the second. The fact that
the second split value is greater than the first one may seem to be somewhat controversial. Indeed, the first split is supposed to be the best, so that
it is the first value that ought to be maximum. Nevertheless, this opinion
is wrong: if the first split was at w4 = 1.7 that would generate just 0.28 of
impurity value, less than the optimal 0.33 at w4 = 0.6. Why? Because the
first taxon has not been extracted yet and would grossly contribute to a
higher confusion (see the top part in Table 6.17). ∎

6.5 Mapping Clusters to Knowledge

6.5.1 Mapping a Cluster to Category

To externally validate a cluster is to successfully map it to a pre-defined category, which is represented, on the entity set, by a subset of entities falling in

the category. For example, a cluster can be a set of genes or proteins and the category may bear a functional meaning such as "metabolism" or "nucleotide repair." Another example: a set of documents retrieved by a search engine (cluster) versus the desired set.

Comparisons between subsets may be of interest in related contexts when the subsets refer to keywords or neighborhoods, among other examples.

To quantitatively assess the extent of similarity between two subsets. F and G, one usually takes the size of the overlap

$$a = |F \cap G| \tag{6.10}$$

and relates it to the size of either set, $p(G/F) = |F \cap G|/|F|$ or $p(F/G) = |F \cap G|/|G|$. When F is the cluster and G, pre-specified category, the former is referred to as the precision (how well the category matches the cluster) and the former, the recall (how well the cluster matches the category). If, for example, a search engine retrieves cluster F of a hundred documents when set G corresponding to query contains just 15 documents so that the overlap size is $F \cap G = 10$, then $p(G/F) = 10/100 = 0.1$ and $p(F/G) = 10/15 = 2/3$. This asymmetry leads the research community to invent symmetric measures of similarity.

Until recently, the most popular among symmetric measures has been the so-called Jaccard index $J = |F \cap G|/|F \cup G|$ relating the overlap size to the total number of different elements in the two sets. In the example, there are 105 documents involved altogether because $|F| = 100, |G| = 15$, and $|G \cap F| = 10$ so that only five documents in G are not accounted in F. Therefore, $J = 10/105 < 0.1$, which is somewhat counterintuitive: one would expect that the symmetric similarity value falls somewhere between the asymmetric conditional frequencies, but this one falls out. In the information retrieval and machine-learning communities, another symmetric measure is becoming increasingly popular, the so-called F-measure which is the harmonic mean of the precision and recall, $F = 2p(F/G)p(G/F)/[p(F/G) + p(G/F)]$. At the example, $F = 2 * 0.1 * 2/3/(0.1 + 2/3) = 4/23 = 0.17$ which is close to the minimum according to a property of the harmonic mean.

This author adheres to another symmetric measure, the arithmetic mean $m = (p(F/G) + p(G/F))/2$, which emerges in the data recovery clustering context [143]; it is referred to as Maryland bridge index, mbi, in [142]. At the example, $m = (1/10 + 2/3)/2 = 23/60 = 0.38$.

There are two intuitively obvious cases [142], at which the similarity measures can be tried to see how they fare against the common sense.

1. *Overlap of equal-sized subsets.* Consider a case at which the sizes of F and G are the same and their overlap is about half of the elements in each of them, so that $p(F/G) = 0.5$ and $p(G/F) = 0.5$. In this case, Jaccard's coefficient is $J = 50/(100 + 50) = 0.33$, whereas one would expect the similarity score to be $1/2$. This is exactly the value

TABLE 6.18

Four-Fold Table

Set	G	$\bar{G}$	Total
F	a	b	$a+b$
$\bar{F}$	c	d	$c+d$
Total	$a+c$	$b+d$	$a+b+c+d=N$

of F-measure and m. The Jaccard index obviously undervalues the similarity in this case.

2. *One subset being part of the other.* Consider a situation in which F is a small part of G so that, for example, $|F| = 0.2|G|$. In this case, $p(F/G) = 0.2$ and $p(G/F) = 1$. The Jaccard index $J = 0.2$, the arithmetic average $m = (1 + 0.2)/2 = 0.6$ and $F = 2 * 1 * 0.2/(1 + 0.2) = 0.33$. The latter follows a known property of the harmonic mean that its value is always much closer to the minimum rather than to the maximum. The common sense tells us that, most likely, F and G are highly related. For example, if these correspond to the contents of two genomes, then the cause of F being part of G would be, most likely, an evolutionary development at which the same organism, under some conditions, underwent a significant loss of genetic material. This may happen, for example, if that became a parasite, thus taking much of its building materials from the host. Similarly, for two text documents considered as bags of words, the relation $F \subset G$ would mean that F is just a brief version of G. In both cases, the user would prefer to have a higher estimate of similarity, which is provided by the m only.

When sets F and G are subsets of a larger set I, it can be of interest to compare the conditional frequencies $p(F/G)$ and $p(G/F)$ with their identifiable analogues on the total set, $p(F)$ and $p(G)$. This can be achieved with the so-called four-fold table presented in Table 6.18. It is a contingency table cross classifying F and its complement $\bar{F} = I - F$ with G and its complement $\bar{G} = I - G$. Its interior entries are cardinalities of intersections of corresponding sets and its marginal entries are cardinalities of the sets themselves.

A category which is over-represented in F so that $p(G/F) = a/(a + b)$ is much higher than $p(G) = (a + c)/N$ is obviously a characteristic that can be used to interpret cluster F. This approach is heavily relied upon when using ontologies to interpret clusters [173].

6.5.2 Mapping between Partitions

Given a cluster partition S and an expertly partition T, similarity between them can be expressed using their cross classification, that is, the contingency table presented in Table 6.19. Rows $k = 1, \ldots, K$ correspond to clusters in

TABLE 6.19

A Contingency Table or Cross Classification of Two Partitions,
$S = \{S_1, \ldots, S_K\}$ and $T = \{T_1, \ldots, T_L\}$ on the Entity Set I

Cluster	1	2	...	L	Total
1	N_{11}	N_{12}	...	N_{1L}	N_{1+}
2	N_{21}	N_{22}	...	N_{2L}	N_{2+}
...	...	...	....	...	...
K	N_{K1}	N_{K2}	...	N_{KL}	N_{K+}
Total	N_{+1}	N_{+2}	...	N_{+L}	N

partition $S = \{S_1, \ldots, S_K\}$ and columns $l = 1, \ldots, L$ to clusters in partition $T = \{T_1, \ldots, T_L\}$, and the (k, l)-th entry, to the number of co-occurrences of S_k and T_l, denoted by N_{kl}. That means $N_{kl} = R_k \cap S_l$. Summary frequencies of row and column categories, N_{k+} and N_{+l}, respectively, are presented in Table 6.19 on the margins titled "Total."

In clustering, partitions are traditionally compared via representing them by clique graphs or, almost equivalently, by binary equivalence relation matrices. Given a partition $S = \{S_1, \ldots, S_K\}$ on I, its clique graph Γ_S has I as its vertice set; edges connect any two entities that are in the same cluster $i, j \in S_k$ for some $k = 1, \ldots, K$. The matrix r_S is defined so that $r_{ij} = 1$ for any i, j belonging to the same class S_k for some $k = 1, \ldots, K$; otherwise, $r_{ij} = 0$. To correspond to graph Γ_S exactly, the matrix must be modified by removing the diagonal and one of the halves separated by the diagonal, either that under or above the diagonal, because of their symmetry.

Let us consider graph Γ_S as the set of its edges. Then, obviously, the cardinality of Γ_S is $|\Gamma_S| = \sum_{k=1}^K \binom{2}{N_{k+}}$ where $\binom{2}{N_{k+}} = N_{k+}(N_{k+} - 1)/2$ is the number of edges in the clique of Γ_S corresponding to cluster S_k. In terms of the corresponding binary matrices, this would be the number of $r_{ij} = 1$ for $i, j \in S_k$, that is, N_k^2 standing for $\binom{2}{N_{k+}}$ here. With a little arithmetic, this can be transformed to:

$$|\Gamma_S| = \left(\sum_{k=1}^K N_{k+}^2 - N \right) / 2 \qquad (6.11)$$

To compare graphs Γ_S and Γ_T as edge sets, one can invoke the four-fold table utilized on p. 243 for comparing sets (see Table 6.18). Table 6.20 presents it in the current context [5].

The elements of Table 6.20 are the cardinalities of the edge sets and their intersections. The only cardinalities of interest in what follows are $|\Gamma_S|$, $|\Gamma_T|$, and a. The first is presented in Equation 6.11 and a similar formula holds for the second, $|\Gamma_T| = (\sum_{l=1}^L N_{+l}^2 - N)/2$. The third, a, is similar too because it is the cardinality of the intersection of graphs Γ_S and Γ_T, which is itself a clique

TABLE 6.20

Four-Fold Table for Partition Graphs

Graph	Γ_T	$\bar{\Gamma}_T$	Total
Γ_S	a	b	$|\Gamma_S|$
$\bar{\Gamma}_S$	c	d	$c+d$
Total	$|\Gamma_T|$	$b+d$	$a+b+c+d = \binom{2}{N}$

graph corresponding to the intersection of partitions S and T. Thus,

$$a = \left(\sum_{k=1}^{K} \sum_{l=1}^{L} N_{kl}^2 - N \right) /2 \qquad (6.12)$$

which is a partition analogue to the overlap set similarity measure (6.10). It is of interest because it is proportional (up to the subtracted N which is due to the specifics of graphs) to the inner product of matrices r_S and r_T, which is frequently used as a measure of similarity between partitions on its own (see Section 7.5).

Among other popular measures of proximity between partitions are partition graph analogues to the mismatch and match coefficients m and s, to Jaccard coefficient J and geometric mean g on p. 242. These analogues are referred to as distance (mismatch), Rand, Jaccard, and Fowlkes–Mallows coefficients, respectively, and can be presented with the following formulas:

$$M = (|\Gamma_S| + |\Gamma_T| - 2a)/\binom{2}{N} \qquad (6.13)$$

$$\text{Rand} = 1 - (|\Gamma_S| + |\Gamma_T| - 2a)/\binom{2}{N} = 1 - M \qquad (6.14)$$

$$J = \frac{a}{|\Gamma_S| + |\Gamma_T| - a} \qquad (6.15)$$

and

$$\text{FM} = \frac{a}{\sqrt{|\Gamma_S||\Gamma_T|}} \qquad (6.16)$$

It should be noted that in the original formulation of FM by Fowlkes and Mallow [51] no N is subtracted from the sum of squares in $|\Gamma_S|$ and $|\Gamma_T|$. The Rand coefficient [172] is frequently used in the standardized form proposed in [81] as ARI. This equals to

$$ARI = \frac{a - \alpha}{\Gamma_S/2 + \Gamma_T/2 - \alpha} \qquad (6.17)$$

where $\alpha = |\Gamma_S||\Gamma_T|/\binom{2}{N}$.

It should be noted also that those partition-to-partition coefficients that are linear with respect to elements of the four-fold table are equal, up to the constant N, to the corresponding between-set measures averaged according to the bivariate distribution of the cross classification of S and T. This is true for the intersection a (6.12), distance M (6.13), and Rand (6.14).

In particular, intersection a (6.12) averages the values N_{kl} themselves. Distance M (6.13) and Rand (6.14) average values $N_{k+} + N_{+l} - 2N_{kl}$ and $N - N_{k+} - N_{+l} + 2N_{kl}$ respectively, as related to the set-theoretic difference between S_k and T_l. This suggests a potential extension of the partition similarity indexes by averaging other, non-linear, measures of set similarity such as the mean m or even the Quetelet coefficients. The averaged Quetelet coefficients, G^2 and Q^2, are akin to traditional contingency measures, which are traditionally labeled as deliberately attached to the case of statistical independence. However, the material of Sections 2.2.3 and 6.3.4 demonstrates that the measures can be utilized just as summary association measures in their own right with no relation to the statistical independence framework.

Example 6.17: Distance and Chi-Squared According to a Model Confusion Data

Consider the following contingency table (Table 6.21) that expresses the idea that binary features S and T significantly overlap and differ by only a small proportion ϵ of the total contents of their classes. If, for instance, $\epsilon = 0.1$, then the overlap is 90%; the smaller the ϵ, the greater the overlap.

Let us analyze how this translates into values of the contingency coefficients described above, in particular, the distance and chi-squared. The distance, or mismatch coefficient, is defined by Formulas 6.13, 6.12, and 6.11 so that it equals the sum of the marginal frequencies squared minus the doubled sum of squares of the contingency elements. To normalize, for the sake of convenience, we divide all items by the total $4n^2$, which is $2n$ squared, rather than by the binomial coefficient $\binom{2}{2n}$. This leads to the distance value

$$M = (4n^2 - 2(2(1 - \epsilon)^2 + 2\epsilon^2)n^2)/(4n^2) = 2\epsilon(1 - \epsilon)$$

To calculate the chi-squared, we use Formula 2.12 from Section 2.2.3. According to this formula, χ^2 is the total of contingency entries squared

TABLE 6.21

Model Confusion Data

Attribute	T	$\bar{T}$	Total
S	$(1 - \epsilon)n$	ϵn	n
$\bar{S}$	ϵn	$(1 - \epsilon)n$	n
Total	n	n	$2n$

and related to both column and row marginal frequencies, minus one:

$$\chi^2 = 2(1 - \epsilon)^2 + 2\epsilon^2 - 1 = 1 - 4\epsilon(1 - \epsilon)$$

These formulas imply that for this model dataset, $\chi^2 = 1 - 2M$.

For example, at $\epsilon = 0.05$ and $\epsilon = 0.1$, $M = 0.095$ and $M = 0.18$, respectively. The respective values of χ^2 are 0.64 and 0.81. This may give an operational meaning to otherwise incomprehensible intermediate values of M and χ^2. ∎

6.5.2.1 Match-Based Similarity versus Quetelet Association

We can distinguish between two types of subset-to-subset and respective partition-to-partition association measures:

M: Match between subsets/partitions measured by their overlap a in Equations 6.10, 6.12, and 6.14, and

C: Conditional probability and Quetelet indexes such as Equations 2.10 and 2.12 and 2.13.

Considered as they are, in terms of co-occurrences, these two types are so different that, to the author's knowledge, they have never been considered simultaneously. There is an opinion that, in the subset-to-subset setting, the former type applies to measuring similarity between entities whereas the latter type to measuring association between features.

However, there exists a framework, that of data recovery models for entity-to-entity similarity matrices, see for example, Section 7.5, in which these two types represent the same measure of correlation applied in two slightly different settings. In this framework, each partition $S = \{S_1, \ldots, S_K\}$ of the entity set I, or a nominal variable whose categories correspond to classes S_k ($k = 1, \ldots, K$), can be represented by a similarity matrix $s = (s_{ij})$ between $i, j \in I$ where $s_{ij} = 0$ for i and j from different classes and a positive real when i and j are from the same class. Consider two definitions:

M: $s_{ij} = 1$ if $i, j \in S_k$ for some $k = 1, \ldots, K$;

C: $s_{ij} = 1/N_k$ if $i, j \in S_k$ for some $k = 1, \ldots, K$ where N_k is the number of entities in S_k.

The assumption is that all similarities within S are the same, in the case M, or inversely proportional to the class size so that the larger the class the smaller the similarity, in the case C. The latter reflects the convention that the more frequent is a phenomenon the less information value in it.

Matrix $s = (s_{ij})$ corresponding to a partition S has dimension $N \times N$. Its average $\bar{s}$ is $\bar{s} = G(S) = 1 - \sum_k p_k^2$, Gini index, in the case M, or $\bar{s} = 1/N$, in the case C.

Let now S and T be two partitions on I. Corresponding similarity matrices $s = (s_{ij})$ and $t = (t_{ij})$, as well as their centered versions $s - \bar{s}$ and $t - \bar{t}$, can be considered as $N \times N$ vectors. Then the inner product of these vectors, in the case M, is equal to $\sum_{k=1}^{K} \sum_{l=1}^{L} N_{kl}^2$ featured in Equation 6.12, whereas the inner product of their centered versions in the case C is the Pearson chi-squared. This shows how the two types of partition-to-partition measures can be related to the two types of entity-to-entity similarities.

6.5.2.2 Average Distance in a Set of Partitions

In some situations, especially when the same clustering algorithm has been applied many times at different initial settings or subsamples, there can be many partitions to compare.

A good many-to-many dis/similarity measure can be produced by averaging a pair-wise dis/similarity between partitions. Let us denote given partitions on I by $S^1, S^2, \ldots, S^m$ where $m > 2$. Then the average distance $M(\{S^t\})$ will be defined by formula:

$$M(\{S^t\}) = \frac{1}{m^2} \sum_{u,w=1}^{m} M(S^u, S^w) \tag{6.18}$$

where $M(S^u, S^w)$ is defined by Formula 6.13.

An interesting measure was suggested in [156] based on the consensus partition matrix which is an entity-to-entity similarity matrix defined by

$$\mu(i,j) = \frac{\sum_{t=1}^{m} s_{ij}^t}{m^*(i,j)} \tag{6.19}$$

where s^t is the binary relation matrix corresponding to S_t with $s_{ij}^t = 1$ when i and j belong to the same class of S^t and $s_{ij}^t = 0$, otherwise, and $m^*(i,j)$ denotes the number of partitions S^t at which both i and j are present. The latter denotation concerns the case when (some of) partitions S^t have been found not necessarily at the entire entity set I but at its sampled parts.

In the situation in which all m partitions coincide, all the consensus values $\mu(i,j)$ are binary being either 1 or 0 depending on whether i and j belong to the same class or not, which means that the distribution of values $\mu(i,j)$ is bimodal in this case. The further away partitions from the coincidence, the further away the distribution of $\mu(i,j)$ from the bimodal. Thus, the authors of [156] suggest watching for the distribution of $\mu(i,j)$, its shape and the area under the empirical cumulative distribution,

$$A(\{S^t\}) = \sum_{l=2}^{L} (\mu_l - \mu_{l-1}) F(\mu_l) \tag{6.20}$$

where $\mu_1, \mu_2, \ldots, \mu_L$ is the sorted set of different entries $\mu(i,j)$ in the consensus matrix and $F(\mu)$ is the proportion of entries $\mu(i,j)$ that are smaller than μ.

In fact, the average distance (6.18) also characterizes the distribution of values $\mu(i,j)$. To see that, let us assume, for the sake of simplicity, that all partitions S^t are defined on the entire set I and denote the average value in the matrix by $\bar{\mu}$ and the variance by σ_μ^2.

Statement 6.1. The average distance $M(\{S^t\})$ (6.18) is proportional to $\bar{\mu}(1 - \bar{\mu}) - \sigma_\mu^2$.

Proof: According to the definition, $M(\{S^t\}) = \sum_{u,w=1}^{m} \sum_{i,j \in I} (s_{ij}^u - s_{ij}^w)^2 / m^2$. This can be rewritten as $M(\{S^t\}) = \sum_{i,j \in I} \sum_{u,w=1}^{m} (s_{ij}^u + s_{ij}^w - 2s_{ij}^u s_{ij}^w)/m^2$. Applying the operations to individual items, we obtain $M(\{S^t\}) = \sum_{i,j \in I} (\sum_{w=1}^{m} \mu(i,j)/m + \sum_{u=1}^{m} \mu(i,j)/m - 2\mu(i,j)\mu(i,j)) = 2\sum_{i,j \in I} (\mu(i,j) - \mu(i,j)^2)$. The proof of the statement follows then from the definition of the variance of the matrix, q.e.d.

As proven in Statement 2.1, p. 46, given a probability of smaller than the average values, the variance of a variable reaches its maximum when the variable is binary. The value $\bar{\mu}$ in this case can be assigned the meaning of being such a probability. Then the formula in Statement 6.1 shows that the average distance M, in fact, measures the difference between the maximum and observed values of the variance of the average partition matrix. Either this difference or the variance itself can be used as an index of similarity between the observed distribution of values $\mu(i,j)$ and the ideal case at which all of them coincide.

6.5.3 External Tree

When a hierarchical tree obtained by clustering has substantive meaning, it also can be used for interpretation of clusters. To illustrate this, let us consider the evolutionary trees built in the previous section and see how one can employ them to reconstruct evolutionary histories of individual genes.

Example 6.18: Using an Evolutionary Tree to Reconstruct the History of a Gene

After an evolutionary tree has been built or supplied externally (see Figure 4.5), one may consider the observed presence–absence patterns. Such a pattern represents a cluster of extant species, that is, tree leaves, that contain a given gene represented by the corresponding COG. Therefore, the issue to be addressed is to interpret such a cluster in the evolutionary perspective. That is, an evolutionary history of the gene should be reconstructed involving the node at which the gene has been gained and major events, the inheritance and loss, that have led to the current presence–absence pattern. Of course, the principle of parsimony, the

TABLE 6.22

Gene Profiles

No	COG	y	a	o	m	p	k	z	q	v	d	r	b	c	e	f	g	s	j
16	COG1709	0	1	1	1	1	1	1	0	0	0	0	0	0	0	1	0	0	0
31	COG1514	0	1	0	1	1	1	1	1	1	1	0	1	0	1	1	0	0	1
32	COG0017	1	1	1	1	1	1	1	0	0	1	0	1	1	1	0	1	1	0

Note: Presence–absence profiles of three COGs over 18 genomes.

Occam's razor, should be followed so that the number of the explanatory events is minimized. To illustrate this line of development, let us consider three COGs in Table 6.22: one from the data in Table 1.3 and two not used for building the tree, COG1514 2'-5' RNA ligase and COG0017 Aspartyl-tRNA synthetase.

Given a gene presence–absence profile, let us consider, for the sake of simplicity, that each of the events of emergence, loss and horizontal transfer of the gene is assigned the same penalty weight while events of inheritance along the tree are not penalized as conforming to the tree structure. Then, COG1709 (the first line of Table 6.22) can be thought of as having emerged in the ancestor of all archaea (node corresponding to cluster *aompkz* and then horizontally transferred to species *f*, which amounts to two penalized events). No other scenario for this COG has the same or smaller number of events.

The reconstructed history of COG1514 (the second line of Table 6.22) is more complicated. Let us reconstruct it on the right-hand tree of Figure 4.5. Since this COG is represented in all non-singleton clusters, it might have emerged in the root of the tree. Then, to conform to its profile, it must be lost at *y, o, s, g, c,* and *r,* which amounts to seven penalized events, one emergence and six losses, altogether. However, there exist better scenarios with six penalized events each. One such scenario assumes that the gene emerged in the node of cluster archaea *aompkz* and then horizontally transferred to cluster *vqbj* and singletons *f, e, d,* thus leading to one emergence event, four horizontal transfers, and one loss, at *o*. Obviously, the profile data by themselves give no clue to the node of emergence. In principle, the gene in question might have emerged in cluster *vqbj* or even in one of the singletons. To resolve this uncertainty, one needs additional data on evolution of the given gene, but this is beyond the scope of this analysis.   ∎

6.5.4 External Taxonomy

The approach of the previous section will be extended here to any taxonomy, that is, a core ontology relation structuring concepts involved in an ontology according to a "generalization" relation such as "*a* is a *b*" or "*a* is part of *b*" or the like. The concept of ontology as a computationally feasible environment for knowledge representation and maintenance has sprung out rather recently [173]. The term refers, first of all, to a set of concepts and relations between them. A taxonomy such as the classification of living organisms in biology, ACM Classification of Computing Subjects (ACM-CCS) [1] and more recently a set of taxonomies comprising the SNOMED CT, the "Systematized Nomenclature of Medicine Clinical Terms" [192] are examples of ontologies.

Specifically, the ACM-CCS taxonomy [1] comprises 11 major partitions (first-level subjects) such as *B. Hardware, D. Software, E. Data, G. Mathematics of Computing, H. Information Systems*, and so on. These are subdivided into 81 second-level subjects. For example, item *I. Computing Methodologies* consists of eight subjects including *I.1 Symbolic and Algebraic Manipulation, I.2 Artificial Intelligence, I.5 Pattern Recognition* and so on. They are further subdivided into third-layer topics as, for instance, *I.5 Pattern Recognition* which is represented by seven topics including *I.5.3 Clustering, I.5.4 Applications*, and so on.

According to [141], most research efforts on computationally handling ontologies may be considered as falling in one of the three areas: (a) developing platforms and languages for ontology representation such as OWL language, (b) integrating ontologies and (c) using them for various purposes. Most efforts in (c) are devoted to building rules for ontological reasoning and querying utilizing the inheritance relation supplied by the ontology's taxonomy in the presence of different data models (e.g. [11,193]). These do not attempt at approximate representations but just utilize additional possibilities supplied by the ontology relations. Another type of ontology usage is in using its taxonomy nodes for interpretation of data-mining results such as association rules and clusters (see, e.g., [173]). Our approach naturally falls within this category. We assume a domain taxonomy, such as ACM-CCS, has been built. What we want to do is to use the taxonomy for representing of a query set, a cluster, comprised of a set of topics corresponding to leaves of the taxonomy by nodes of the taxonomy's higher ranks. The representation should approximate a query topic cluster in a "natural" way, at a cost of some "small" discrepancies between the query set and the taxonomy structure. This sets this approach apart from other work on queries to ontologies that rely on purely logical approaches [11,193].

Situations at which a "low-cost" generalization can be of interest may relate to visualization or generalization or interpretation or any combination of them. Let us discuss these, in brief:

- *Visualization* For computational purposes, the author assumes a view [140] that visualization necessarily involves a ground image, the structure of which should be well known to the viewer. This can be a Cartesian plane, a geography map, or a genealogy tree, or a scheme of London's Tube. Then visualization of a data set is such a mapping of the data on the ground image that translates important features of the data into visible relations over the ground image. For a research department, the following elements can be taken for the visualization:

- *Ground image:* a tree of the ACM-CCS taxonomy of Computer Science, the ground image,
- *Entities to visualize:* the set of CS research subjects being developed by members of the department, and
- *Mapping:* representation of the research on the taxonomy tree.

There can be various management goals achieved with this such as, for example [152]:

- Positioning of a department within the ACM-CCS taxonomy
- Analyzing and planning the structure of research in the organization
- Finding nodes of excellence, nodes of failure and nodes needing improvement for the organization
- Discovering research elements that poorly match the structure of AMS-CCS taxonomy
- Planning of research and investment
- Integrating data of different organizations in a region, or on the national level, for the purposes of regional planning and management

- *Generalization*   Let us specify the term, generalization, as deriving a general conception or principle from particulars. Consider two examples of query sets of research topics [141]. One can think of them as those being under development by different project groups in a research center:

Query topic set $S1$ consists of four items from the ACM-CCS taxonomy:

F.1 Computation by abstract devices

F.3 Logics and meaning of programs

F.4 Mathematical logic and formal languages

D.3 Programming languages

Query topic set $S2$ consists of three items:

C.1 Processor architectures

C.2 Computer-communication networks

C.3 Special-purpose and application-based systems

The topics are labeled according to ACM-CCS coding. They fall in higher rank subjects: D—Software, F—Theory of computation and C—Computer Systems Organization. If one asks whether the groups' research can be further generalized according to the ACM-CCS nomenclature, one should probably ask first, how many more topics fall under D, F, and C headings in ACM-CCS. Then upon learning that D, F, and C have only 4, 5, and 4 meaningful children in the taxonomy, respectively*, one should claim that, obviously, group S1 conducts research in F—Theory of computation, whereas group S2's research is in C—Computer Systems Organization. This is a meaningful example of generalization. Basically, topic sets are generalized to their least common ancestors in the taxonomy tree if this involves not too many *gaps* and *offshoots*.

Gaps are nodes, like F.2 or C.4 and C.5, that fall under the generalizing subjects but do not belong to the query topic sets. Offshoots are nodes like D.3 that do belong to the topic set—S1 in our example, but they fall through when generalizing.

- *Interpretation* According to the common view, to interpret something means to explain the meaning of it in understandable terms. It is assumed that the set of concepts forming the nodes of a taxonomy, together with the corresponding relation, "to be part of," forms a language which is to be used for interpretation. What are the items to interpret? This can be, for example, a text. Or this can be a concept that is not part of the taxonomy. Let us illustrate these.

For the former, consider the text of a research paper, P, that covers a number of research subjects, exactly those listed in the crisp query set S1 above. Assume that one has been able to assign them the following fuzzy membership values describing the paper in terms of the taxonomy:

F.1 Computation by abstract devices—0.30

F.3 Logics and meaning of programs—0.20

F.4 Mathematical logic and formal languages—0.30

D.3 Programming languages—0.20

The membership values can be derived, for example, by analysing the summaries of papers referring to P according to Google Scholar.

For the latter, let us consider a concept C that does not belong to the set of concepts listed in the ACM-CCS nomenclature. Yet the analysis of Internet documents may have provided us with a fuzzy list of ACM-CCS subjects coinciding with that listed in the query set S2, together with the following membership values:

C.1 Processor architectures—0.7

C.2 Computer-communication networks—0.2

---

* In ACM-CCS, each subject has topics General and Miscellaneous, which are meaningless in this context.

C.3 Special-purpose and application-based systems—0.1 summing to unity.

Then one would imagine that paper P and concept C can be interpreted as falling in higher nodes F and C of the ACM-CCS taxonomy, respectively. These nodes then provide the interpretations sought.

6.5.5 Lifting Method

This method has been presented by the author and his colleagues in a number of papers. This text follows [152] and [147].

Assume that there are a number of concepts in an area of research or practice that are structured according to the relation "a is part of b" into a taxonomy, that is a rooted hierarchy T. The set of its leaves is denoted by I. The relation between taxonomy nodes is conventionally expressed in genealogical terms so that each node $t \in T$ is said to be the parent of the nodes immediately descending from t in T, its children. Each interior node $t \in T$ corresponds to a concept that generalizes the concepts corresponding to the subset of leaves $I(t)$ descending from t, viz. the leaves of the subtree $T(t)$ rooted at t, which will be referred to as the leaf-cluster of t.

A fuzzy set on I is a mapping u of I to the non-negative real numbers assigning a membership value $u(i) \geq 0$ to each $i \in I$. The set $S_u \subset I$, where $S_u = \{i : u(i) > 0\}$, is referred to as the support of u.

Given a taxonomy T and a fuzzy set u on I, one can think that u is a, possibly noisy, projection of a high rank concept to the leaves I. Under this assumption, there should exist a "head subject" h among the interior nodes of the tree T that more or less comprehensively (up to small errors) covers S_u. Two types of possible errors are gaps and offshoots as illustrated in Figure 6.5.

A gap is a maximal node g in the subtree $T(h)$ rooted at h such that $I(g)$ is disjoint from S_u. The maximality of g means that $I(parent(g))$, the leaf-cluster of g's parent, does overlap S_u. A gap under the head subject h can be interpreted as a loss of the concept h by the topic set u. In contrast, establishing a node h as a head concept can be technically referred to as a gain.

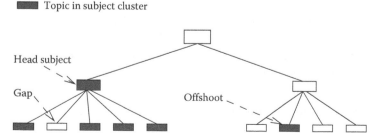

FIGURE 6.5
Three types of features in lifting a topic set within taxonomy.

An offshoot is a leaf $i \in S_u$ which is not covered by h, that is, $i \notin I(h)$.

Since no taxonomy perfectly reflects all of the real-world phenomena, some topic sets u may refer to general concepts that are not captured in T. In this case, two or more, rather than just one, head subjects are needed to cover them. This motivates the following definition.

The pair (T, u) will be referred to as an interpretation query. Consider a set H of nodes of T that covers the support S_u; that is, each $i \in S_u$ either belongs to H or is a descendant of a node in H, viz. $S_u \subseteq \cup_{h \in H} I(h)$. This set H is a possible result of the query (T, u). Nodes in H will be referred to as head subjects if they are interior nodes of T or offshoots if they are leaves. A node $g \in T$ is a gap for H if it is a gap for some $h \in H$. Of all the possible results H, those bearing the minimum penalty are of interest only. A minimum penalty result sometimes is referred to as a parsimonious one.

Any penalty value $p(H)$ associated with a set of head subjects H should penalize the head subjects, offshoots and gaps commensurate with the weighting of nodes in H determined from the membership values in the topic set u. We assign the head penalty to be *head*, offshoot penalty, *off*, and the gap penalty, *gap*.

To take into account the u membership values, we need to combine them to nodes of higher rank in T. In order to define appropriate membership values for interior nodes of tree T, we assume the following normalization condition:

$$\sum_{i \in I} u^2(i) = 1 \tag{6.21}$$

This is not that odd after all. The equation expresses just Euclidean normalization coming from the additive spectral approach taken by Mirkin and Nascimento [151].

A crisp set $S \subseteq I$ can be considered as a fuzzy set with the non-zero membership values defined according to the normalization principle so that $u(i) = 0$ if $i \notin S$, and $u(i) = 1/\sqrt{(|S|)}$.

For each interior node $t \in T$, its membership weight is defined as $u(t) = \sqrt{\sum_{i \in I(t)} u(i)^2}$. The weight of a gap is zero. The membership weight of the root is 1.

We now define the notion of pruned tree. Pruning the tree T at t results in the tree remaining after deleting all descendants of t. The definitions in this section are consistent in that the weights of the remaining nodes are unchanged by any sequence of successive prunings. Note, however, that the sum of the weights assigned to the leaves in a pruned tree is typically less than that in the original tree.

We consider that weight $u(t)$ of node t influences not only its own contribution, but also contributions of those gaps that are children of t. Therefore, the contribution to the penalty value of each of the gaps g of a head subject $h \in T$ is weighted according to the membership weight of its parent, as defined by $\gamma(g) = u(parent(g))$. Let us denote by $\Gamma(h)$ the set of all gaps below h. The gap contribution of h is defined as $\gamma(h) = \sum_{g \in \Gamma(h)} \gamma(g)$. For a crisp query set S with no condition, (N), this is just the number of gaps in $\Gamma(h)$.

To distinguish between proper head subjects and offshoots in H, we denote the set of leaves and interior nodes in H as H^- and H^+, respectively.

Then penalty function $p(H)$ for the tree T is defined by:

$$p(H) = head \times \sum_{h \in H^+} u(h) + gap \times \sum_{h \in H^+} \gamma(h) + off \times \sum_{h \in H^-} u(h) \qquad (6.22)$$

The problem is to find such a set H that minimizes the penalty—this will be the result of the query (T, u).

A preliminary step is to prune the tree T of irrelevant nodes. Then all interior nodes $t \in T$ are annotated by extending the leaf membership values. Those nodes in the pruned tree that have a zero weight are gaps; they are assigned with a γ-value which is the u-weight of its parent. This can be accomplished as follows:

a. Label with 0 all nodes t whose clusters $I(t)$ do not overlap S_u. Then remove from T all nodes that are children of 0-labeled nodes since they cannot be gaps. We note that all the elements of S_u are in the leaf set of the pruned tree, and all the other leaves of the pruned tree are labeled 0.

b. The membership vector u is extended to all nodes of the pruned tree according to the rule above.

c. Recall that $\Gamma(t)$ is the set of gaps, that is, the 0-labeled nodes of the pruned tree, and $\gamma(t) = \sum_{g \in \Gamma(t)} u(parent(g))$. We compute $\gamma(t)$ by recursively assigning $\Gamma(t)$ as the union of the Γ-sets of its children and $\gamma(t)$ as the sum of the γ-values of its children. For leaf nodes, $\Gamma(t) = \oslash$ and $\gamma(t) = 0$ if $t \in S_u$. Otherwise, that is, if t is a gap node (or, equivalently, if t is labeled 0), $\Gamma(t) = t$ and $\gamma(t) = u(parent(t))$.

Let us assume that the tree T has already been pruned and annotated as described above. The algorithm proceeds recursively from the leaves to the root. For each node t, we compute two sets, $H(t)$ and $L(t)$, containing those nodes at which gains and losses of head subjects occur. The respective penalty is computed as $p(t)$.

I   *Initialization*
    At each leaf $i \in I$: If $u(i) > 0$, define $H(i) = i$, $L(i) = \oslash$ and $p(i) = off \times u(i)$.
    If $u(i) = 0$, define $H(i) = \oslash$, $L(i) = \oslash$ and $p(i) = 0$.

II *Recursion*
    Consider a node $t \in T$ having a set of children W, with each child $w \in W$ assigned a pair $H(w)$, $L(w)$ and associated penalty $p(w)$. One of the following two cases must be chosen:

a. The head subject has been gained at t, so the sets $H(w)$ and $L(w)$ at its children $w \in W$ are not relevant. Then $H(t)$, $L(t)$ and $p(t)$ are defined by: $H(t) = t$;
$L(t) = \Gamma(t)$;
$p(t) = head \times u(t) + gap \times \gamma(t)$.

b. The head subject has not been gained at t, so at t we combine the H- and L-sets as follows: $H(t) = \bigcup_{w \in W} H(w)$, $L(t) = \bigcup_{w \in W} L(w)$ and $p(t) = \sum_{w \in W} p(w)$.

Choose whichever of (a) and (b) has the smaller value of $p(t)$.

III *Output*: Accept the values at the root:
$H(root)$—the heads and offshoots, $L(root)$—the gaps, $p(root)$—the penalty.

These give a full account of the parsimonious scenario obtained for the query (T, u).

Example 6.19 [147]:

Consider the tree in Figure 6.6a along with a fuzzy set comprised of four leaves, A.1, A.2, B.1, B.2 and membership values normalized according to the quadratic option Q. Figure 6.6b presents the pruned version of the tree, at which the nodes are annotated by membership values, in thousandth; and the gap weights are put in parentheses at the higher rank nodes.

Table 6.23 presents a number of different membership values and penalty systems leading to different query outcomes at the taxonomy presented in Figure 6.6a. The membership values are not normalized to give the reader a more convenient reading. The query illustrated in Figure 6.6 and Tables 6.24 and 6.25 is number 2 in Table 6.23.

The Tables 6.24 and 6.25 represent sequential recursion steps of the lifting algorithm at the nodes A, B (Table 6.24) and the root (Table 6.25).

TABLE 6.23

Different Queries and Corresponding Lifting Outputs for the Illustrative Taxonomy in Figure 6.6a

#	$u(A1)$	$u(A2)$	$u(B1)$	$u(B2)$	off	gap	H(Root)	L(Root)	P(Ro)
1	0.8	0.5	0.1	0.01	0.9	0.2	$\{A1, A2, B1, B2\}$	$\{\}$	1.34
2	0.1	0.01	0.8	0.5	0.9	0.2	$\{A1, A2, B\}$	$\{B3\}$	1.30
3	0.8	0.5	0.1	0.1	0.9	0.2	$\{A1, A2, B\}$	$\{B3\}$	1.40
4	0.1	0.01	0.8	0.5	0.9	0.1	$\{A1, A2, B\}$	$\{B3\}$	1.20
5	0.8	0.5	0.1	0.01	0.9	0.1	$\{A, B1, B2\}$	$\{A3, A4\}$	1.30
6	0.8	0.5	0.1	0.1	1.1	0.2	$\{A, B\}$	$\{A3, A4, B3\}$	1.56
7	0.8	0.5	0.1	0.1	0.9	0.1	$\{Root\}$	$\{A3, A4, B3, C\}$	1.31
8	0.1	0.1	0.8	0.5	0.9	0.1	$\{Root\}$	$\{A3, A4, B3, C\}$	1.23

Note: Membership values are further normalized according to option Q.

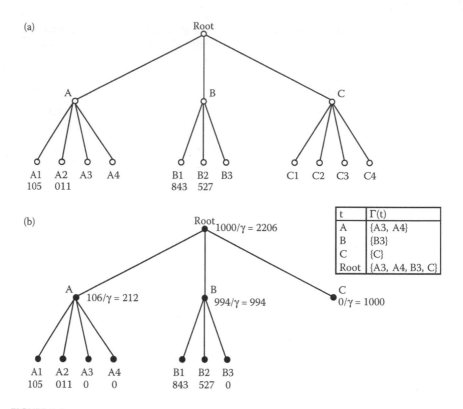

FIGURE 6.6
Illustrative taxonomy *T* with a fuzzy topic set membership values assigned to leaves, in thousandth (a); part (b) presents *T* after pruning and annotating with membership and gap values.

The Table 6.25 shows that No Gain solution at the root relates to a smaller penalty, 1.297, so that the query results in the lifting to node B, leaving A1 and A2 as offshoots.

Changing the gap penalty to *gap* = 0.1 leaves the result the same (case number 4), B as the only genuine head subject, which is no wonder because the major membership values are concentrated at B. If the membership values are changed to shift the weight to A, A becomes the head subject indeed—see case 5. Yet such an outcome is not a certainty, as case 3 clearly demonstrates—a very small increase in membership values towards B, still leaving the bulk of membership with A, makes B the head subject again, against the odds—just because of a slight asymmetry in the structure of A and B nodes. Cases 6 and 7 show that the same membership values as in case 3, but with somewhat changed gap or offshoot penalty can lead to a different output: both A and B, or even the root, as the head subjects.   ∎

Some real-world examples are given in [141,152,153].

TABLE 6.24

Recursion Lifting Steps at Nodes A and B of the Pruned Tree for the Query
in Figure 6.6

le	H(le)	L(le)	p(le)	t		H(t)	L(t)	p(t)
A1	{A1}	⊘	$0.9 \times 0.105 = 0.095$		Gain	{A}	{A3, A4}	0.148
A2	{A2}	⊘	$0.9 \times 0.0105 = 0.009$	A				
A3	⊘	⊘	0		No gain	{A1, A2}	{A3, A4}	**0.104**
A4	⊘	⊘	0					
B1	{B1}	⊘	$0.9 \times 0.843 = 0.759$		Gain	{B}	{B3}	1.193
B2	{B2}	⊘	$0.9 \times 0.527 = 0.474$	B				
B3	⊘	⊘	0		No gain	{B1, B2}	{B3}	1.233
C	⊘	⊘	0					

Note: Penalty values are *head* = 1, *off* = 0.9, *gap* = 0.2. The minimum penalty is highlighted in bold.

TABLE 6.25

Recursion Lifting Steps at the Root for the Query in Figure 6.6 Following
the Results Presented in Table 6.24

t	H(t)	L(t)	p(t)	t		H(Root)	L(Root)	p(Root)
A	{A1, A2}	{A3, A4}	0.104	Root	Gain	{Root}	{A3, A4, B3, C}	1.441
B	{B}	{B3}	1.193		No gain	{A1, A2, B}	{A3, A4, B3}	**1.297**
C	{}	{C}	1.000					

Note: Penalty values are *head* = 1, *off* = 0.9, *gap* = 0.2. The minimum penalty is highlighted in bold.

The lifting method should be a useful addition to the methods for interpreting thematic clusters produced by various data analysis tools. Unlike the methods based on mapping them within individual taxonomy nodes with the goal of finding overrepresented features, the interpretation capabilities of this method come from an interplay between the topology of the taxonomy tree, the membership values and the penalty weights for the head subjects and associated gap and offshoot events. As our illustrative example shows, the tree topology can substantially contribute to the query outcome.

On the other hand, the definition of the penalty weights remains of an issue in the method. Potentially, this issue can be overcome if statistics are accumulated from applications of this approach for interpreting various concepts or papers. Then each node can be assigned with frequencies of gain and loss events so that the criterion of maximum parsimony can be changed for the criterion of maximum likelihood. Under usual independence assumptions, the maximum likelihood criterion would inherit the additive structure of the minimum penalty criterion (6.22). Then the recursions of the lifting algorithm will remain valid, with respective changes in the criterion of course.

6.6 Overall Assessment

The chapter addresses the issue of validation and interpretation of clusters as two sides of the same problem. The statistics approach leads to testing how good is the fit between data and clusters. The knowledge discovery approach would rather focus on the fit between the clusters and domain knowledge. These two approaches are highly incompatible which probably contributed to the current state of the art which is rather embryonic.

The two approaches to the issues of validation and interpretation are operationalized as internal validation and external validation.

On the former side, some indexes of relationship between the data and clusters are reviewed. Then rules for cluster interpretation are given including cluster representative, cluster tendency, and comprehensive description of clusters. These are based mostly on contributions to data scatter that can be computed within the data recovery framework.

On the latter side, decision tree techniques are described, as well as measures of correspondence between subsets and partitions. A recent "lifting" method for interpretation, generalization and visualization of clusters in a taxonomy of the domain knowledge is described.

7

Least-Squares Data Recovery
Clustering Models

Main subjects covered

1. What is the data recovery approach?
2. Clustering as data summarization
3. A data recovery model and method for principal component analysis PCA
4. A data recovery model and data scatter decomposition for K-Means and anomalous cluster clustering
5. Equivalent reformulations of the square error clustering criterion
6. A data recovery model and data scatter decompositions for cluster hierarchies
7. Divisive and agglomerative data recovery hierarchical clustering
8. Feature weighting in a data recovery model using Minkowski's metric
9. Laplacian transformation and spectral clustering
10. Additive clustering
11. Data recovery models for combined consensus and ensemble consensus partitioning
12. One-by-one data recovery clustering methods
13. Statistical correlation and association measures as contributions to the data scatter

---

Key Concepts

Alternating optimization A method of optimization applicable when the function to be optimized depends on two or more groups of variables. The method works iteratively by optimizing the function over a group of variables having the other groups specified. K-Means is a method of alternating minimization of the square error clustering criterion.

Anomalous cluster A cluster which is furthest away from the reference point. Iterated anomalous clustering implements the one-by-one separating strategy of principal component analysis in the data recovery model for K-Means, and thus allows for extending K-Means to its intelligent version, iK-Means, mitigating the need in defining an initial setting.

Attraction coefficient A measure of attraction of an entity to a cluster derived within the data recovery approach. This is equal to the entity's average similarity to the cluster minus half of the average within cluster similarity. In K-Means and Ward clustering, attractions of entities to their clusters are greater than to other clusters. Moreover, in K-Means, entities are always positively attracted to their clusters and negatively to other clusters.

Contingency data A data table whose rows and columns correspond to two sets of categories and entries are counts or proportions of observations at the intersection of a row and column categories (co-occurrence counts). Because of their summability across the table, contingency data are analyzed after having been transformed into relative Quetelet coefficients. The data scatter of the transformed data is measured by the chi-square contingency coefficient.

Correspondence factor analysis A PCA-like method for visualization of co-occurrence values in a contingency table by displaying both row and column items as points in the same 2D or 3D space.

Data recovery models in clustering A data recovery clustering model includes a rule which restores a value, according to a cluster structure, to every data entry. Therefore, every datum is represented as the sum of a cluster-recovered value and a residual. This provides for a built-in goodness-of-fit principle: the smaller the residuals, the better the cluster structure.

Data scatter decomposition A decomposition of the data scatter in two parts, that explained by the cluster structure and that remaining unexplained. Such a decomposition provides for both an explicit cluster criterion and interpretation aids. The explained part in K-Means and Ward clustering is always the sum of contributions of individual clusters or splits.

Feature weighting The distances between entity points, and thus clusters, much depend on feature weights. In the cluster data recovery approach, feature weights can be explicitly included and optimized. A K-Means-like system involves alternating minimization of the criterion with respect to three groups of variables: clusters, centroids and feature weights.

Linear regression A method for the analysis of correlation between two quantitative features x and y in which y is approximated by a linear transformation $ax + b$ of x, where a and b are constant. This setting is a genuine ground on which the coefficients of correlation and determination can be defined and explained.

Minkowski metric clustering An extension of the data recovery clustering criterion from the squares of the absolute values of the residuals to an arbitrary positive power (Minkowski's p). This makes the problem of finding centers harder than usual but allows, instead, to get rid of irrelevant features.

One-by-one clustering A method in clustering in which clusters or splits are taken one by one. In this text, all such methods exploit the additive structure of data recovery clustering models, which is analogous to that of the model of principal component analysis.

PDDP A strategy for divisive clustering based on projection of the data points to the first principal component direction.

Principal component analysis (PCA) A method for approximation of a data matrix with a small number of hidden factors, referred to as principal components, such that data entries are expressed as linear combinations of the hidden factor scores. It appears the principal components can be determined with the singular value decomposition (SVD) of the data matrix.

Reference point A point in the feature space serving as the space origin. The anomalous pattern is sought starting from an entity furthest from the reference point.

Split versus separation Difference between two perspectives: cluster-versus-the-rest and cluster-versus-the-whole, reflected in different coefficients attached to the distance between centroids of split parts. The former perspective is taken into account in the Ward-like divisive clustering methods, the latter in the anomalous pattern clustering.

Ward-like divisive clustering A divisive clustering method using Ward distance as the splitting criterion. The method can be considered an implementation of the one-by-one PCA strategy within the data recovery clustering.

7.1 Statistics Modeling as Data Recovery

7.1.1 Data Recovery Equation

The data recovery approach is a cornerstone of contemporary thinking in statistics and data analysis. It is based on the assumption that the observed data reflect a regular structure in the phenomenon of which they inform. The regular structure A, if known, would produce data $F(A)$ that should coincide with the observed data Y up to small residuals which are due to possible flaws in any or all of the following three aspects: (a) sampling entities, (b) selecting features and tools for their measurements, and (c) modeling the phenomenon in question. Each of these can drastically affect results. However, so far only the simplest of the aspects, (a), has been addressed by introduction of probabilities to study the reliability of statistical inference in data analysis. In this text, we are not concerned with these issues. We are concerned with the underlying equation:

$$\text{Observed data } Y = \text{recovered data } F(A) + \text{residuals } E \qquad (*)$$

The quality of the model A is assessed according to the level of residuals E: the smaller the residuals, the better the model. Since quantitative models involve unknown coefficients and parameters, this naturally leads to the idea of fitting these parameters to data in such a way that the residuals become as small as possible. To put this idea as a minimization problem, one needs to combine the multiple residuals in an aggregate criterion. In particular, the so-called principle of maximum likelihood has been developed in statistics. When the data can be modeled as a random sample from a multivariate Gaussian distribution, this principle leads to the so-called least-squares criterion, the sum of squared residuals to be minimized. In the data analysis or data-mining framework, the data do not necessarily come from a probabilistic population. Moreover, analysis of the mechanism of data generation is not of primary concern here. One needs only to see if there are any patterns in the data as they are. In this case, the principle of maximum likelihood may be not applicable. Still, the sum of squared residuals criterion can be used in the context of data mining as a measure of the incompatibility between the data and the model. It provides for nice geometric properties and leads to provably reasonable cluster solutions. It also leads to useful decompositions of the data scatter into the sum of explained and unexplained parts. To show the working of model $(*)$ along with the least-squares principle, let us introduce four examples covering important methods in data analysis: (a) averaging, (b) linear regression, (c) principal component analysis, and (d) correspondence analysis [140]. These are also useful for introduction of data analysis concepts that are used throughout this text.

7.1.2 Averaging

Let a series of real numbers, $x_1, \ldots, x_N$, have been obtained to represent the same unknown value a. Equation $(*)$ then becomes

$$x_i = a + e_i$$

with e_i being the residual for $i = 1, \ldots, N$. To minimize the sum of squares $L(a) = \sum_i e_i^2 = \sum_i (x_i - a)^2$ as a function of a, one may utilize the first-order optimality condition, $dL/da = 0$, that is, $dL/da = -2(\sum_i x_i - Na) = 0$. Therefore, the least-squares solution is the average $a = \bar{x} = \sum_i x_i/N$. By substituting this for a in $L(a)$, one obtains $L(\bar{x}) = \sum_i x_i^2 - N\bar{x}^2$. The last expression gives, in fact, a decomposition of the data scatter, the sum of data entries squared $T(x) = \sum_i x_i^2$, into the explained and unexplained parts, $T(x) = N\bar{x} + L(\bar{x})$. The averaged unexplained value $L(\bar{x})/N$ is the well-known variance s_x^2 of the series, and its square root, $s_x = \sqrt{L(\bar{x})/N}$, the standard deviation. It appears thus that the average minimizes the standard deviation s_x of observations from a.

7.1.3 Linear Regression

Let a series of pairs of reals $(x_1, y_1), \ldots, (x_N, y_N)$ such as IQ scores for x_i and marks in Mathematics for y_i for individuals i ($i = 1, \ldots, N$) have been collected. The linear regression model assumes that y-values are effected by x-values according to a linear equation $y = ax + b$ where a and b are constants, referred to as the slope and intercept, respectively. To fit the values of a and b to data, the conventional thinking considers that only y-values are to be explained by model ($*$), thus leading to equations

$$y_i = ax_i + b + e_i$$

The least-squares criterion for minimizing the residuals in this case is a function of two unknown coefficients, $L = L(a, b) = \sum_i e_i^2 = \sum_i (y_i - ax_i - b)^2$. The first-order optimality conditions lead to $a = \sum_i (x_i - \bar{x})(y_i - \bar{y})/s_x^2$ and $b = \bar{y} - a\bar{x}$. The linear regression $y = ax + b$ with a and b fitted to the data can be used for analysis and prediction of y, given x, if L is small.

A symmetric function of features x and y, the correlation coefficient, has been defined as $\rho = \sum_i (x_i - \bar{x})(y_i - \bar{y})/[Ns_x s_y]$ where s_x and s_y are standard deviations of x and y, respectively. The optimal a thus can be expressed through ρ as $a = \rho s_y / s_x$. By putting the optimal a and b into $L(a, b)$, the minimum L can be expressed as $L = Ns_y^2(1 - \rho^2)$.

According to the above formulas, the correlation coefficient ρ in the data recovery paradigm has the following properties:

1. Its square, the so-called determinacy coefficient ρ^2, expresses the relative decrease of the variance of y after its linear relation to x has been taken into account.

2. The values of ρ fall within the interval between -1 and 1. The closer ρ is to either 1 or -1, the smaller are the residuals in equation ($*$). For instance, at $\rho = 0.9$, the unexplained variance of y accounts for $1 - \rho^2 = 19\%$ of its original variance.

3. The slope a is proportional to ρ so that a is positive or negative depending on the sign of correlation coefficient. When $\rho = 0$, the slope is 0 too; the variables y and x are referred to as non-correlated, in this case, which means that there is no linear relation between them, though another functional relation, such as a quadratic one, may exist.

The case of $\rho = 0$ geometrically means that the centered versions of feature vectors $x = (x_i)$ and $y = (y_i)$ are mutually orthogonal.

4. With the data standardized by z-scoring transformation as

$$x' = \frac{x - \bar{x}}{s_x} \quad \text{and} \quad y' = \frac{y - \bar{y}}{s_y} \tag{7.1}$$

the variances are unities, thus leading to simpler formulas: $a' = \rho = <x', y'>/N$, $b' = 0$, $L' = N(1 - \rho^2)$ and $N = N\rho^2 + L'$ where N happens to be equal to the scatter of y'.

5. The value of ρ does not change under linear transformations of scales of x and y.

7.1.4 Principal Component Analysis

Principal component analysis* is a major tool for approximating observed data with model data formed by a few "hidden" factors. Observed data such as marks of students $i \in I$ at subjects labeled by $v = 1, \ldots, V$ constitute a data matrix $X = (x_{iv})$. Assume that each mark x_{iv} reflects a hidden ability z_i $(i \in I)$ up to a coefficient c_v, due to subject v's specifics. The principal component analysis model, as presented in the textbook [140], suggests that the student i's score over subject v reflects the product of the mutual impact of student and subject, $z_i c_v$. Then equation (*) can be formulated as

$$x_{iv} = c_v z_i + e_{iv} \tag{7.2}$$

The least-squares criterion is $L^2 = \sum_{i \in I} \sum_{v \in V} (x_{iv} - c_v z_i)^2$ and the first-order optimality conditions lead to equations $\sum_v x_{iv} c_v = z_i \sum_v c_v^2$ and $\sum_i x_{iv} z_i = c_v \sum_i z_i^2$ for all $v = 1, \ldots, V$ and $i \in I$. Sums $\sum_v c_v^2$ and $\sum_i z_i^2$ are squared norms of vectors c and z with the norms being defined as $||c|| = \sqrt{\sum_v c_v^2}$ and $||z|| = \sqrt{\sum_i z_i^2}$. Let us multiply the first of the optimality conditions by z_i and sum the obtained equations over $i \in I$. This would lead to

$$\sum_i \sum_v z_i x_{iv} c_v = \sum_i z_i^2 \sum_v c_v^2$$

which is, in matrix notation, $z^T X c = ||z||^2 ||c||^2$, or equivalently,

$$\frac{z^T X c}{||z|| ||c||} = ||z|| ||c|| \tag{7.3}$$

---

* Sections 7.1.4 and 7.1.5 can be understood in full only if introductory concepts of the linear algebra are known, including the concepts of matrix and matrix rank.

A vector whose norm is unity is referred to as a normed vector; to make $z = (z_i)$ normed, z has to be divided by its norm: vector $z^* = z/||z||$ is the normed version of z. Let us denote by μ the product $\mu = ||z||||c||$ and by z^* and c^* the normed versions of the least-squares solution c, z. Then the equations above can be rewritten, in matrix algebra notation, as $Xc^* = \mu z^*$, $X^T z^* = \mu c^*$ and $\mu = z^{*T} X c^*$. These are quite remarkable equations expressing the fact that optimal vectors c and z are linear combinations of, in respect, rows and columns of matrix X. These expressions define an inherent property of matrix X, its singular value and vectors, where μ is a singular value of matrix X and c^* and z^* are the normed singular vectors corresponding to μ. It is well known that the number of non-zero singular values of a matrix is equal to its rank and, moreover, the singular vectors c^* corresponding to different singular values are mutually orthogonal, as well as the vectors z^* [61]. In our case, μ must be the maximum singular value of X. Indeed, consider the least-squares criterion $L^2 = \sum_{i,v}(x_{iv} - c_v z_i)^2 = \sum_{i,v} x_{iv}^2 - 2\sum_{i,v} x_{iv} c_v z_i + \sum_{i,v} z_i^2 c_v^2$. Recalling that $\sum_{i,v} x_{iv}^2 = T(X)$ is the data scatter, this can be rewritten as

$$L^2 = T(X) - 2z^T Xc + ||z||^2 ||c||^2$$

or, using Equation 7.3, as

$$L^2 = T(X) - \left(\frac{z^T Xc}{||z||||c||}\right)^2 \tag{7.4}$$

Expression on the right, $(z^T Xc/(||z||||c||))^2$ is equal to μ^2 according to Equation 7.3, which means that the least-squares criterion is equivalently represented by the requirement that μ is maximized. Therefore, the least-squares criterion leads to the maximum singular value and corresponding singular vectors as the only solution. Two statements proven above deserve to be stated explicitly.

First, the found solution provides for the decomposition of the data scatter $T(X) = \sum_{i,l} x_{il}^2$

$$T(X) = \mu^2 + L^2 \tag{7.5}$$

in two parts, that explained by the model, μ^2, and that remaining unexplained, L^2.

Second, the maximum singular value μ can be found by maximizing the following singular quotient:

$$\mu(z, c) = \frac{z^T Xc}{||z||||c||} \tag{7.6}$$

with respect to all possible vectors z and c. The normed versions of the maximizers are the corresponding singular vectors.

To keep up with the model (7.2), vectors z^* and c^* must be rescaled to contain μ in their product, which is done with formulas $c = \sqrt{\mu}c^*$ and $z = \sqrt{\mu}z^*$.[*] The talent score vector z is referred to as the principal component and the corresponding c as the loading vector.

Generalizing the one-factor model (7.2), one may assume a small number K of different hidden "talent" factors $z_1,\ldots, z_K$ forming an unknown $N \times K$ score matrix Z_K with corresponding K loading vectors $c_1, \ldots, c_K$ forming rows of a $K \times V$ loading matrix C_K so that Equation 7.2 is extended to

$$x_{iv} = c_{1v}z_{i1} + \cdots + c_{Kv}z_{iK} + e_{iv} \tag{7.7}$$

for all $i \in I$ and $v \in V$, or, in the matrix algebra notation,

$$X = Z_K C_K + E \tag{7.8}$$

We are interested in finding the least-squares solution to Equations 7.7 and 7.8, that is, matrices Z_K and C_K minimizing the sum of squared elements of the residual matrix E. It is not difficult to see that the solution, if exists, may not be unique. Indeed, any solution Z_K, C_K can be transformed to $Z = Z_K F^T$ and $C = FC_K$ with any $K \times K$ matrix F satisfying equation $F^T = F^{-1}$, that is, being a "rotation" matrix [44,61]. Obviously $ZC = Z_K C_K$, that is, the rotated solution Z, C corresponds to the same residual matrix E and thus the same value of the least squares criterion L^2. This shows that the least-squares solution is defined only up to the linear subspace of the space of N-dimensional vectors, whose base is formed by the columns of matrix Z_K.

The optimal linear subspace can be specified in terms of the so-called singular value decomposition (SVD) of matrix X. Let us recall that the SVD of $N \times V$ matrix X of rank r amounts to equation

$$X = ZMC^T = \mu_1 z_1^* c_1^{*T} + \mu_2 z_2^* c_2^{*T} + \cdots + \mu_r z_r^* c_r^{*T}$$

where Z is $N \times r$ matrix of mutually orthogonal normed N-dimensional column-vectors z_k^*, C is $r \times V$ matrix of mutually orthogonal normed V-dimensional column-vectors c_k^*, and M is a diagonal $r \times r$ matrix with positive singular values μ_k on the main diagonal such that z_k^* and c_k^* are normed singular vectors of X corresponding to its singular value μ_k. These are defined by equations $Xc_k^* = \mu_k z_k^*$ and $X^T z_k^* = \mu_k c_k^*$, $\mu_k = z_k^{*T} X c_k^*$, $k = 1,\ldots,r$ [61]. The SVD decomposition is proven to be unique when singular values μ_k are all different.

A least-squares solution to model (7.8) can now be derived from matrices Z_k^* and C_k^* of the K singular vectors z_k^* and c_k^* corresponding to the K greatest singular values μ_k, $k = 1,\ldots,K$. (The indices reflect a conventional assumption that the singular values have been placed in the descending order, $\mu_1 > \mu_2 > \cdots > \mu_r > 0$. Another conventional assumption is that all

---

[*] Amazingly, no other rescaling factors α and β such that $\alpha\beta = \mu$ are possible, see [140].

the singular values are different, which is not of an issue if X, as it happens, is a real-world data matrix.) Let us denote the diagonal matrix of the first K singular values by M_K. Then a solution to the problem is determined by rescaling the normed singular vectors with formulas $Z_K = Z_K^* \sqrt{M_K}$ defining the principal components, and $C_K = \sqrt{M_K} C_K^*$ defining their loadings.

Since singular vectors z corresponding to different singular values are mutually orthogonal, the factors can be found one by one as solutions to the one-factor model (7.2) above applied to the so-called residual data matrix: after a factor z and loadings c are found, X must be substituted by the matrix of residuals, $x_{iv} \leftarrow x_{iv} - c_v z_i$. After k subtractions, the maximum singular value of the residual matrix corresponds to the $(k + 1)$th largest singular value of the original matrix X ($k = 1, \ldots, r - 1$). Repeating the process K times, one obtains the K first principal components and loading vectors.

Each kth component additively contributes μ_k^2 to the data scatter $T(X)$ ($k = 1, \ldots, K$) so that Equation 7.5 in the general case extends to

$$T(X) = \mu_1^2 + \cdots + \mu_K^2 + L^2(Z_K, C_K) \tag{7.9}$$

At a given K, the minimum value of $L^2(Z_K, C_K) = \sum_{i,v} e_{iv}^2$ is equal to the sum of the $r - K$ smallest singular values squared, $L^2(Z_K, C_K) = \sum_{k=K+1}^{r} \mu_k^2$, whereas the K largest singular values and corresponding singular vectors define the factor space solving the least-squares fitting problem for Equation 7.5.

Computation of the singular values and vectors can be performed by computing eigenvalues and eigenvectors of its derivative $|I| \times |I|$ matrix XX^T or $|V| \times |V|$ matrix $X^T X$. The former can be derived by putting expression $c^* = X^T z^*/\mu$ for c^* in the formula $Xc^* = \mu z^*$ leading to $XX^T z^* = \mu^2 z^*$. This equation means, according to the conventional definition, that the normed z^* is an eigenvector (sometimes referred to as a latent vector) of the square matrix $A = XX^T$ corresponding to its eigenvalue $\lambda = \mu^2$. Then the corresponding normed singular vector c^* can be determined from the equation $X^T z^* = \mu c^*$ as $c^* = X^T z^*/\mu$.

Thus, for XX^T, its eigenvalue decomposition rather than SVD must be sought, because singular vectors z of X are eigenvectors of XX^T corresponding to the eigenvalues $\lambda = \mu^2$. Consider the square symmetric $N \times N$ matrix $A = XX^T$. As is well known, its rank is equal to the rank of X, that is, r. This means that A has exactly r eigenvalues $\lambda_1 > \lambda_2 > \cdots \lambda_r > 0$ which are positive, because they are squares of the corresponding positive singular values μ_k of X. In fact, the number of eigenvalues of a square matrix is equal to its dimension, N, in this case. If $r < N$, which is true if the number of entities exceeds that of features, then the other $N - r$ eigenvalues are all zeros. A square matrix with all its eigenvalues being non-negative is what is referred to as a positive semi-definite matrix; a defining property of a positive semi-definite A is that, for any vector z, the product $z^T A z \geq 0$. The eigenvalue decomposition of A is usually referred to as its spectral decomposition because the set of all the eigenvalues is usually referred to as its spectrum. The spectral decomposition can be expressed as

$$A = Z\Lambda Z^{\mathrm{T}} = \lambda_1 z_1^* z_1^{*\mathrm{T}} + \lambda_2 z_2^* z_2^{*\mathrm{T}} + \cdots + \lambda_r z_r^* z_r^{*\mathrm{T}} \tag{7.10}$$

where $Z = (z_k^*)$ is an $N \times r$ matrix whose columns are the normed eigenvectors z_k^* ($k = 1, \ldots, r$); note, that these very vectors are singular vectors for the matrix X from which A is derived as $A = XX^{\mathrm{T}}$.

A procedure for sequential extraction of the eigenvectors can be described as follows. Given a matrix $A = (a_{ij})$, let us find a normed vector $z = (z_i)$ and its intensity λ in such a way that equations

$$a_{ij} = \lambda z_i z_j + e_{ij}$$

hold, up to as small residuals e_{ij} as possible. The least-squares solution to this problem must maximize an analogue to the singular quotient $\mu(z, c)$ above

$$\lambda(z) = \frac{z^T A z}{z^T z} \tag{7.11}$$

which is usually referred to as the Rayleigh quotient; its maximum is the maximum eigenvalue of A while the maximizer, the corresponding eigenvector. The data scatter, in this case, $T(A) = \sum_{i,j} a_{ij}^2$, is decomposed now as $T(A) = \lambda^2 + \sum_{i,j} e_{ij}^2$, which is somewhat different from the decomposition of the original data scatter $T(X)$ in Equation 7.9 which sums eigenvalues themselves rather than their squares. After the maximum eigenvalue is found, the next eigenvalue is found as the maximum eigenvalue of the residual covariance matrix $A - \lambda z z^{\mathrm{T}}$. The following eigenvalues and eigenvectors are found similarly by sequentially subtracting them from the residual matrix. Equation 7.9 can be reformulated in terms of the Rayleigh quotients (7.11):

$$T(X) = \frac{z_1^{\mathrm{T}} A z_1}{z_1^{\mathrm{T}} z_1} + \cdots + \frac{z_K^{\mathrm{T}} A z_K}{z_K^{\mathrm{T}} z_K} + L^2(Z_K, C_K) \tag{7.12}$$

since $\mu_k^2 = \lambda_k$ for all $k = 1, \ldots, K$.

The other derived matrix, $X^{\mathrm{T}} X$, has similar properties except that its eigenvectors now are the right normed singular vectors c_k^*, corresponding to the same eigenvalues $\lambda_k = \mu_k^2$, so that its spectral decomposition is $X^{\mathrm{T}} X = C\Lambda C^{\mathrm{T}}$ where $C = (c_k^*)$ is a $|V| \times r$ matrix of the eigenvectors c_k^*. Under the assumption that X has been standardized by subtraction of the feature grand means, matrix $X^{\mathrm{T}} X$ related to the number of entities N is what is called feature-to-feature covariance coefficients matrix, see p. 49. These are correlation coefficients if the features have been standardized by z-scoring, that is, further normalized by dividing over the feature's standard deviations. It should be noted that in many texts the method of principal components is explained by using only this matrix $B = X^{\mathrm{T}} X / N$, without any reference to the base Equations 7.2 or 7.7, see for instance [44,88,103]. Indeed, finding the maximum eigenvalue and the corresponding eigenvector of $B = X^T X / N$ can be interpreted as finding such a linear combination of the original variables that takes into account the maximum share of the data scatter. In this, the fact that the principal components are linear combinations of variables is an assumption of the method, not a corollary, which it is with models (7.2) and (7.7). The procedure of sequential extraction of the eigenvectors of matrix $A = XX^T$ described above can be applied to matrix $B = X^T X / N$ as well.

The singular value decomposition is frequently used as a data visualization tool on its own (see, for instance, [220]). Especially interesting results can be seen when entities are naturally mapped onto a visual image such as a geographic map; Cavalli-Sforza has interpreted as many as seven principal components in this way [23]. There are also popular data visualization techniques such as correspondence analysis [12,111], Latent semantic analysis [31] and Eigenface [209] that heavily rely on SVD. The former will be reviewed in the next section following the presentation in [138].

7.1.5 Correspondence Factor Analysis

Correspondence analysis (CA) is a method for visually displaying both row and column categories of a contingency table $P = (p_{ij})$, $i \in I, j \in J$, in such a way that distances between the presenting points reflect the pattern of co-occurrences in P. There have been several equivalent approaches developed for introducing the method (see, e.g., Benzécri [12]). Here, we introduce CA in terms that are similar to those of PCA above.

To be specific, let us concentrate on the problem of finding just two "underlying" factors, $u_1 = \{(v_1(i)), (w_1(j))\}$ and $u_2 = \{(v_2(i)), (w_2(j))\}$, with $I \cup J$ as their domain, such that each row $i \in I$ is displayed as point $u(i) = (v_1(i), v_2(i))$ and each column $j \in J$ as point $u(j) = (w_1(j), w_2(j))$ on the plane as shown in Figure 7.1. The coordinate row-vectors, v_l, and column-vectors, w_l, constituting u_l ($l = 1, 2$) are calculated to approximate the relative Quetelet coefficients $q_{ij} = p_{ij}/(p_{i+}p_{+j}) - 1$ according to equations:

$$q_{ij} = \mu_1 v_1(i) w_1(j) + \mu_2 v_2(i) w_2(j) + e_{ij} \tag{7.13}$$

where μ_1 and μ_2 are positive reals, by minimizing the weighted least-squares criterion

$$E^2 = \sum_{i \in I} \sum_{j \in J} p_{i+} p_{+j} e_{ij}^2 \tag{7.14}$$

with regard to μ_l, v_l, w_l, subject to conditions of weighted ortho-normality:

$$\sum_{i \in I} p_{i+} v_l(i) v_{l'}(i) = \sum_{j \in J} p_{+j} w_l(j) w_{l'}(j) = \begin{cases} 1, & l = l' \\ 0, & l \neq l' \end{cases} \tag{7.15}$$

where $l, l' = 1, 2$.

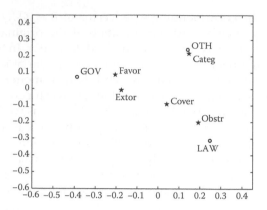

FIGURE 7.1

CA display for the rows and columns of Table 7.1 represented by circles and pentagrams, respectively.

The weighted criterion E^2 is equivalent to the unweighted least-squares criterion L applied to the matrix with entries $a_{ij} = q_{ij}\sqrt{p_{i+}p_{+j}} = (p_{ij} - p_{i+}p_{+j})/\sqrt{p_{i+}p_{+j}}$. This implies that the factors are determined by the singular-value decomposition of matrix $A = (a_{ij})$. More explicitly, the optimal values μ_l and row-vectors $f_l = (v_l(i)\sqrt{p_{i+}})$ and column-vectors $g_l = (w_l(j)\sqrt{p_{+j}})$ are the maximal singular values and corresponding singular vectors of matrix A, defined by equations $Ag_l = \mu_l f_l$, $f_l A = \mu_l g_l$.

These equations, rewritten in terms of v_l and w_l, are considered to justify the joint display: the row points appear to be averaged column points and, vice versa, the column points appear to be averaged versions of the row points. The mutual location of the row points is considered as justified by the fact that between-row-point squared Euclidean distances $d^2(u(i), u(i'))$ approximate chi-square distances between corresponding rows of the contingency table $\chi^2(i, i') = \sum_{j \in J} p_{+j}(q_{ij} - q_{i'j})^2$. Here, $u(i) = (v_1(i), v_2(i))$ for v_1 and v_2 rescaled in such a way that their norms are equal to μ_1 and μ_2, respectively. To see it, one needs to derive first that the weighted averages $\sum_{i \in I} p_{i+}v_i$ and $\sum_{j \in J} p_{+j}w_j$ are equal to zero. Then, it will easily follow that the singularity equations for f and g are equivalent to equations $\sum_{j \in J} p(j/i)w_j = \mu v_i$ and $\sum_{i \in I} p(i/j)v_i = \mu w_j$ where $p(j/i) = p_{ij}/p_{i+}$ and $p(i/j) = p_{ij}/p_{+j}$ are conditional probabilities defined by the contingency table P. These latter equations define elements of v as weighted averages of elements of w, up to the factor μ, and vice versa.

The values μ_l^2 are latent values of matrix $A^\mathsf{T}A$. As is known, the sum of all latent values of a matrix is equal to its trace, defined as the sum of diagonal entries, that is, $Tr(A^\mathsf{T}A) = \sum_{t=1}^r \mu_t^2$ where r is the rank of A. On the other hand, direct calculation shows that the sum of diagonal entries of $A^\mathsf{T}A$ is

$$Tr(A^\mathsf{T}A) = \sum_{i,j} (p_{ij} - p_{i+}p_{+j})^2 / (p_{i+}p_{+j}) = X^2 \qquad (7.16)$$

Thus,

$$X^2 = \mu_1^2 + \mu_2^2 + E^2 \qquad (7.17)$$

which can be seen as a decomposition of the contingency data scatter, measured by X^2, into contributions of the individual factors, μ_l^2, and unexplained residuals, E^2. (Here, $l = 1, 2$, but, actually, the number of factors sought can be raised up to the rank of matrix A.) In a common situation, the first two latent values account for a major part of X^2, thus justifying the use of the plane of the first two factors for visualization of the interrelations between I and J.

Therefore, CA is similar to PCA except for the following:

1. CA applies to contingency data in such a way that the relative Quetelet coefficients are modeled rather than the original frequencies.

2. Rows and columns are weighted by the marginal frequencies; these are used in both the least-squares criterion and orthogonality constraints.

3. Both rows and columns are visualized on the same display so that geometric distances between the representations reflect chi-square distances between row or column conditional frequency profiles.

4. The data scatter is measured by the Pearson chi-square association coefficient.

As shown in [12] (see also [111]), CA better reproduces the visual shapes of contingency data than the standard PCA.

Example 7.1: Contingency Table for the Synopsis of Bribery Data and Its Visualization

Let us build on results of the interpretation of the Bribery data set over its clustering results on p. 223 in Section 6.3.3: all features of the Bribery data in Table 1.13 are well represented by two variables only: the branch at which the corrupt service occurs and the type of the service, features X and III in Table 1.12, respectively. Let us take a look at the cross-classification of these features (Table 7.1) and visualize it with the method of correspondence analysis.

In Figure 7.1, one can see which columns are attracted to which rows: Change of category to Other branch, Favors and Extortion to Government and Cover-up and Obstruction of justice to Law Enforcement. This is compatible with the conclusions drawn on p. 223. A feature of this display is that the branches constitute a triangle within which the services are— this property follows from the fact that the column-points are convex combinations of the row points. ∎

TABLE 7.1

Bribery

	Type of Service					
Branch	ObstrJus	Favors	Extort	CategCh	Cover-up	Total
Government	0	8	7	0	3	18
LawEnforc	14	1	3	2	9	29
Other	1	1	0	5	1	8
Total	15	10	10	7	13	55

Note: Cross-classification of features Branch (X) and type of service (III) from Table 1.12.

7.1.6 Data Summarization versus Learning in Data Recovery

Data analysis involves two major activities: summarization and correlation [140]. In a correlation problem, there is a target feature or a set of target features that are to be related to other features in such a way that the target feature can be predicted from values of the other features. Such is the linear regression problem considered above. In a summarization problem, such as the averaging or the principal component analysis, all the features available are considered target features so that those to be constructed as a summary can be considered as "hidden input features."

In the data recovery context, the structure of a summarization problem may be likened to that of a correlation problem if a rule is provided to predict all the original features from the summary. That is, the original data in the summarization problem act as the target data in the correlation problem. This implies that there should be two rules involved in a summarization problem: one for building the summary, and the other to recover the data from the summary.

However, unlike in the correlation problem, the feedback rule must be pre-specified in the summarization problem so that the focus is on building a summarization rule rather than on using the summary for prediction. In the machine-learning literature, the issue of data summarization has not been given that attention it deserves; this is why the problem is usually considered somewhat simplistically as just deriving a summary from data without any feedback (see Figure 7.2c).

Considering the structure of a summarization problem as a data recovery problem, one should rely on existence of a rule providing the feedback from a summary back to the data and consider the summarization process as more or less similar to that of the correlation learning process (see Figure 7.2b). This makes it possible to use the same criterion of minimization of the difference between the original data and the output so that the less the difference, the better.

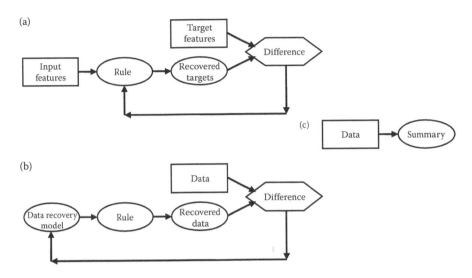

FIGURE 7.2
Diagrams for data recovery correlation (a) versus summarization (b); in (c), summarization without data recovery is illustrated. Rectangles are for observed data, ovals for computational constructions and hexagons for comparison of the data with the recovered data.

7.2 K-Means as a Data Recovery Method

7.2.1 Clustering Equation and Data Scatter Decomposition

In K-Means, a clustering is represented by a partition $S = \{S_k\}$ of the entity set I consisting of K cluster lists S_k that do not overlap and cover all entities, that is, $S_k \cap S_l \neq \emptyset$ if $k \neq l$ and $\cup_{k=1}^{K} S_k = I$. The latter condition can be relaxed as described in Sections 7.2.6 and 3.3. The lists are frequently referred to as cluster contents. The number of elements in S_k is frequently referred to as the cluster's size or cardinality and denoted by N_k. Centroids of clusters are vectors $c_k = (c_{kv})$ representing cluster "prototypes" or "standard" points.

Given a partition S and set of centroids $c = \{c_k\}$ resulting from K-Means, the original data can be recovered in such a way that any data entry y_{iv} (where $i \in I$ denotes an entity and $v \in V$ a category or quantitative feature) is represented by the corresponding centroid value c_{kv}, up to a residual, $e_{iv} = y_{iv} - c_{kv}$ where $i \in S_k$. In this way, clustering (S, c) leads to the data recovery model described by equations

$$y_{iv} = c_{kv} + e_{iv}, \quad i \in S_k, \ k = 1, \ldots, K \tag{7.18}$$

It is this model, in all its over-simplicity, that stands behind K-Means clustering. Let us see how this can happen.

Multiplying Equations 7.18 by themselves and summing up the results, it is not difficult to derive the following equation:

$$\sum_{i \in I} \sum_{v \in V} y_{iv}^2 = \sum_{v \in V} \sum_{k=1}^{K} N_k c_{kv}^2 + \sum_{i \in I} \sum_{v \in V} e_{iv}^2 \qquad (7.19)$$

The derivation is based on the assumption that c_{kv} is the average of within cluster values y_{iv}, $i \in S_k$, so that $\sum_{i \in S_k} c_{kv} y_{iv} = N_k c_{kv}^2$. Note that the right-hand term in Equation 7.19, $\sum_{i \in I} \sum_{v \in V} e_{iv}^2 = \sum_{k=1}^{K} \sum_{i \in S_k} \sum_{v \in V} (y_{iv} - c_{kv})^2 = \sum_{k=1}^{K} \sum_{i \in S_k} d(y_i, c_k)$, is the K-Means square error criterion $W(S, c)$ (3.2), Equation 7.19 can be rewritten as

$$T(Y) = B(S, c) + W(S, c) \qquad (7.20)$$

where $T(Y) = \sum_{i,v} y_{iv}^2$ is the data scatter, $W(S, c) = \sum_{k=1}^{M} \sum_{i \in S_k} d(y_i, c_k)$ square-error clustering criterion and $B(S, c)$ the middle term in decomposition (7.19):

$$B(S, c) = \sum_{v \in V} \sum_{k=1}^{K} c_{kv}^2 N_k \qquad (7.21)$$

Equation 7.19, or its equivalent (7.20), is well-known in the analysis of variance; its parts $B(S, c)$ and $W(S, c)$ are referred to as between-group and within-group variance in statistics. In the context of model (7.18), these, however, denote the explained and unexplained parts of the data scatter, respectively. The square error criterion of K-Means, therefore, minimizes the unexplained part of the data scatter, or, equivalently, maximizes the explained part, $B(S, c)$ (7.21). In other words, this criterion expresses the idea of approximation of data Y by the K-Means partitioning as expressed in Equation 7.18.

Equation 7.20 can be rewritten in terms of distances d since $T(Y) = \sum_{i=1}^{N} d(y_i, 0)$ and $B(S, c) = \sum_{k=1}^{K} N_k d(c_k, 0)$ according to the definition of the Euclidean distance squared:

$$\sum_{i=1}^{N} d(y_i, 0) = \sum_{k=1}^{K} N_k d(c_k, 0) + \sum_{k=1}^{K} \sum_{i=1}^{N} d(y_i, c_k) \qquad (7.22)$$

7.2.2 Contributions of Clusters, Features and Entities

According to Equations 7.20 and 7.21, each individual cluster $k = 1, \ldots, K$ additively contributes

$$B(S_k, c_k) = \sum_{v \in V} c_{kv}^2 N_k \qquad (7.23)$$

to $B(S, c)$. In its turn, any individual cluster's contribution is the sum of cluster-specific feature contributions $B_{kv} = c_{kv}^2 N_k$. Following from the preliminary standardization of data by subtracting features' grand means, the contribution B_{kv} is proportional to the squared difference between variable v's grand mean c_v and its within cluster mean c_{kv}: the larger the difference, the greater the contribution. This nicely fits into our intuition: the farther away is the cluster from the grand mean on a feature range, the more useful should be the feature in separating the cluster from the rest.

To find contributions of individual entities to the explained part of the data scatter, one needs yet another reformulation of Equation 7.23. Let us refer to the definition $c_{kv} = \sum_{i \in S_k} y_{iv}/N_k$ and put it into $c_{kv}^2 N_k$ "halfway." This then becomes $c_{kv}^2 N_k = \sum_{i \in S_k} y_{iv}c_{kv}$ leading to $B(S_k, c_k)$ reformulated as the summary inner product:

$$B(S_k, c_k) = \sum_{v \in V} \sum_{i \in S_k} y_{iv}c_{kv} = \sum_{i \in S_k} <y_i, c_k> \tag{7.24}$$

thus suggesting that the contribution of entity $i \in S_k$ to the explained part of the data scatter is $<y_i, c_k>$, the inner product between the entity point and the cluster's centroid, as follows from Equation 6.2. This may give further insights into the scatter decomposition to highlight contributions of individual entities (see an example in Table 6.10 on p. 222).

These and related interpretation aids are suggested for use in Section 6.3.2 as a non-conventional but informative tool.

7.2.3 Correlation Ratio as Contribution

To measure statistical association between a quantitative feature v and partition $S = \{S_1, \ldots, S_K\}$, the so-called correlation ratio η^2 has been defined in statistics:

$$\eta^2(S, v) = \frac{\sigma_v^2 - \sum_{k=1}^{K} p_k \sigma_{kv}^2}{\sigma_v^2} \tag{7.25}$$

where $\sigma_v^2 = \sum_{i \in I}(x_{iv} - c_v)^2/N$ and $\sigma_{kv}^2 = \sum_{i \in S_k}(x_{iv} - c_{kv})^2/N_k$ are the variance and within-cluster variance of variable v, respectively, before data preprocessing and $p_k = N_k/N$, see Section 2.2.2. Actually, the correlation ratio expresses the proportion of the variance of v explained by the within-cluster averages and, in this sense, is analogous to the determinacy coefficient in the model of linear regression. The correlation ratio ranges between 0 and 1, and it is equal to 1 only when all the within-class variances are zero. The greater the within-category variances, the smaller the correlation ratio.

Assuming that the raw data x_{iv} have been standardized to $y_{iv} = (x_{iv} - c_v)/b_v$ by shifting the origins to c_v and rescaling results by dividing them over b_v, it is not difficult to prove that the total contribution

$B(S,v) = \sum_{k=1}^{K} B_{kv} = \sum_{k=1}^{K} c_{kv}^2 N_k$ of a quantitative feature v to the cluster-explained part of the data scatter in Equation 7.21 equals to $N\eta^2(S,v)\sigma_v^2/b_v^2$ [139].

The cluster to variable contribution $N\eta^2(S,v)\sigma_v^2/b_v^2$ becomes plain $N\eta^2(S,v)$ when the variable has been normalized with b_v being its standard deviation, the option which hides the shape of the variable distribution. Otherwise, with b_v being the range r_v, the contribution should be considered a partition-to-feature association coefficient on its own:

$$\mu^2(S,v) = \eta^2(S,v)\sigma_v^2/r_v^2 \tag{7.26}$$

7.2.4 Partition Contingency Coefficients

Consider now the case of a nominal variable l represented by its set of categories V_l. The summary contribution $B(S,l) = \sum_{v \in V_l} \sum_k B_{vk}$ of nominal feature l to partition S, according to decomposition (7.19), appears to have something to do with association coefficients in contingency tables considered in Section 2.2.3.

To analyze the case, let us initially derive frequency-based reformulations of centroids for binary variables $v \in V_l$. Let us recall that a categorical variable l and cluster-based partition S, when cross-classified, form contingency table p_{kv}, whose marginal frequencies are p_{k+} and p_{+v}, $k = 1, \ldots, K, v \in V_l$.

For any three-stage pre-processed column $v \in V_l$ and cluster S_k in the clustering (S,c), its within-cluster average is equal to

$$c_{kv} = \left[\frac{p_{kv}}{p_{k+}} - a_v \right] / \left[b_v b_v' \right] \tag{7.27}$$

where $b_v' = \sqrt{|V_l|}$. Indeed, within-cluster average in this case equals $c_{kv} = p_{kv}/p_{k+}$, the proportion of v in cluster S_k. The mean a_v of binary attribute $v \in V_l$ is the proportion of ones in it, that is, the frequency of the corresponding category, $a_v = p_{+v}$.

This implies that

$$B(S,l) = N \sum_{k=1}^{K} p_{k+} \sum_{v \in V_l} (p_{kv}/p_{k+} - p_{+v})^2 / |V_l| b_v^2 \tag{7.28}$$

which can be transformed, with little arithmetic, into

$$B(S,l) = \frac{N}{|V_l|} \sum_{k=1}^{K} \sum_{v \in V_l} \frac{(p_{kv} - p_{k+}p_{+v})^2}{p_{k+}b_v^2} \tag{7.29}$$

where b_v and $|V_l|$ stand for the second stage, scaling, and the third stage, rescaling, during data pre-processing, respectively. The items summarized in Equation 7.29 can be further specified depending on scaling coefficients b_v as

1. $\dfrac{(p_{kv} - p_k p_v)^2}{p_k}$ if $b_v = 1$, the range.

2. $\dfrac{(p_{kv} - p_k p_v)^2}{p_k p_v (1 - p_v)}$ if $b_v = \sqrt{p_v(1 - p_v)}$, Bernoulli standard deviation.

3. $\dfrac{(p_{kv} - p_k p_v)^2}{p_k p_v}$ if $b_u = \sqrt{p_u}$, Poisson's standard deviation.

These lead to the following statement.

Statement 7.1. The contribution $B(S, l)$ of a nominal variable l and partition S to the explained part of the data scatter, depending on the standardizing coefficients, is proportional to contingency coefficient G^2 (2.11) or $Q^2 = X^2$ (2.12) if scaling coefficients b_v are taken to be $b_v = 1$ or $b_v = \sqrt{p_{+v}}$, respectively, and rescaling coefficients $b'_v = 1$. It is further divided by the number of categories $|V_l|$ if rescaling coefficients are $b'_v = \sqrt{|V_l|}$ for $v \in V_l$.

Two well-known normalizations of the Pearson chi-square contingency coefficient are due to Tchouproff, $T = X^2/\sqrt{(K-1)(|V_l|-1)}$, and Cramer, $C = X^2/\min(K-1, |V_l|-1)$. They both are symmetric over the numbers of categories and clusters. Statement 7.1 implies one more, asymmetric, normalization of X^2, $M = X^2/|V_l|$, as a meaningful part of the data scatter in the clustering problem.

When the chi-square contingency coefficient or related indexes are applied in the traditional statistics context, the presence of zeros in a contingency table becomes an issue because it bluntly contradicts the hypothesis of statistical independence. In the context of data recovery clustering, zeros are not different from any other numbers and create no problems at all because the coefficients are measures of contributions and bear no probabilistic meaning in this context.

7.2.5 Equivalent Reformulations of the Least-Squares Clustering Criterion

Expression $B(S, c)$ in Equation 7.21 is an equivalent reformulation of the least-squares clustering criterion $W(S, c)$ with the only difference that $B(S, c)$ is to be maximized rather than minimized. Merging its formula with that in Equation 7.22, one can easily see that in fact it is the sum of the squared Euclidean distance from 0 to all the centroids c_k weighted by cluster cardinalities N_k:

$$B(S, c) = \sum_{k=1}^{K} \sum_{v \in V} c_{kv}^2 N_k = \sum_{k=1}^{K} N_k d(0, c_k) \qquad (7.30)$$

To maximize this criterion, the clusters should be as far away from 0 as possible. This idea is partly implemented in the Anomalous Pattern one-by-one clustering method described in Section 3.2.3 and later in Section 7.2.6. An AP batch version is to be developed yet.

Let us note that the internal sum $\sum_{v \in V} c_{kv}^2$ in Equation 7.44 is in fact the inner square $\langle c_k, c_k \rangle$ and substitute one within cluster average by its definition: $\langle c_k, c_k \rangle = \langle c_k, \sum_{i \in S_k} y_{kv}/N_k \rangle = \sum_{i \in S_k} \langle c_k, y_i \rangle / N_k$. This implies that

$$B(S,c) = \sum_{k=1}^{K} \sum_{i \in S_k} < y_i, c_k > \qquad (7.31)$$

This expression shows that the K-Means criterion that minimizes the within-cluster distances to centroids is equivalent to the criterion (7.31) for maximization of the within-cluster inner products with centroids. These two total to the data scatter (7.20), which does not depend on the clustering. Note, however, that the distance-based criterion makes sense at any set of centroids whereas the inner product-based criterion makes sense only when the centroids are within-cluster averages. Also, the distance does not depend on the location of the origin whereas the inner product does. There is a claim in the literature that the inner product is much beneficial at larger sizes of the feature set if the data are pre-normalized in such a way that the rows, corresponding to entities, are normed so that each has its Euclidean norm equal to unity (note the change of the empasis in normalization here, from features to entities), see, for example, [54]. By further substituting $c_k = \sum_{j \in S_k} y_j / N_k$ in Equation 7.31, we arrive at equation

$$B(S,c) = \sum_{k=1}^{K} \sum_{i,j \in S_k} < y_i, y_j > /N_k \qquad (7.32)$$

expressing the criterion in terms of entity-to-entity similarities $a_{ij} = < y_i, y_j >$, with centroids c_k present implicitly. Criterion (7.32) is the total within-cluster semi-averaged similarity that should be maximized to minimize the least-squares clustering criterion.

There should be two comments made on this.

First, $a_{ij} = \sum_v y_{iv} y_{jv}$ is the sum of the entity-to-entity similarities over individual features, $a_{ij} = \sum_v a_{v,ij}$, where $a_{v,ij} = y_{iv} y_{jv}$ scores the level of similarity between entities due to a single feature v. Assuming that y_{iv} has been pre-processed by subtracting a reference value such as grand mean $m_v = \sum_{i \in I} y_{iv}/|I|$, one can see that $a_{v,ij}$ is negative if i and j differ over m_v so that y_{iv} and y_{jv} lie on v-axis at the different sides of zero. In contrast, $a_{v,ij}$ is positive if y_{iv} and y_{jv} lie on v-axis at the same side, either positive or negative. The greater the distance from 0 to y_{iv} and y_{jv}, the greater the value of $a_{v,ij}$. If the distribution is similar to Gaussian, most entities fall near grand mean, which is 0 after the normalization, so that most similarities are quite small.

Second, obviously matrix $A = (a_{ij})$ can be expressed through the original data matrix Y as $A = YY^T$ so that criterion (7.32) can be rewritten, in a matrix form, as

$$B(S, c) = \sum_{k=1}^{K} \frac{s_k^T A s_k}{s_k^T s_k} \tag{7.33}$$

where $s_k = (s_{ik})$ is a binary membership vector for cluster S_k so that $s_{ik} = 1$ if $i \in S_k$ and $s_{ik} = 0$, otherwise. Indeed, obviously, $s_k^T s_k = N_k$ and $\sum_{i,j \in S_k} a_{ij} = s_k^T A s_k$ so that Equation 7.33 easily follows from Equation 7.32. This is much similar to formula (7.12). The only difference is that the Rayleigh quotients in Equation 7.12 have the global maximizers as their arguments, whereas the arguments of those in Equation 7.33 are intended to be maximizers under the constraint that only the binary 0/1 vectors are admissible. The similarity follows from the similarity of the underlying data recovery models.

Indeed, the clustering model in Equation 7.18 can be rewritten in the format of the PCA equation (7.7)

$$y_{iv} = c_{1v} s_{i1} + \cdots + c_{Kv} s_{iK} + e_{iv} \tag{7.34}$$

or, in the matrix algebra notation,

$$Y = S_K C_K + E \tag{7.35}$$

where S_K is an $N \times K$ binary matrix, k-th column of which is the cluster membership vector $s_k = (s_{ik})$ where $s_{ik} = 1$ if $i \in S_k$ or $s_{ik} = 0$, otherwise. Matrix C_K consists of columns c_k which are the within cluster means, according to the least-squares criterion. That is, clusters in model (7.34) correspond to factors in model (7.7), and centroids, to the loading vectors. Equation 7.34 is equivalent to that in Equation 7.18 indeed, because the clusters S_k do not mutually overlap, so that the summation in Equation 7.34 is somewhat fictitious—at each $i \in I$, only one of $s_{i1}, s_{i2}, \ldots, s_{iK}$ is not zero.

Therefore, the contributions of individual clusters, $\sum_v c_{kv}^2 N_k$, are akin to the contributions of individual principal components, μ_k^2, both expressing the maximized values of the corresponding Rayleigh quotients, albeit over different domains.

One more formulation extends the K-Means criterion to the situation of monothetic clustering in which individual features are utilized for clustering rather than just distances over all the features. This comes from the analysis of the contributions of feature-cluster pairs to the data scatter in the preceding section. Consider a measure of association $\zeta(S, v)$ between a partition S of the entity set I and a feature v, $v \in V$. Consider, for the sake of simplicity, that all features are quantitative and have been standardized by z-scoring. Then

$\zeta(S, v)$ is the correlation ratio η^2 defined by formula (7.25). Then maximizing the summary association $\zeta(S) = \sum_{v=1}^{V} \eta(S, v)^2$ is equivalent to minimization of the K-Means least-squares criterion $W(S, c)$ defined by using the squared Euclidean distances. A very similar equivalent consensus criterion can be formulated for the case when feature set consists of categorical or even mixed scale features (see more on this in [140]).

Therefore, there are at least five different expressions for the least-squares partitioning criterion, each potentially leading to a different set of techniques for optimizing it:

1. The original formulation as "clusters around centroids" leading to K-Means method for the method of alternating minimization of the criterion.

2. Reformulation as "similarity clusters" in Equation 7.32 leading to algorithms of entity exchange and addition described further in Section 7.4.2.

3. Reformulation as "binary-valued principal components" in Equation 7.34 leading to several perspectives. One of them, advocated by the author, is the "one-principal-cluster-at-a-time" approach leading to the methods of AP and iK-Means as explained in Section 7.2.6. Another approach would be in utilizing the principal components themselves, as solutions to a relaxed form of the model, for building a cluster over each of them. This did not work in our preliminary experiments conducted a while ago.

4. Reformulation in terms of partition-to-feature association measures. On the one hand, this leads to a conceptual clustering approach described in Section 7.3.4.3. On the other hand, this is akin to the consensus clustering approaches described in Section 7.5.

5. Reformulation as "faraway clusters" maximizing criterion (7.30), a method for simultaneously finding anomalous clusters is yet to be developed.

7.2.6 Principal Cluster Analysis: Anomalous Pattern Clustering Method

Let us pick on the formulation of clustering model in Equation 7.18 in the format of the PCA equation in Equation 7.34:

$$y_{iv} = c_{1v}s_{i1} + \cdots + c_{Kv}s_{iK} + e_{iv}$$

In this equation, s_k is a binary cluster membership vector $s_k = (s_{ik})$ so that $s_{ik} = 1$ if $i \in S_k$ and $s_{ik} = 0$, otherwise ($k = 1, \ldots, K$). Vector $c_k = (c_{kv})$ is M-dimensional centroid of kth cluster. The problem is, given a pre-processed matrix $Y = (y_{iv})$, to find vectors s_k, c_k minimizing the sum of squared residuals

$\sum_{i\in I, v\in V} e_{iv}^2$. Unlike the SVD problem, this one is computationally intensive, because of the constraint that the membership vector entries are to be zero-one binary.

As a practically reasonable way to go, let us use a simplified "greedy" optimization strategy of the PCA. Let us find just one cluster S with centroid c to minimize the least squares one-cluster criterion

$$L = \sum_{i\in I}\sum_{v\in V}(y_{iv} - c_v s_i)^2 \tag{7.36}$$

After finding a cluster S together with its centroid c, how could one proceed to find the other clusters? One approach would be to follow the PCA one-by-one procedure straightforwardly and apply the same criterion to the residual data $y_{iv}^1 = y_{iv} - c_v s_i$ for finding the second "principal" cluster. Unfortunately, in contrast to the PCA case, the optimal second cluster membership vector s' is not necessarily orthogonal to the first membership vector s. That is, the second cluster may overlap the first one, because binary vectors are orthogonal if and only if corresponding clusters do not overlap. This would undermine the approach as applicable for building a partition. Therefore, we are left with a different option: after the first cluster is found, its elements are to be removed from I, so that the next cluster is built on the remaining subset of I. Continuing the process in this way, after a finite number of steps, no more unclustered entities will remain. "Overlapping" versions of the approach are considered in [32].

Let us now take a closer look at the problem of minimization of one cluster criterion (7.36). Given a binary membership vector s and the corresponding cluster $S = \{i : s_i = 1\}$, the criterion can be rewritten as

$$L = \sum_{i\in S}\sum_{v\in V}(y_{iv} - c_v)^2 + \sum_{i\notin S}\sum_{v\in V}y_{iv}^2 = \sum_{i\in S}d(y_i, c) + \sum_{i\notin S}d(y_i, 0) \tag{7.37}$$

because $c_v s_i$ is equal to either c_v if $i \in S$ or 0 if $i \notin S$.

Criterion (7.37) is a version of K-Means criterion at $K = 2$, at which one centroid, 0, is pre-specified and unmoved. The optimal c is the vector of within-cluster feature means. At the start of the alternating minimization procedure for criterion (7.37), we can reasonably restrict ourselves, in a "greedy-wise" fashion, with singleton clusters S only. In this case, the minimum of Equation 7.37 is reached, obviously, at the entity y_i which is the farthest away from 0. Therefore, the initial centroid c can be reasonably put at the farthest away y_i. Then the cluster minimizing L in Equation 7.37 would represent an anomalous pattern. An alternating minimization procedure can therefore be formulated, as presented in the box.

ANOMALOUS PATTERN (AP) CLUSTERING

1. *Pre-processing.* Specify a reference point $a = (a_1, \ldots, a_n)$ (this can be the grand mean) and standardize the original data table with formula (2.22) at which shift parameters a_k are the reference point coordinates. (In this way, the space origin is shifted into a.)

2. *Initial setting.* Put a tentative centroid, c, at that entity which is the the farthest away from the origin, 0, that is, the reference point.

3. *Cluster update.* Determine cluster list S around c, versus the only other "centroid" 0, with the minimum distance rule so that y_i is assigned to S if $d(y_i, c) < d(y_i, 0)$.

4. *Centroid update.* Calculate the within S mean c' and check whether it differs from the previous centroid c. If c' and c do differ, update the centroid by assigning $c \leftarrow c'$ and go to Step 3. Otherwise, go to 5.

5. *Output.* Output list S and centroid c with accompanying interpretation aids (as advised in Section 6.3) as the AP.

The convergence of AP procedure follows from two facts: (i) the criterion L value decreases at each iteration and (ii) the number of iterations cannot be infinite because the number of possible clusters S is finite.

An iterative application of AP method to the subsets of yet unclustered entities has been referred to as PCA [135], but the current term, AP, better captures the intuitive meaning of the method.

Obviously, each of the AP clusters (S_k, c_k) contributes to the data scatter the corresponding item $\sum_{v \in V} c_{kv}^2 |S_k|$ from the complementary criterion $B(S, c)$. This can be used for deciding whether the process of AP generation should be halted before the set I is fully clustered.

Chiang and Mirkin [26] and Amorim and Mirkin [4] have shown that centroids of the largest AP clusters can be efficiently used for initialization of K-Means. The default option is using all the non-singleton AP clusters for the purpose.

7.2.7 Weighting Variables in K-Means Model and Minkowski Metric

The contents of this section follows [4]. Let us consider some clustering structure (S, c) where S is a partition of the entity set I and c the sets of cluster centroids. Assume that there is a system of feature weights w_v ($v \in V$) that are non-negative and sum to unity, that is, $w_v \geq 0$ and $\sum_{v \in V} w_v = 1$. Then the data recovery equations in Equation 7.18 can be modified, by multiplying them by w_v, to the following:

$$w_v y_{iv} = w_v c_{kv} + w_v e_{iv}, \quad i \in S_k, \ k = 1, \ldots, K \tag{7.38}$$

Considering the weights as feature rescaling factors to be found yet, Equations 7.38 can be rewritten as

$$y'_{iv} = c'_{kv} + e'_{iv}, \quad i \in S_k, \ k = 1, \ldots, K \qquad (7.39)$$

where $y'_{iv} = w_v y_{iv}$ are the rescaled feature values so that $c'_{iv} = w_v c_{iv}$ are appropriately rescaled cluster centroids. The rescaled residuals e'_{iv} are not of much interest here since they are to be minimized with a non-linear criterion anyway. For a real $\beta > 0$, consider a Minkowski metric criterion to be minimized:

$$W_\beta(S, c, w) = \sum_{i=1}^{N} \sum_{k=1}^{K} \sum_{v \in V} w_v^\beta |y_{iv} - c_{kv}|^\beta = \sum_{i=1}^{N} \sum_{k=1}^{K} d_\beta(y'_i, c'_k) \qquad (7.40)$$

where d_β is β power of the Minkowski's distance defined for any $x = (x_v)$ and $z = (z_v)$ as

$$d_\beta(x, z) = \sum_{v \in V} |x_v - z_v|^\beta \qquad (7.41)$$

The case of $\beta = 2$ corresponds to the least squares K-Means criterion considered so far.

On the one hand, criterion (7.40) is but the ordinary K-Means criterion of the summary distances between the entities and their cluster's centroids, just the distance is the β power of Minkowski's distance. On the other hand, this criterion involves three groups of variables to be found rather than two: the unknown clusters and centroids of K-Means criterion now are supplemented by the unknown feature weights.

Minkowski metric K-Means criterion in Equation 7.40 can be minimized alternatingly in iterations over the three groups of variables: clusters, centroids and weights. Each iteration would involve the following three steps:

i. Given centroids c_k and weights w_v, update the cluster assignment of entities by using the Minimum distance rule with distance defined as the β power of Minkowski β-metric in Equation 7.41.

ii. Given clusters S_k and weights w_v, update centroid $c'_k = (c'_{kv})$ of each cluster S_k as its Minkowski center so that, at each v, c'_{kv} is minimizing an item in Minkowski's distance β power,

$$d_{kv}(c) = \sum_{i \in S_k} |y_{iv} - c|^\beta \qquad (7.42)$$

(see an algorithm for that on page 287): $c'_{kv} = w_v c$. (Indeed, since criterion (7.40) is the sum of independent items of the form of Equation 7.42, optimal centroid components c_{kv} are found by minimizing Equation 7.42 for the corresponding $v \in V$ and $k = 1, \ldots, K$).

iii. Given clusters S_k and centroids c_k, update weights according to formula

$$w_v = \frac{1}{\sum_{u \in V} [D_{v\beta}/D_{u\beta}]^{1/\beta-1}} \qquad (7.43)$$

where $D_{v\beta} = \sum_{k=1}^{K} \sum_{i \in S_k} |y_{iv} - c_{kv}|^\beta$.

To start the computation, a set of centroids is selected and equal weights are assigned to the features. Then iterations of the alternating minimization are to be run, each as a sequence of steps (i), (ii) and (iii) above. Let us refer to this algorithm as MWK-Means. An exact formulation of MWK-Means, in both cluster-specific and cluster-independent versions, can be found in Section 3.4.2.

To prove that formula (7.43) gives the optimal weights indeed, let us use the first-order optimality condition for the problem of minimization of W_β in Equation 7.40 with respect to w_v constrained by the condition that $\sum_{v \in V} w_v = 1$; there will be no need in invoking the condition that the weights must be non-negative. Let us rewrite the Minkowski criterion using $D_{v\beta}$ from (7.43): $W_\beta(S, c, w) = \sum_{v \in V} w_v^\beta D_{v\beta}$. To minimize this with regard to the constraint, take the Lagrange function

$$L = \sum_{v \in V} w_v^\beta D_{v\beta} + \lambda \left(1 - \sum_{v \in V} w_v\right)$$

and its partial derivative with respect to w_v: $\partial L/\partial w_v = \beta w_v^{\beta-1} D_{v\beta} - \lambda$. By equating this to zero, one can easily derive that $\lambda^{1/\beta-1} = (\beta D_{v\beta})^{1/\beta-1} w_v$. With little algebraic manipulation, this leads to Equation 7.43 indeed. Note that Equation 7.43 warrants that the weights cannot be negative, which is a bonus implied by the additive structure of the criterion. Of course, Equation 7.43 is not applicable when $D_{u\beta} = 0$ for some feature $u \in V$.

Another issue of the MWK-Means algorithm is finding Minkowski centers: their components are minimizers of the summary Minkowski distances (7.42). The problem is to find a real value c minimizing the summary distance (7.42) for a given set of real values representing a feature within a cluster. Let us restrict ourselves to the case of $\beta \geq 1$. The summary distance $d_{kv}(c)$ in Equation 7.42 is a convex function of c at $\beta \geq 1$. Then a steepest descent procedure can be applied to find the global minimizer. As it is well known, at $\beta = 1$, the median minimizes the distance $d_{kv}(c)$ in Equation 7.42, so that further on only $\beta > 1$ are considered. For the sake of convenience let us drop off the indexes at $d_{kv}(c)$. Assume that the y-values are sorted in the ascending order so that $y_1 \leq y_2 \leq \cdots \leq y_n$. Let us first prove that the optimal c must be between the minimum, y_1, and the maximum, y_n, of the range. Indeed, if, on the contrary, the minimum is reached outside of the interval, say at $c > y_n$, then $d(y_n) < d(c)$ because $|y_i - y_n| < |y_i - c|$ for all $i = 1, 2, \ldots, n$; and the same holds for the βth powers of those. This contradiction proves the statement. To prove the convexity, consider any c in the interval between y_1 and y_n. Distance function (7.42) then can be rewritten as

$$d(c) = \sum_{i \in I^+} (c - y_i)^\beta + \sum_{i \in I^-} (y_i - c)^\beta$$

where I^+ is set of those indices $i = 1, 2, \ldots, n$ for which $c > y_i$, and I^- is set of such i that $c \le y_i$. Then the first derivative of $d(c)$ is equal to

$$d'(c) = \beta \left[\sum_{i \in I^+} (c - y_i)^{\beta-1} - \sum_{i \in I^-} (y_i - c)^{\beta-1} \right]$$

and the second derivative, to

$$d'(c) = \beta(\beta - 1) \left[\sum_{i \in I^+} (c - y_i)^{\beta-2} + \sum_{i \in I^-} (y_i - c)^{\beta-2} \right]$$

The latter expression is positive for each c value, provided that $\beta > 1$, which proves that $d(c)$ is convex indeed.

The convexity leads to one more useful property: assume that $d(y_{i*})$ is minimum among all $d(y_i)$ values ($i = 1, 2, \ldots, n$). Then the point of minimum $y*$ lies in the vicinity of y_{i*}, that is, between y_{i*-1} and y_{i*+1}.

These properties justify the following steepest descent algorithm for finding Minkowski center of a set $\{y_i\}$ of values $y_1 \le y_2 \le \cdots \le y_n$ at $\beta > 1$.

FINDING MINKOWSKI CENTER

1. Initialize with $c_0 = y_{i*}$, the minimizer of $d(c)$ on the set $\{y_i\}$, and a positive learning rate λ that can be taken, say, as 10% of the range $y_n - y_1$.

2. Compute $c_0 - \lambda d'(c_0)$ and take it as c_1 if it falls within the interval (y_{i*-1}, y_{i*+1}). Otherwise, decrease λ a bit, say, by 10%, and repeat the step.

3. Test whether c_1 and c_0 coincide up to a pre-specified precision threshold. If yes, halt the process and output c_1 as the optimal value of c. If not, move on.

4. Test whether $d(c_1) < d(c_0)$. If yes, set $c_0 = c_1$ and $d(c_0) = d(c_1)$, and go to step 2. If not, decrease λ a bit, say by 10%, and go to step 2 without changing c_0.

AP clustering algorithm remains applicable with Minkowski metric if centroids are determined as Minkowski centers. Similarly, the reference point of the anomalous pattern algorithm should be defined as the Minkowski center of the entity set. The MWK-Means algorithm initialized with centroids of anomalous clusters found by using Minkowski metric and the found features' weights is referred to as iMWK-Means.

In relation to $D_{v\beta} = 0$ in formula (7.43), according to our experiments, adding a very small constant to the denominator alleviates the issue [4]. This applies only at the initialization stage of the algorithm; MWK-Means remains then unchanged.

7.3 Data Recovery Models for Hierarchical Clustering

7.3.1 Data Recovery Models with Cluster Hierarchies

To formulate supporting models for agglomerative and divisive clustering, one needs to explicitly define the concept of cluster hierarchy. A set S of subsets $S_w \subseteq I$ is called nested if for every two subsets S_w and $S_{w'}$ from S either one of them is part of the other or they do not overlap at all. Given a nested set S, $S_w \in S$ is referred to as a child of $S_{w'} \in S$ if $S_w \subset S_{w'}$ and no other subset $S_{w''} \in S$ exists such that $S_w \subset S_{w''} \subset S_{w'}$. A subset $S_w \in S$ is referred to as a terminal node or leaf if S_w has no children in S. A nested set S is referred to as a cluster hierarchy over I if any non-terminal subset $S_w \in S$ has two children $S_{w'}, S_{w''} \in S$ covering it entirely so that $S_{w'} \cup S_{w''} = S_w$. The subsets $S_w \in S$ are referred to as clusters.

Two types of cluster hierarchy are of interest in relation to agglomerative and divisive clustering algorithms: those S containing singleton clusters $\{i\}$ for all $i \in I$ and those containing set I itself as a cluster. The former will be referred to as the cluster down-hierarchy and the latter, the cluster up-hierarchy. An up-hierarchy can be thought of as resulting from a divisive clustering algorithm, see Figure 4.1b and a down-hierarchy from an agglomerative clustering algorithm, see Figure 4.1c. A complete result of a clustering algorithm of either type can be represented by a cluster tree, that is, a cluster hierarchy which is both down- and up-hierarchy, see Figure 4.1a.

For an up-cluster hierarchy S, let us denote the set of its leaves by $L(S)$; it is obviously a partition of I as illustrated in Figure 4.1b. Similarly, for a cluster down-hierarchy S, let us denote its set of maximal clusters by $M(S)$; this is also a partition of I.

Given an up- or down-hierarchy S over set I and a pre-processed data matrix $Y = (y_{iv})$, let us, for any feature v, denote the average value of y_{iv} within $S_w \in S$ by c_{wv}.

Given a cluster up-hierarchy S, let us consider its leaf partition $L(S)$. For any data entry y_{iv} and a leaf cluster $S_{w^*} \in L(S)$ containing it, the model underlying K-Means suggests that y_{iv} is equal to c_{w^*v} up to the residual $e_{iv} = y_{iv} - c_{w^*v}$.

Obviously, $e_{iv} = 0$ if S_{w^*} is a singleton consisting of just one entity i. To extend this to the S, let us denote the set of all non-singleton clusters containing i by S_i and add to and subtract from the equation the averages of feature v within each $S_w \in S_i$. This leads us to the following equation:

$$y_{iv} = \sum_{S_w \in S_i} (c_{w1,v} - c_{wv}) + e_{iv} \tag{7.44}$$

where S_{w1} is a child of S_w that runs through S_i. Obviously, all $e_{iv} = 0$ in Equation 7.44 if S is a full cluster hierarchy, that is, a cluster tree.

7.3.2 Covariances, Variances and Data Scatter Decomposed

Model in Equation 7.44 is not just a trivial extension of the K-Means model to the case of cluster up-hierarchies, in spite of the fact that the only "real" item in the sum in Equation 7.44 is c_{w^*v} where S_{w^*} is the leaf cluster containing i. Indeed, the equation leads to the following decomposition.

Statement 7.2. For every two feature columns $v, u \in V$ in the pre-processed data matrix Y, their inner product can be decomposed over the cluster up-hierarchy S as follows:

$$\langle y_v, y_u \rangle = \sum_w \frac{N_{w1}N_{w2}}{N_w} (c_{w1,v} - c_{w2,v})(c_{w1,u} - c_{w2,u}) + < e_v, e_u > \tag{7.45}$$

where w runs over all interior clusters $S_w \in S$ with children denoted by S_{w1} and S_{w2}; N_w, N_{w1}, N_{w2} being their respective cardinalities.

Proof: The proof follows from equation $N_w c_w = N_{w1} c_{w1} + N_{w2} c_{w2}$ which relates the centroid of an interior cluster S_w with those of its children. By putting this into Equation 7.44, one can arrive at Equation 7.45 by multiplying the decomposition for y_{iv} by that for y_{iu} and summing up results over all $i \in I$, q.e.d.

Given a cluster down-hierarchy S, model (7.44) remains valid, with e_{iv} redefined as $e_{iv} = c_{w^\#i}$ where $S_{w^\#} \in M(S)$ is the maximal cluster containing i. The summation in Equation 7.44 still runs over all $S_w \in S$ containing i, so that in the end the equation may be reduced to the definition $e_{hi} = c_{w^\#i}$. Yet taken as they are, the equations lead to the same formula for decomposition of inner products between feature columns.

Statement 7.3. Statement 7.2 is also true if S is a cluster down-hierarchy, with residuals redefined accordingly.

These lead to a decomposition described in the following statement.

Statement 7.4. Given a cluster down- or up-hierarchy **S**, the data scatter can be decomposed as follows:

$$\sum_{i \in I} \sum_{v \in V} y_{iv}^2 = \sum_{w} \frac{N_{w1} N_{w2}}{N_w} \sum_{v \in V} (c_{w1,v} - c_{w2,v})^2 + \sum_{i \in I} \sum_{v \in V} e_{iv}^2 \tag{7.46}$$

where w runs through all nonterminal clusters $S_w \in \mathbf{S}$.

Proof: To prove Equation 7.46, set $u = v$ in Equations 7.45 and sum them up over all feature columns $v \in V$. This also produces, on the left-hand side, the sum of all inner products $< y_v, y_v >$, which is obviously the data scatter, and on the right side, exactly the right side of Equation 7.46, q.e.d.

Note that the central sum in Equation 7.46 is but the squared Euclidean distance between centroids of clusters S_{w1} and S_{w2}, which leads to the following reformulation:

$$T(Y) = \sum_{w} \frac{N_{w1} N_{w2}}{N_w} d(c_{w1}, c_{w2}) + \sum_{i \in I} \sum_{v \in V} e_{iv}^2 \tag{7.47}$$

Further reformulations easily follow from the definitions of e_{iv} in cluster up- or down-hierarchies.

In particular, if **S** is a cluster down-hierarchy then the residual part $\sum_i \sum_v e_{iv}^2$ of the data scatter decomposition in Equation 7.46 is equal to the complementary K-Means criterion $B(S, c)$ where $S = M(\mathbf{S})$ is the set of maximal clusters in **S** and c the set of their centroids. That means that for any cluster down-hierarchy **S** with $S = M(\mathbf{S})$:

$$T(Y) = \sum_{w} \frac{N_{w1} N_{w2}}{N_w} d(c_{w1}, c_{w2}) + B(S, c) \tag{7.48}$$

Similarly, if **S** is a cluster up-hierarchy, the residual part is equal to the original K-Means square error criterion $W(S, c)$ where $S = L(\mathbf{S})$ is the set of leaf clusters in **S** with c being their centroids. That means that for any cluster up-hierarchy **S** with $S = L(\mathbf{S})$:

$$T(Y) = \sum_{w} \frac{N_{w1} N_{w2}}{N_w} d(c_{w1}, c_{w2}) + W(S, c) \tag{7.49}$$

These decompositions explain what is going on in Ward clustering in terms of the underlying data recovery model. Every merging step in agglomerative clustering or every splitting step in divisive clustering adds (or subtracts) the Ward distance

$$\lambda_w = dw(S_{w1}, S_{w2}) = \frac{N_{w1} N_{w2}}{N_w} d(c_{w1}, c_{w2}) \tag{7.50}$$

to (or, from) the central sum in Equation 7.47 by reducing the other part, $B(S, c)$ or $W(S, c)$, respectively. The central part appears to be that explained by the up-hierarchy in divisive clustering and that unexplained by the down-hierarchy in agglomerative clustering. Thus, the Ward distance (7.50) must be as large as possible in divisive clustering and as small as possible in agglomerative clustering. q.e.d.

7.3.3 Split Base Vectors and Matrix Equations for the Data Recovery Model

In fact, the model (7.44) can be presented with the same equations as the PCA model in Equations 7.7 and 7.8 and K-Means partitioning model in Equations 7.34 and 7.35. The only difference is in the N-dimensional scoring vectors: they are arbitrary in PCA and binary in K-Means. Now they are three-valued to correspond to the interior nodes of a cluster hierarchy.

Consider any non-terminal cluster of a cluster hierarchy $S_w \in \mathbf{S}$ and its children S_{w1} and S_{w2}. Define a vector ϕ_w so that its components are zeros outside of the cluster, and they have one value in S_{w1} and another in S_{w2}. Specifically, define: $\phi_{iw} = 0$ if $i \notin S_w$, $\phi_{iw} = a_w$ for $i \in S_{w1}$, and $\phi_{iw} = -b_w$ for $i \in S_{w2}$ where a_w and b_w are positive reals satisfying the conditions that vector ϕ_w must be centered and normed. These two conditions lead to a_w and b_w uniquely defined as $a_w = \sqrt{1/N_{w1} - 1/N_w} = \sqrt{N_{w2}/N_w N_{w1}}$ and $b_w = \sqrt{1/N_{w2} - 1/N_w} = \sqrt{N_{w1}/N_w N_{w2}}$ [135]. Thus defined, vector ϕ_w will be referred to as a split base vector. Note that the assignment of plus and minus signs in a split base vector is arbitrary and can be associated with an ordering of the nodes specified by a drawing of the hierarchy such as in Figure 4.1. Having such a drawing specified, one may also specify a rule, say, that, for any split, the subcluster on the right gets a minus while the subcluster on the left, plus.

It is not difficult to prove that the split base vectors ϕ_w are mutually orthogonal and, therefore, form an orthonormal base (see Section 4.1 and [135]). In the case when $\mathbf{S}$ is a cluster full hierarchy, that is, both up and down-hierarchy, the set of all split base vectors ϕ_w is a base of the space of all N-dimensional centered vectors [135]: indeed, the dimensionality of the space, $N - 1$, coincides with the number of split base vectors.

Now, define an $N \times m$ matrix Φ where N is the number of entities in I and m the number of nonterminal clusters in $\mathbf{S}$. Matrix Φ has split vectors ϕ_w as its columns.

By using matrix Φ, Equation 7.44 can be equivalently rewritten as

$$Y = \Phi A + E \tag{7.51}$$

where A is an $m \times M$ matrix with entries $a_{wv} = \sqrt{N_{w1} N_{w2}/N_w}(c_{w1,v} - c_{w2,v})$, or in the matrix entry format,

$$y_{iv} = \sum_w \phi_w(i) a_{wv} + e_{iv} \tag{7.52}$$

Multiplying Equation 7.51 by Y^T on the left, one arrives at the matrix equation $Y^T Y = A^T A + E^T E$, since: (i) $\Phi^T \Phi$ is the identity matrix and (ii) $\Phi^T E$ the zero matrix. These latter equations are matrix expressions of the conditions that all the split vectors ϕ_w are (i) normed and mutually orthogonal and (ii) centered. This matrix equation is a matrix reformulation for the decomposition of covariances (7.45). This derivation gives another proof to Statement 7.2.

Moreover, each contribution λ_w (7.50) to the data scatter in Equation 7.49 can be expressed as the Rayleigh quotient of the matrix YY^T and the split vector ϕ_w:

$$\lambda_w = \frac{\phi_w^T YY^T \phi_w}{\phi_w^T \phi_w} \tag{7.53}$$

Indeed, because of the definition, $\phi_w^T YY^T \phi_w = \sum_{i,j \in I} <y_i, y_j> \phi_{iw} \phi_{jw} = \left(\sum_{i \in I} \phi_{iw} y_i, \sum_{j \in I} \phi_{jw} y_j\right)$. Because $\phi_{iw} = 0$ outside of cluster S_w, the sum $\sum_{i \in I} \phi_{iw} y_i$ can be further represented as $\sum_{i \in I} \phi_{iw} y_i = \sqrt{N_{w2}/(N_w N_{w1})} \sum_{i \in S_{w1}} y_i - \sqrt{N_{w1}/(N_w N_{w2})} \sum_{i \in S_{w2}} y_i$ where S_{w1} and S_{w2} are children of cluster S_w in hierarchy **S**. Since $\sum_{i \in S_{w1}} y_i = N_{w1} c_1$ and $\sum_{i \in S_{w2}} y_i = N_{w2} c_2$, this can be further transformed to $\sum_{i \in S_{w1}} y_i = \sqrt{N_{w2} N_{w1}/N_w}(c_{w1} - c_{w2})$. This proves the statement.

The data recovery model for K-Means partitioning leads to decomposition (7.20) of the data scatter and the data recovery model for a cluster up-hierarchy (the divisive Ward-like clustering), to decomposition (7.49). By comparing these two equations, one can conclude that the split contributions λ_w sum to $B(S, c)$, that is, they give an alternative way of explaining clusters in S: by splits leading to S as the leaf set $L(\mathbf{S})$ of a cluster hierarchy rather than by clusters in S themselves. Obviously, decomposition (7.49) holds for any up-hierarchy leading to S.

7.3.4 Divisive Partitioning: Four Splitting Algorithms

As explained in Section 7.1.4, the PCA model assumes a small number K of hidden factors, the principal components, $z_1, z_2, \ldots, z_K$, to underlie the observed matrix Y so that $Y = ZA + E$ where Z is an $N \times K$ matrix whose columns are the principal components, A is a $K \times M$ matrix of so-called factor loadings and E is a matrix of the residuals that are to be least-square minimized with respect to the sought Z and A. Each consecutive principal component z_k additively contributes μ_k^2 to the explained part of the data scatter $T(Y)$ ($k = 1, \ldots, K$). Here μ_k is a singular value of Y or, equivalently, μ_k^2 is a latent value of both feature-to-feature 'covariance' related matrix $Y^T Y$ and entity-to-entity similarity matrix YY^T. The principal components can be found one by one in such a way that the PCA model with $K = 1$ is iteratively applied to the residual data matrix Y from which principal components found at previous steps have been subtracted. The first iteration yields the component corresponding to the maximum singular value; the second yields the next largest singular value, and so on.

The data recovery models in Equation 7.18, for partitioning, and Equation 7.44, for hierarchical clustering, have the same format: equation $Y = ZA + E$ at which the residuals E are to be minimized over unknown Z and A. In the hierarchical clustering problem, Z is a matrix Φ of split vectors ϕ_w corresponding to non-terminal clusters S_w of a cluster down- or up-hierarchy to be found. In the partitioning problem, Z is a matrix of the binary membership vectors z_k defined by clusters S_k. The vectors z_k are mutually orthogonal, because clusters S_k are not to overlap.

Therefore, the PCA strategy of extracting columns of Z one by one, explained in Section 7.1.4, can be applied in each of these situations. The PCA-like model in Equation 7.52 can be fitted by using the one-by-one approach to build a (sub)-optimal cluster up-hierarchy.

Starting from a standardized data matrix Y and corresponding universal cluster $S_0 = I$ of the cardinality $N_0 = |I|$, split the cluster in two parts maximizing the Rayleigh quotient λ_0 over all the possible three-valued split vectors. Since the Rayleigh Quotient (7.53) in this case is equal to Ward distance, $\lambda_0 = N_1 N_2 / N_0 d(c_1, c_2)$, a split is found by maximizing the distance. After split parts S_1 and S_2 are found, their contributions $T(S_1) = \sum_{i \in S_1} d(y_i, c_1)$ and $T(S_2) = \sum_{i \in S_2} d(y_i, c_2)$ are computed. Then a stop-condition is tested. If the condition does not hold, the same splitting process applies to the cluster with the maximum contribution, according to the principle of the one-by-one optimal splitting, and so forth.

The problem of splitting a cluster to maximize the Ward distance has not received much attention; it seems rather computationally intensive even in spite of the fact that only hyperplane separable splits are admitted. This is why only suboptimal procedures have been proposed so far. We present four of them:

1. Alternating minimization: Bisecting K-Means, or 2-splitting
2. Relaxing the constraints: Principal direction division
3. One-feature splitting: Conceptual clustering
4. Separating a cluster: Split by Separation

7.3.4.1 Bisecting K-Means, or 2-Splitting

In this approach, the fact that maximizing the Ward distance is equivalent to minimizing K-Means criterion at $K = 2$ is utilized. The fact easily follows from the definition of Ward distance: this is the difference between two K-Means criteria, that for the entire cluster to be split minus the value at the two-split partition. The statement follows from the fact that the former is constant here. A computational proof is given in the next section. Therefore, one can apply the alternating minimization of the 2-Means criterion, which is 2-Means, or 2-splitting, or Bisecting K-Means, as it has become known starting from paper [199]. Paper [199] experimentally approved the method, albeit referred to as just a heuristic rather than a locally optimal least-squares hierarchical clustering procedure (as it was introduced in [135]).

The Bisecting K-Means, aka 2-Means splitting, can be specified as follows.

2-MEANS SPLITTING OR BISECTING K-MEANS

1. *Initial setting.* Given $S_w \subseteq I$, specify initial seeds of split parts, $c_{w1} = y_1$ and $c_{w2} = y_2$.

2. *Batch 2-Means.* Apply 2-Means algorithm to S_w with initial seeds specified at step 1 and the squared Euclidean distance (2.17).

3. *Output* results: (a) split parts S_{w1} and S_{w2}; (b) their centroids c_{w1} and c_{w2} along with their heights, $h(S_1)$ and $h(S_2)$; (c) contribution of the split, that is, Ward distance between S_{w1} and S_{w2}.

In spite of the fact that Euclidean squared distance d, not Ward distance dw is used in splitting, the algorithm in fact goes in line with Ward agglomeration.

The only issue remaining to be covered is the choice of the two initial centroid seeds. Beyond a set of random starts, one idea would be to begin with two entities that are furthest away in the cluster. Another idea would be to use a version of iK-Means, by starting from centroids of two maximum-sized Anomalous pattern clusters. In our experiments, the latter works much better than the former, probably because the Anomalous Pattern seeds do relate to relatively dense parts of the entity "cloud," whereas the maximally distant seeds may fail on this count.

7.3.4.2 Principal Direction Division

This approach, advocated in [19] under the title Principal Direction Divisive Partitioning (PDDP), can be considered as making use of the expression of the Ward distance as a Rayleigh quotient (7.53) for the matrix YY^T.

By relaxing the constraint that only the special three-valued split vectors ϕ are admitted, one arrives at the problem of maximizing the Rayleigh quotient over any vector, that is, finding the very first principal component of that part of the data set that relates to the cluster under consideration. By doing this, all the components related to the entities outside of the cluster are made zero, so that the only problem remaining is to approximate the principal component scores by the special values. In [19], a simple approximation rule is taken. The entities are divided according to the sign of the principal component score. Those with a negative score go into one split part, and those with a positive score, to another.

Of course, the result much depends on the location of the zero point. The zero is put into the cluster's centroid in the PDDP [19]. After this is specified, the method requires no further adjustment of parameters, unlike many other methods, 2-splitting included.

As mentioned in [180], the results of PDDP and Bisecting K-Means are typically very similar. This should be obvious in the light of the description above. Both methods optimize the same criterion, though somewhat differently. Bisecting K-Means involves the constraints that are relaxed in PDDP. The former, however, can only approximate the global minimum with a suboptimal solution, whereas the latter does find the global minimum, for an approximate problem, and then approximates it with a cluster solution.

It should be pointed out, though, that splitting according to the sign, plus or minus, is not necessarily a best clustering approximation of the optimal scoring vector. There can be different splitting criteria at the PDDP method. That originally proposed in [19] is the sign-based splitting of the principal component (PDDP-Sign). Another approach would rather take that split base vector, which is the nearest to the principal component (PDDP Projection).

Yet one more version could be proposed: use the principal direction vector to sort all its values and then take such a split of the sorted order at which the value of Rayleigh quotient (7.53) at the corresponding vector ϕ is maximum (PDDP-RQ). Unlike the PDDP-Projection result, PDDP-RQ result does not depend on the scale of the principal component scoring vector.

One more version was proposed in [204]: the authors first build a density function for the entity factor scores and then split the sorted set at the place of the deepest minimum of the density function. The density function is built according to the Parzen's recipe [188]. Each entity i is presented with the so-called Gaussian kernel

$$f(x, z_i) = \exp\left(-\frac{(x - z_i)^2}{2h^2}\right) \Big/ [\sqrt{2\pi}h]$$

where z_i is the principal component score of entity $i \in I$ and h, the so-called Parzen's window characterizing the width of the 'bell' shape of the function. According to an advice, its value is specifed as $h = \sigma(4/(3n))^{1/5}$ with σ being an estimate of the standard deviation of the density [204]. The total density function is defined as the average of individual density functions $f(x) = \sum_{i \in I} f(x, z_i)/N$. This version can be referred to as PDDP-MD (Minimum of the Density function). This function is used not only for selecting the split, but also for stopping the division process, when no minimum exists on the function so that it is either monotone or concave.

Therefore, four splitting principles can be considered: two, PDDP-Sign and PDDP-MD, based on heuristic considerations and two, PDDP-Projection and PDDP-RQ, inspired by the least-squares framework involving split base vectors. PDDP-MD performs quite well if no noisy objects are added to a Gaussian clusters sample [100].

7.3.4.3 Conceptual Clustering

This method differs from the rest because of the way the features are taken into account. Each split is made by using just only one of the features. If the feature, x, is quantitative or rank, then a value, a, is taken from within its range and the cluster under consideration is split so that entities at which $x > a$ go into one cluster while those satisfying the opposite predicate, $x \leq a$, go to the other part. The number of all possible a-values is infinite of course, but those a values that do matter are not too many—just a sorted list of all the values of x on the entities under consideration. This yet can be too many in the case of a large data set. Then a pre-specified number of the admissible a-values can be defined by uniformly dividing the range of x in that number of bins. If x is categorical, then the two parts are formed according to predicates $x = a$ and $x \neq a$. The choice of x and a is made according to a criterion. In our case, that should be the maximum of Ward distance, of course—the more so, that the criterion is rather universal, as will be seen later in this section. The following formalizes the description.

CONCEPTUAL CLUSTERING WITH BINARY SPLITS

1. *Initialization.* Define an up-hierarchy to consist of just one universal cluster, the entire entity set I.

2. *Evaluation.* Run a cycle over all the leaf clusters k, variables v and their values a, and compute the value of the criterion at each of them.

3. *Split.* Select the triplet (k, v, a) that has received the highest score and do the binary split of cluster k in two parts defined by conditions "$y_v > a$" and "$y_v \leq a$." Make a record of the split.

4. *Halt.* Check the stop-condition. If it is true, stop computation, and output the obtained hierarchy along with other interpretation aids. Otherwise go to 2.

When the scoring function is the Ward distance, the stop-condition here can be borrowed from that on p. 142.

Originally, the method was developed for categorical features only. A goodness-of-split criterion is utilized to decide what class S in a hierarchy to split and by what variable. To define such a criterion, various approaches can be used. Let us consider two popular approaches that are compatible with the data recovery framework.

1. *Impurity.* Denote the set of entities to be split by $J \subseteq I$ and a qualitative variable on it by l, with p_v denoting frequencies of its categories

$v \in V_l$ in J. Let us measure the dispersion of l on J with the *Gini coefficient*, $G(l) = 1 - \sum_{v \in V_l} p_v^2 = \sum_{v \in V_l} p_v(1 - p_v)$ defined in Section 2.1.3. If J is divided in clusters $\{S_k\}$, this cross classifies l so that any category $v \in V_l$ may co-occur with any S_k; let us denote the frequency of the co-occurrence by p_{kv}. Then the Gini coefficient for l over S_k will be $G(l/k) = 1 - \sum_{v \in V_l} p(v/k)^2$ where $p(v/k) = p_{kv}/p_k$ with p_k denoting proportion of entities of J in S_k. The average change of the Gini coefficient after the split of J into $S_k, k = 1, \ldots, K$, can be expressed as the difference, $\Delta(l, S) = G(l) - \sum_k p_k G(l/k)$. It is this expression which is referred to as the impurity function in [20]. It also equals the summary Quetelet index G^2 according to the analysis on p. 58. The greater the $\Delta(l, S)$, the better the split S.

2. *Category utility.* Consider a partition $S = \{S_k\}$ ($k = 1, \ldots, K$) on J and a set of categorical features $l \in L$ with categories $v \in V_l$. The category utility function scores partition S against the set of variables according to formula [49]:

$$u(S) = \frac{1}{K} \sum_{k=1}^{K} p_k \left[\sum_l \sum_{v \in V_l} P(l = v|S_k)^2 - \sum_l \sum_{v \in V_l} P(l = v)^2 \right] \quad (7.54)$$

The term in square brackets is the increase in the expected proportion of attribute values that can be correctly predicted given a class, S_k, over the expected number of attribute values that could be correctly predicted without using the knowledge of the class. The assumed prediction strategy follows a probability-matching approach. According to this approach, category v is predicted with the frequency reflecting its probability, $p(v/k)$ within S_k, and $p_k = N_k/N$ when information of the class is not provided. Factors p_k weigh classes S_k according to their sizes, and the division by K takes into account the difference in partition sizes: the smaller, the better.

Either of these functions can be applied to a partition S to decide which of its clusters is to be split and how. They are closely related to each other as well as to the contributions B_{kv} in Section 6.3.2. The relation can be stated as follows.

Statement 7.5. The impurity function $\Delta(l, S)$ equals the summary contribution $B(l, S) = \sum_{v \in V_l} \sum_{k=1}^{K} B_{kv}$ with scaling factors $b_v = 1$ for all $v \in V_l$. The category utility function $u(S)$ is the sum of impurity functions over all features $l \in L$ related to the number of clusters K, $u(S) = \sum_l \Delta(l, S)/K$.

Proof: Indeed, according to the definition of impurity function, $\Delta(l, S) = G(l) - \sum_k p_k G(l/k) = 1 - \sum_{v \in V_l} p_v^2 - \sum_k p_k(1 - \sum_{v \in V_l} p(v/k)^2) = \sum_k \sum_v p_{kv}/p_k - \sum_{v \in V_l} p_v^2 = B(l, S)$. To prove the second part of the statement, let us note that

$P(l = v|S_k) = p(v/k)$ and $P(l = v) = p_v$. This obviously implies that $u(S) = \sum_l \Delta(l, S)/K$, which proves the statement.

The summary impurity function, or the category utility function multiplied by K, is exactly the summary contribution of variables l to the explained part of the data scatter, $B(S, c)$, that is, the complement of the K-Means square error clustering criterion to the data scatter, when the data pre-processing has been done with all $b_v = 1$ ($v \in V_l$) [138]. In brief, maximizing the category utility function is equivalent to minimizing the K-Means square error criterion divided by the number of clusters with the data standardized as described above. This function, therefore, can be used as another stopping criterion to use both the contribution and the number of clusters: stop the splits when the category utility goes down. Unfortunately, this approach does not always work, as can be seen in the examples in Section 4.4.

It should be pointed out that conceptual clustering is akin to the so-called decision tree techniques in machine learning (see Section 6.4.4). The difference comes from the goal, not method. Here the goal is to accurately reproduce all the features used for building the tree, whereas decision trees in machine learning are built to accurately reproduce a target feature which does not belong to the feature set. Yet the method, basically, is the same (see [140] for more on this). Different decision tree programmes use different goodness-of-split criteria. Among those utilized for categorical variables are: the chi-squared in CHAID [66], the entropy in C4.5 [171] and the so-called twoing rule in CART [20].

7.3.4.4 Separating a Cluster

The Ward distance in Equation 4.1 admits a different formulation:

$$dw(S_{w1}, S_{w2}) = \frac{N_w N_{w1}}{N_{w2}} d(c_{w1}, c_w) \tag{7.55}$$

in which the center of S_{w2} is changed for the center of the parental cluster S_w together with the change in the numeric factor at the distance. Formula 7.55 expresses Ward distance through one of the split parts only. The proof easily follows from equation $N_w c_w = N_{w1} c_{w1} + N_{w2} c_{w2}$ and the definition of the squared Euclidean distance d. Indeed, this implies $c_{w2} = (N_w c_w - N_{w1} c_{w1})/N_{w2}$.

Therefore, $c_{w1} - c_{w2} = c_{w1} - N_w c_w/N_{w2} + N_{w1} c_{w1}/N_{w2} = N_w(c_{w1} - c_w)/N_{w2}$ since $1 + N_{w1}/N_{w2} = N_w/N_{w2}$. This implies that $dw = N_{w1} N_{w2} <c_{w1} - c_{w2}, c_{w1} - c_{w2}>/N_w = N_{w1} N_{w2} < N_w(c_{w1} - c_w)/N_{w2}, N_w(c_{w1} - c_w)/N_{w2}>/N_w = N_{w1} N_w < c_{w1} - c_w, c_{w1} - c_w > /N_{w2}$, which proves Equation 7.55.

To maximize Equation 7.55, one needs to keep track of just one cluster S_{w1} because c_w and N_w are pre-specified by S_w and do not depend on splitting.

Bisecting K-Means can work with criterion (7.55) in the same way as it does with criterion (7.50) except for the initialization. Let us formulate an incremental version, akin to the incremental K-Means algorithm in Section 3.1.3 by exploiting the asymmetry in Equation 7.55. According to this approach, cluster S_{w2} and its center c_{w2} are updated incrementally by considering one entity at a time. Given an entity, denote $z = 1$ if the entity has been added to S_{w1} and $z = -1$ if the entity has been removed, that is, added to S_{w2}. Then the new value of Ward distance (7.55) after the move is $N_w(N_{w1} + z)d(c'_{w1}, c_w)/(N_{w2} - z)$ where c'_{w1} is the updated centroid of S_{w1}. Relating this to dw (7.55), we can see that the value of Ward distance increases if the ratio is greater than 1, that is, if

$$\frac{d(c_w, c_{w1})}{d(c_w, c'_{w1})} < \frac{N_{w1}N_{w2} + zN_{w2}}{N_{w1}N_{w2} - zN_{w1}} \tag{7.56}$$

and it decreases otherwise. This leads to the following incremental splitting algorithm.

SEPARATION OF S_{W1}

1. *Initial setting.* Given $S_w \subseteq I$ and its centroid c_w, specify its split part S_{w1} to consist of an entity y_1, which is the furthest from c_w, and set $c_{w1} = y_1$, $N_{w1} = 1$ and $N_{w2} = N_w - 1$.

2. *Next move.* Take a next entity y_i; this can be that nearest to c_{w1}.

3. *Update.* Check the inequality (7.56) with y_i added to S_{w1} if $y_i \notin S_{w1}$ or removed from S_{w1}, otherwise. If Equation 7.56 holds, change the state of y_i with respect to S_{w1} accordingly, recalculate $c_{w1} = c'_{w1}$, N_{w1}, N_{w2} and go to step 2.

4. *Output* results: split parts S_{w1} and $S_{w2} = S_w - S_{w1}$; their centroids c_{w1} and c_{w2}; their heights, $h_1 = W(S_{w1}, c_{w1})$ and $h_2 = W(S_{w2}, c_{w2})$; and the contribution of the split, that is, Ward distance $dw(S_{w1}, S_{w2})$.

To specify the initial seed, different strategies can be used: (a) random selection or (b) taking the centroid of the Anomalous pattern found with the AP algorithm from Section 7.2.6. These strategies are similar to those suggested for the 2-Means splitting algorithm, and so are their properties.

7.3.5 Organizing an Up-Hierarchy: To Split or Not to Split

To build an up-hierarchy with a splitting process, one needs to specify two more rules: (a) a rule for deciding what cluster is to be split next and (b) a stopping rule to decide at what point no splitting is to be done anymore.

These are rather simple in the context of the data recovery model for binary hierarchies (7.44) and (7.51). According to this, that cluster is to be split that leads to a greater value of the Ward distance or, which is the same, the Rayleigh quotient (7.53). This rule is followed in the conceptual clustering because nothing better is available. Yet at clustering per se, trying each leaf cluster of an up-hierarchy already having been built would make the computation drastically more intensive. As a proxy, the choice can be made over the cluster's height, the overall variance. This may make sense because in general, the larger the contribution of a cluster, the larger the contribution of its split.

As to the stopping rule, the data recovery perspective suggests using the contribution to the data scatter: either that accumulated or that of a single cluster or both. For example, a cluster should not be split anymore if it accounts 4% of the data scatter or less and the splits should stop after the cumulative contribution of the splits has reached 80% of the data scatter. An issue with this approach is that the contribution so far has received no clear interpretation.

This is why an intuitive approach by Tasoulis et al. [204] seems rather promising. They propose that a cluster should not be split anymore if the density function of the principal direction scores has no minima, which may happen if the function is concave or monotone. This rule has been supported in [204] with successful computational experiments. Kovaleva and Mirkin [100] further extended the rule to the situation of Ward-like divisive clustering, in which random directions are taken rather than the principal one. A cluster is not to be split anymore if at least one-third of the projections show a density function with no minima. This threshold has been derived as an approximate estimate of the probability that a random direction would show a separation when there is one [100]. It appears that the random projections can be competitive when noise is present in data (see Table 4.6 on p. 150).

7.3.6 A Straightforward Proof of the Equivalence between Bisecting K-Means and Ward Criteria

On the first glance, the Ward criterion for dividing an entity set in two clusters has nothing to do with that of K-Means. Given $S_w \subseteq I$, the former is to maximize λ_w in Equation 7.50 over all splits of S_w in two parts, while the K-Means criterion, with $K = 2$, in the corresponding denotations is

$$W(S_{w1}, S_{w2}, c_{w1}, c_{w2}) = \sum_{i \in S_{w1}} d(y_i, c_{w1}) + \sum_{i \in S_{w2}} d(y_i, c_{w2}) \qquad (7.57)$$

Criterion (7.57) is supposed to be minimized over all possible partitions of S_w in two clusters, S_{w1} and S_{w2}. According to Equation 7.22, this can be equivalently reformulated as the problem of maximization of

$$B(S_{w1}, S_{w2}, c_{w1}, c_{w2}) = N_{w1}d(c_{w1}, 0) + N_{w2}d(c_{w2}, 0) \qquad (7.58)$$

over all partitions S_{w1}, S_{w2} of S_w.

Now we are ready to prove that criteria (7.57) and (7.50) are equivalent.

Statement 7.6. Maximizing Ward criterion (7.50) is equivalent to minimizing 2-Means criterion (7.57).

Proof: To see if there is any relation between Equation 7.50 and Equation 7.58, let us consider an equation relating the two centroids with the total gravity center, c_w, in the entity set S_w under consideration:

$$N_{w1}c_{w1} + N_{w2}c_{w2} = (N_{w1} + N_{w2})c_w \tag{7.59}$$

The equation holds because the same sum of all the entity points stands on both sides of it.

Let us assume $c_w = 0$. This should not cause any trouble because the split criterion (7.50) depends only on the difference between c_{w1} and c_{w2}, which does not depend on c_w. Indeed, if $c_w \neq 0$, then we can shift all entity points in S_w by subtracting c_w from each of them, thus defining $y'_{iv} = y_{iv} - c_{wv}$. With the shifted data, the averages are obviously, $c'_w = 0$, $c'_{w1} = c_{w1} - c_w$, and $c'_{w2} = c_{w2} - c_w$, which does not change the difference between centroids. This does change the principal components, though. Making $c_w = 0$ is a base of the PDDP method described in the previous section and, in fact, is consistent with one-by-one extracting splits found by maximizing Equation 7.50 according to Equation 7.52 in Section 7.3.3.

With $c_w = 0$, Equation 7.59 implies $c_{w1} = (-N_{w2}/N_{w1})c_{w2}$ and $c_{w2} = (-N_{w1}/N_{w2})c_{w1}$. Based on these, the inner product (c_{w1}, c_{w2}) can be presented as either $(c_{w1}, c_{w2}) = (-N_{w2}/N_{w1})(c_{w1}, c_{w1})$ or $(c_{w1}, c_{w2}) = (-N_{w1}/N_{w2})(c_{w2}, c_{w2})$. By substituting these instead of (c_{w1}, c_{w2}) in decomposition $d(c_{w1}, c_{w2}) = [(c_{w1}, c_{w1}) - (c_{w1}, c_{w2})] + [(c_{w2}, c_{w2}) - (c_{w1}, c_{w2})]$ one can see that λ_w (7.50) is equal to:

$$\lambda_w = \frac{N_{w1}N_{w2}}{N_w}[N_w(c_{w1}, c_{w1})/N_{w2}] + [N_w(c_{w2}, c_{w2})/N_{w1}]$$

By removing redundant items, this leads to equation:

$$\lambda_w = B(S_{w1}, S_{w2}, c_{w1}, c_{w2})$$

which completes the proof. q.e.d.

7.3.7 Anomalous Pattern versus Splitting

A question related to the Anomalous Pattern method is whether any relation exists between its criterion (7.9) and Ward criterion λ_w, especially in its form

(7.55) used in the Separation algorithm. Both methods separate a cluster from a body of entities with a square error criterion.

To analyze the situation at a split, let us denote the entity cluster under consideration by J and the separated cluster, by S_1, with its centroid c_1. Consider also a pre-specified point c in J, which is the centroid of J in Ward clustering and a reference point in the Anomalous pattern clustering. The other subset, $J - S_1$, will be denoted by S_2. Then the Anomalous pattern criterion can be expressed as

$$W = \sum_{i \in S_1} d(y_i, c_1) + \sum_{i \in S_2} d(y_i, c)$$

and the Ward splitting criterion as

$$\lambda_w = \frac{N_J N_1}{N_2} d(c_1, c)$$

where N_J, N_1 and N_2 are the sizes (cardinalities) of J, S_1 and S_2, respectively.

To analyze the relationship between these criteria, let us add to and subtract from W the complementary part $\sum_{i \in S_1} d(y_i, c)$. Then $W = T(Y, c) + \sum_{i \in S_1} d(y_i, c_1) - \sum_{i \in S_1} d(y_i, c)$ where $T(Y, c)$ denotes the sum of squared Euclidean distances from all points in J to c, that is, the scatter of the data around c, an item which is constant with respect to the cluster S_1 being separated. By noting that $d(y_i, c) = (y_i, y_i) + (c, c) - 2(y_i, c)$ and $d(y_i, c_1) = (y_i, y_i) + (c_1, c_1) - 2(y_i, c_1)$, the last expression can be equivalently rewritten as $W = T(Y, c) - N_1 d(c_1, c)$. The following is proven:

Statement 7.7. Both separating methods, Anomalous pattern and Separation, maximize the weighted distance $\alpha d(c_1, c)$ between centroid c_1 of the cluster being separated and a pre-specified point c. The difference between the methods is that:

1. The weight is $\alpha = N_1$ in the Anomalous pattern and $\alpha = N_1/N_2$ in the Splitting by separation.
2. Point c is the user-specified reference point, in the former, and the unvaried centroid of set J being split, in the latter.

The difference (2) disappears if the reference point has been set at the centroid of J. The difference (1) is fundamental: as proven in Section 7.4.2 further on, both sets, S_1 and S_2, are tight clusters at Separation (Statement 7.9), whereas only one of them, S_1, is to be tight as the Anomalous pattern (Statement 7.8): the rest, $J - S_1$, may be a set of rather disconnected entities. Intuitively, if there is a range of uniformly distributed unidimensional values, then the Splitting would divide the set in two halves, whereas the Anomalous pattern would pick up a fragment of about one fourth of the range on one side of that. This shows once again the effect of the size coefficients in the cluster criteria.

7.4 Data Recovery Models for Similarity Clustering

7.4.1 Cut, Normalized Cut, and Spectral Clustering

Given a symmetric similarity matrix $A = (a_{ij})$ on set I, let us try to split I in two parts, S_1 and S_2, according to the compactness principle. The principle states that the similarity between S_1 and S_2 is minimum, whereas the within-cluster similarity is at its maximum.

There are several heuristic criteria that can be used to formalize the idea of compactness. Given a cluster $S \subseteq I$, the within cluster similarity can be scored by the following criteria:

i. Summary similarity

$$a(S, S) = \sum_{i,j \in S} a_{ij} \qquad (7.60)$$

ii. Average similarity

$$a(S) = \frac{\sum_{i,j \in S} a_{ij}}{|S|\delta(S)} \qquad (7.61)$$

where $\delta(S) = |S|$, the number of elements in S, if the diagonal entries a_{ii} are specified in the similarity matrix A, and $\delta(S) = |S| - 1$ if no diagonal entries a_{ii} are supplied in A.

iii. Semi-average similarity

$$A(S) = a(S)|S| = \sum_{i,j \in S} a_{ij}/\delta(S) \qquad (7.62)$$

where $\delta(S)$ is defined above at (ii).

iv. Normalized similarity (tightness)

$$ns(S) = \frac{\sum_{i,j \in S} a_{ij}}{\sum_{i \in S} \sum_{j \in I} a_{ij}} \qquad (7.63)$$

Criteria (i) and (iv) will be discussed in this section, including the data recovery models from which they can be derived. A similar discussion of (ii) and (iii) will be presented in the next section.

The summary similarity criterion straightforwardly satisfies the compactness principle.

What follows is set for a two-cluster partition, but it is true for any partition as well. For $f, g = 1, 2$, let us denote the summary similarity between S_f and S_g by $a(S_f, S_g)$ so that $a(S_f, S_g) = \sum_{i \in S_f} \sum_{j \in S_g} a_{ij}$. Then, of course, $a(S_f, S_f)$ is the summary similarity within S_f and $a(S_1, S_2) = a(S_2, S_1)$ is the summary similarity between S_1 and S_2. Moreover, the sum $a(S_1, S_1) + a(S_2, S_2) +$

$a(S_1, S_2) + a(S_2, S_1)$ is the total sum $a(I, I)$ of all the similarities, which is constant because it does not depend on the split. The total value $a(S_1, S_2)$ of the between cluster similarity is referred to as a cut. A natural clustering criterion, the minimum cut, would correspond then to the maximum of the summary within cluster similarity $a_w(S_1, S_2) = a(S_1, S_1) + a(S_2, S_2)$ because $a_w(S_1, S_2) = a(I, I) - 2a(S_1, S_2)$.

Yet the criterion of minimum cut usually is not considered appropriate for clustering because, at a non-negative A, its minimum would be reached normally on the most unbalanced a partition in which one part is a singleton formed by the entity with the weakest summary links. Such a result cannot be considered a proper aggregation.

However, if the similarities are standardized by subtracting a background part, either a constant threshold or the random interactions as described in Sections 2.5.5 and 5.1, the minimum cut criterion is a consistent tool to explore the data structure. It should be noted that the summary criterion is referred to as the uniform criterion when the background part is just a similarity shift value, and it is referred to as the modularity criterion when the background is (proportional to) the product of marginal probabilities.

The reference to "uniformity" has something to do with the properties of the criterion. Indeed, denote $a_\pi(S_f, S_g)$ the summary value of similarities $a_{ij} - \pi$ for $i \in S_f$ and $j \in S_g$ at which the similarity shift value π has been subtracted. Then the summary within-cluster similarity at the shift π can be expressed through the sum of the original similarities as follows:

$$a_\pi(S_1, S_1) + a_\pi(S_2, S_2) = a(S_1, S_1) + a(S_2, S_2) - \pi(N_1^2 + N_2^2)$$

This means that the uniform within-cluster summary criterion reaches its maximum with a penalty depending on the extent of uniformity of the distribution of entities over the clusters. Indeed, $N_1^2 + N_2^2$ reaches its minimum at $N_1 = N_2$, and maximum, at $N_1 = 1, N_2 = N - 1$. The closer N_1 and N_2 to each other, the greater the summary similarity criterion value. In fact, it can be proven that the greater the shift π, the smaller the index $N_1^2 + N_2^2$ at any optimal partition [135].

Unfortunately, when A-entries can be both positive and negative, the problem of minimum cut becomes much intensive computationally. Therefore, local or approximate algorithms would be a welcome development for the problem. One of such algorithms is AddRemAdd(j) in Section 7.4.2, based on one-by-one addition/removal of individual entities in a greedy-wise manner. Although this is a one-cluster algorithm, collecting a cluster makes the set split in two parts, the cluster and the rest. The only thing that needs to be changed is the criterion: the maximum summary within cluster similarity is to be extended to cover both clusters. Of course, in this case, the operation of addition-removal loses its one-cluster asymmetry and becomes just operation of exchange between the two clusters.

Another approach comes from the matrix spectral decomposition theory. Indeed, define N-dimensional vector $z = (z_i)$ so that $z_i = 1$ if $i \in S_1$ and $z_i = -1$

if $i \in S_2$. Obviously, $z_i^2 = 1$ for any $i \in I$ so that $z^T z = N$ which is constant at any given entity set I. On the other hand, $z^T A z = a(S_1, S_1) + a(S_2, S_2) - 2a(S_1, S_2) = 2(a(S_1, S_1) + a(S_2, S_2)) - a(I, I)$, which means that the summary within-cluster criterion is maximized when $z^T A z$ or $z^T A z / (z^T z)$ is maximized, that is, the problem of finding a minimum cut is equivalent to the problem of maximization of Rayleigh quotient $r(z) = z^T A z / (z^T z)$ with respect to the unknown N-dimensional z whose components are either 1 or -1. Matrix A should be pre-processed by subtraction of a threshold or of the random interactions or using a different transformation. The maximum of the Rayleigh quotient, with respect to arbitrary z, is equal to the maximum eigenvalue of A and it is reached at the corresponding eigenvector. This brings forth the idea that is referred to as spectral clustering: Find the first eigenvector as the best solution and then approximate it with a $(1, -1)$-vector by putting 1 for positive components and -1 for non-positive components—then produce S_1 as the set of entities corresponding to 1, and S_2, corresponding to -1.

One more splitting criterion, the so-called normalized cut, was proposed in [186]. The normalized cut involves the summary similarities $a_{i+} = a(i, I)$. Denote $v(S_1) = \sum_{i \in S_1} a_{i+}$, the volume of S_1 according to A. Obviously, $v(S_1) = a(S_1, S_1) + a(S_1, S_2)$; a similar equation holds for $v(S_2)$. The normalized cut is defined then as

$$nc(S_1, S_2) = \frac{a(S_1, S_2)}{v(S_1)} + \frac{a(S_1, S_2)}{v(S_2)} \tag{7.64}$$

to be minimized. It should be noted that minimized cut, in fact, includes the requirement of maximization of the within-cluster similarities. Indeed consider the normalized within-cluster tightness

$$ns(S_1, S_2) = \frac{a(S_1, S_1)}{v(S_1)} + \frac{a(S_2, S_2)}{v(S_2)} \tag{7.65}$$

These two measures are highly related: $nc(S) + ns(S) = 2$, which can be proven with elementary manipulations. This latter equation warrants that minimizing the normalized cut simultaneously maximizes the normalized tightness. It appears that the criterion of minimizing $nc(S)$ can be expressed in terms of a corresponding Rayleigh quotient for the Laplace transformation of the similarities. Given a (pre-processed) similarity matrix $A = (a_{ij})$, let us denote its row sums, as usual, by $a_{i+} = \sum_{j \in I} a_{ij}$ and introduce diagonal matrix D in which all entries are zero except for diagonal elements (i, i) that hold a_{i+}, $i \in I$. The so-called (normalized) Laplacian is defined as $L = E - D^{-1/2} A D^{-1/2}$ where E is identity matrix and $D^{-1/2}$ is a diagonal matrix with (i, i)th entry equal to $1/\sqrt{a_{i+}}$. That means that L's (i, j)th entry is $\delta_{ij} - a_{ij}/\sqrt{a_{i+} a_{j+}}$ where δ_{ij} is 1 if $i = j$ and 0, otherwise. It is not difficult to prove that $L f_0 = 0$ where $f_0 = (\sqrt{a_{i+}}) = D^{1/2} 1_N$ and 1_N is N-dimensional vector whose all entries are unity. That means that 0 is an eigenvalue of L with f_0 being the corresponding eigenvector.

Moreover, for any N-dimensional $f = (f_i)$, the following equation holds:

$$f^{\mathrm{T}}Lf = 1/2 \sum_{I,j \in I} a_{ij}(f_i/\sqrt{a_{i+}} - f_j/\sqrt{a_{j+}})^2 \qquad (7.66)$$

This equation shows that $f^{\mathrm{T}}Lf \geq 0$ for all f, that is, matrix L is positive semidefinite. It is proven in matrix theory that any positive semidefinite matrix has all its eigenvalues non-negative so that 0 must be L's minimum eigenvalue. Given a partition $S = \{S_1, S_2\}$ of I, define vector s so that $s_i = \sqrt{a_{i+}v(S_1)/v(S_2)}$ for $i \in S_1$ and $s_i = -\sqrt{a_{i+}v(S_2)/v(S_1)}$ for $i \in S_2$. Obviously, the squared norm of this vector is constant, $||s||^2 = \sum_{i \in I} s_i^2 = v(S_2) + v(S_1) = a(I,I) = a_{++}$. Moreover, s is orthogonal to the eigenvector $f_0 = D^{1/2}1_N$ corresponding to the 0 eigenvalue of L. Indeed, the product of i-th components of these vectors has a_{i+} as its factor multiplied by a value which is constant within clusters. Sum of these components over S_1 is $v(S_1)\sqrt{v(S_2)/v(S_1)} = \sqrt{v(S_1)v(S_2)}$ and over S_2, $-v(S_2)\sqrt{v(S_1)/v(S_2)} = -\sqrt{v(S_1)v(S_2)}$. The sum of these two is 0, which proves the statement.

It remains to prove that minimization of the normalized cut criterion (7.64) is equivalent to minimization of the Rayleigh quotient $s^{\mathrm{T}}Ls/s^{\mathrm{T}}s$ for thus defined s. Take a look at the items in the sum on the right in (7.66) at $f = s$. This is equal to 0 for i and j taken from the same set S_1 or S_2. If i and j belong to different classes of S, this is equal to $v(S_1)/v(S_2) + v(S_2)/v(S_1) + 2 = [a_{++} - v(S_2)]/v(S_2) + [a_{++} - v(S_1)]/v(S_1) + 2 = a_{++}/v(S_2) + a_{++}/v(S_1)$. Therefore, $s^{\mathrm{T}}Ls = a_{++}nc(S)$, that is, $nc(S) = s^{\mathrm{T}}Ls/s^{\mathrm{T}}s$ indeed. The normalized cut minimizes the Rayleigh quotient for Laplace matrix L over specially defined vectors s that are orthogonal to the eigenvector $f_0 = (\sqrt{a_{i+}})$ corresponding to the minimum eigenvalue 0 of L. Therefore, one may consider the problem of finding the minimum non-zero eigenvalue for L together with the corresponding eigenvector as a proper relaxation of the normalized cut problem. That means that the spectral clustering approach in this case would be to grab that eigenvector and approximate it with an s-like binary vector. The simplest, though not necessarily optimal, way to do that would be by putting all the plus components to S_1 and all the minus components to S_2.

It remains to define the pseudo-inverse Laplace transformation L^+, Lapin for short, for a symmetric matrix A. Consider all the non-zero eigenvalues $0 < \lambda_1 < \lambda_2 < \cdots < \lambda_r$ of Laplace matrix L and corresponding eigenvectors $f_1, f_2, \ldots, f_r$. The spectral decomposition of L is $L = \sum_{k=1}^{r} \lambda_k f_k f_k^{\mathrm{T}}$. The pseudo-inverse L^+ is defined by leaving the same eigenvectors but reversing the eigenvalues, $L^+ = \sum_{k=1}^{r} \frac{1}{\lambda_k} f_k f_k^{\mathrm{T}}$. The eigenvalues of L^+ are in the reverse order, so that $1/\lambda_1$ is the largest, and what is nice is that the gap between the eigenvalues increases drastically. If, for example, $\lambda_1 = 0.1$ and $\lambda_2 = 0.2$, then $1/\lambda_1 = 10$ and $1/\lambda_2 = 5$: a 50-fold increase of the gap, that is, the difference between the two, from $0.2 - 0.1 = 0.1$ to $10 - 5 = 5$.

7.4.2 Similarity Clustering Induced by K-Means and Ward Criteria

Given a similarity matrix $A = (a_{ij})$, $i, j \in I$, let us define a reasonable cluster tightness index. For the sake of simplicity, the diagonal entries are assumed to be present in A in this section. For any subset $S \subseteq I$, one could try its average within-cluster similarity, $a(S) = \sum_{i,j \in S} a_{ij}/N_S^2$: the greater $a(S)$, the better. Unfortunately, this normally would not work. Indeed, take an $i \in I$ and cluster it with its nearest neighbor j so that a_{ij} is the maximum in ith row of matrix A. For this doublet cluster S, $a(S) = a_{ij}$. If one adds one more entity to S, this next entity normally would not be as close to i as j so that the average similarity would decrease—the more you add, the greater the decline in $a(S)$. That means that maximizing $a(S)$ would not lead to large clusters at all. To include the idea of maximizing the cardinality of cluster S, N_S, let us take the product,

$$A(S) = N_S a(S) = \sum_{i,j \in S} a_{ij}/N_S \qquad (7.67)$$

as the cluster scoring criterion. The index $A(S)$ combines two contradictory criteria, (i) maximize the tightness as measured by $a(S)$ and (ii) maximize the size, measured by N_S. These two should balance each other and, in this way, lead to a relatively tight cluster of a reasonable size.

Indeed, the tightness of the cluster can be mathematically explored by using the following concept. For any entity $i \in I$, let us define its attraction to subset $S \subseteq S$ as the difference between its average similarity to S, $a(i, S) = \sum_{j \in S} a_{ij}/N_S$, and half the average within-cluster similarity:

$$\beta(i, S) = a(i, S) - a(S)/2 \qquad (7.68)$$

The fact that $a(S)$ is equal to the average $a(i, S)$ over all $i \in S$, leads us to expect that normally $a(i, S) \geq a(S)/2$ for the majority of elements $i \in S$, that is, normally $\beta(i, S) \geq 0$ for $i \in S$. It appears that a cluster S maximizing the criterion $A(S)$ is much more than that: S is cohesive internally and separated externally because all its members are positively attracted to S whereas non-members are negatively attracted to S.

Statement 7.8. If S maximizes $A(S)$ (7.67) and $N_S \geq 2$ then $\beta(i, S)$ is not negative for all $i \in S$ and not positive for all $i \notin S$.

Proof: Indeed, if $i \in S$ and $\beta(i, S) < 0$, then $A(S - i) > A(S)$ which contradicts the assumption that S maximizes Equation 7.67. To prove this inequality, let us consider $A(S)N_S = \sum_{i,j \in S} a_{ij}$, the sum of within cluster similarities. Obviously, $A(S)N_S = A(S - i)(N_S - 1) + 2a(i, S)N_S - a_{ii}$. This leads to $[A(S - i) - A(S)](N_S - 1) = a_{ii} + A(S) - 2a(i, S)N_S = a_{ii} - 2\beta(i, S)N_S > 0$ since a_{ii} must be non-negative, which proves the inequality. The fact that $i \notin S$ contradicts $\beta(i, S) > 0$ is proved similarly, q.e.d.

It is nice to note now that the cluster scoring index $A(S)$ has something to do with K-Means criterion. Indeed, recall the inner product similarity matrix $A = YY^T$ and the equivalent expression (7.32) for the K-Means criterion. By substituting a_{ij} for $< y_i, y_j >$ in Equation 7.32 and having in mind the definition (7.67), the following equation for the maximized K-Means criterion is true:

$$B(S,c) = \sum_{k=1}^{K} \sum_{i,j \in S_k} a_{ij}/N_k = \sum_{k=1}^{K} A(S_k) \qquad (7.69)$$

Obviously, criterion (7.69), with the kernel trick meaning that the inner product $<y_i, y_j>$ is changed for a semidefinite positive kernel function $K(y_i, y_j)$, can be applied to any kernel-based similarity measure a_{ij}, not necessarily the inner product.

There can be several approaches to optimization of the criterion (7.69), similar to those in the previous sections:

1. All clusters at once
2. Hierarchical clustering
3. One-by-one clustering

These will be presented in the remainder of the section.

7.4.2.1 All Clusters at Once

Two approaches should be mentioned:

1. *Entity moving:* This approach starts from a pre-specified partition $S = \{S_1, \ldots, S_K\}$ and tries to improve it iteratively. At each iteration, each entity is moved to a cluster of its maximum attraction. More precise computations could be done using formulas that take into account the change of the cluster score when an entity is moved in to or out of the cluster. Specifically, if i is moved out of cluster S_k, the score's change is

$$\Delta(i,k) = \frac{A(S_k)}{N_k - 1} - 2a(i, S_k - i) - a_{ii}/(N_k - 1) \qquad (7.70)$$

whereas the change is

$$\Delta(i,k) = -\frac{A(S_k)}{N_k + 1} + 2a(i, S_k + i) - a_{ii}/(N_k + 1) \qquad (7.71)$$

if i is added to S_k. Here $A(S_k)$ is the cluster's tightness score (7.67) and $a(i, R)$ is the average similarity between i and subset R. The formulas are easy to prove by using elementary algebraic manipulations over the definition in Equation 7.67.

Then the moves of entities can be executed according to a hill-climbing principle. For example, for each $l = 1, 2, \ldots, K$ and $i \in S_l$ find a cluster S_k at which the total increment $\alpha(i, l, k) = \Delta(i, k) + \Delta(i, l)$ is maximum. After this, move that element i at which the maximum of $\alpha(i, l, k)$ is the largest, if it is positive, from S_l to S_k. If the maximum increment is negative, the computation is stopped: a local maximum has been reached. This algorithm is obviously convergent, because at each step the criterion gets a positive increment, and the number of possible moves is finite.

2. *Spectral approach:* Since criterion (7.69) admits an equivalent reformulation in terms of Rayleigh quotients (7.33) for an arbitrary similarity matrix A, it makes a natural point for application of the spectral approach. According to this approach, the constraints of the cluster memberships being 1/0 binary are relaxed, and the first K eigenvectors of A are found as the globally optimal solution, after which the eigenvectors are converted into the binary format, by assigning a cluster S_k membership to the largest components of the k-th eigenvector ($k = 1, \ldots, K$). The author has never heard of any project systematically trying this approach for similarity data.

7.4.2.2 Hierarchical Clustering

For a triplet of clusters S_{w1}, S_{w2}, S_w in which the latter is the union of the former two, $S_w = S_{w1} \cup S_{w2}$, the difference between the scores of criterion (7.69) on the partition $S = S_1, \ldots, S_K$ and that found by the merger $S_w = S_{w1} \cup S_{w2}$ will be equal to $\Delta(S_{w1}, S_{w2}) = A(S_{w1}) + A(S_{w2}) - A(S_w)$. This is the value to be minimized in the agglomerative process, or maximized, in a divisive process.

Since the former does not differ from other agglomerative algorithms, except for the criterion, we concentrate on a splitting process for divisive clustering to maximize $\Delta(S_{w1}, S_{w2})$.

The sum $\sum_{i,j \in S_w} a_{ij}$ in the parent class S_w is equal to the sum of within children summary similarities plus the between children similarity doubled. Thus, by subtracting the former from their counterparts in $\Delta(S_{w1}, S_{w2})$, the criterion can be transformed into

$$\Delta(S_{w1}, S_{w2}) = \frac{N_{w1} N_{w2}}{N_w} (A_{11} + A_{22} - 2A_{12}) \tag{7.72}$$

where

$$A_{12} = \frac{\sum_{i \in S_{w1}} \sum_{j \in S_{w2}} a_{ij}}{N_{w1} N_{w2}}$$

the average similarity between S_{w1} and S_{w2}. A similar definition applies for $A_{21}, A_{11},$ and A_{22}.

Moving an entity i from S_{w2} to S_{w1} leads to the change in Δ equal to

$$\delta(i) = A_{22}\frac{N_{w2}}{N_{w2}-1} - A_{11}\frac{N_{w1}}{N_{w1}+1} + 2(A1(i,w1) - A2(i,w2)) + \alpha a_{ii} \qquad (7.73)$$

where $A1(i,w1) = \sum_{j\in S_{w1}} a_{ij}/(N_{w1}+1)$, $A2(i,w2) = \sum_{j\in S_{w2}} a_{ij}/(N_{w2}-1)$ and $\alpha = N_w/((N_{w1}+1)(N_{w2}-1))$.

A similar value for the change in $\Delta(S_{w1}, S_{w2})$ when $i \in S_{w1}$ is moved into S_{w2} can be obtained from this by swapping symbols 1 and 2 in the indices.

Let us describe a local search algorithm for maximizing criterion $\Delta(S_{w1}, S_{w2})$ in Equation 7.72 by splitting S_w in two parts. At first, a tentative split of S_w is carried out according to the dissimilarities $\lambda(i,j) = (a_{ii} + a_{jj} - 2a_{ij})/2$ that are defined according to the criterion (7.72) applied to individual entities. Then the split parts are updated by exchanging entities between them until the increment δ in Equation 7.73 becomes negative.

SPLITTING BY SIMILARITY

1. *Initial seeds.* Find a pair of entities i^*, j^* maximizing $\lambda(i,j) = (a_{ii} + a_{jj} - 2a_{ij})/2$ over $i, j \in S_w$.

2. *Initial clusters.* Create initial S_{w1} and S_{w2} by putting i^* into the former and j^* into the latter and by assigning each $i \in S_w$ either to S_{w1} or S_{w2}, according to the sign of an analogue to Equation 7.73, $\delta(i) = 3(a_{ii^*} - a_{ij^*}) - (a_{i^*i^*} - a_{j^*j^*})$.

3. *Candidacy marking.* For any $i \in S_w$, calculate $\delta(i)$ and take i^* maximizing $\delta(i)$.

4. *Exchange.* If $\delta(i^*) > 0$, move i^* to the other split part and go to step 3. Otherwise, halt the process and output current clusters S_{w1} and S_{w2} along with corresponding λ_w.

In this method, step 3 is the most challenging computationally: it requires to find a maximum $\delta(i)$ over $i \in I$. To scale the method to large data sizes, one can use not the optimal i but rather any i at which $\delta(i) > 0$.

In spite of its simplicity, the Splitting by similarity produces rather tight clusters, which can be expressed in terms of the measure of attraction $\beta(i, S) = a(i, S) - a(S)/2$ (7.68).

If an entity $i \in S_w$ is moved from split part S_{w2} to S_{w1}, the change in the numbers of elements of individual parts can be characterized by the number $n_{12} = Nw1(N_{w2} - 1)/N_{w2}(N_{w1} + 1)$ or by the similar number n_{21} if an entity is moved in the opposite direction. Obviously, for large N_{w1} and N_{w2}, n_{12} tends to unity. It appears that elements of a split part are more attracted to it than to the other part, up to this quantity.

Statement 7.9. For any $i \in S_{w2}$,

$$n_{12}\beta(i, S_{w1}) < \beta(i, S_{w2}) \tag{7.74}$$

and a symmetric inequality holds for any $i \in S_{w1}$.

Proof: Indeed, $\delta(i) < 0$ for all $i \in S_{w2}$ in Equation 7.73 after the Splitting by similarity algorithm has been applied to S_w. This implies that $(N_{w1}/N_{w1} + 1)\beta(i, S_{w1}) + \alpha < (N_{w2}/N_{w2} - 1)\beta(i, S_{w2})$. The inequality remains true even if α is removed from it because $\alpha > 0$. Dividing the result by $N_{w2}/N_{w2} - 1$ leads to Equation 7.74, q.e.d.

7.4.2.3 One-by-One Clustering

This approach imitates the one-by-one extraction strategy of the Principal component analysis. The only difference is that in PCA, it is a principal component which is extracted, whereas here it is a cluster. After a "principal" cluster is found, it is removed from the entity set. The process stops when no unclustered entities remain.

Since the clusters are not overlapping here, a similarity data scatter decomposition holds similar to that in Equation 7.12:

$$T(A) = \left(\frac{s_1^T A s_1}{s_1^T s_1}\right)^2 + \cdots + \left(\frac{s_K^T A s_K}{s_K^T s_K}\right)^2 + L^2 \tag{7.75}$$

where $T(A) = \sum_{i,j \in I} a_{ij}^2$ and s_k is a binary membership vector for cluster S_k so that $s_{ik} = 1$ if $i \in S_k$ and $s_{ik} = 0$, otherwise. In this way, each cluster contribution, $((s_k^T A s_k)/(s_k^T s_k))^2 = A(S_k)^2$, can be accounted for.

To find a cluster S optimizing $A(S) = \sum_{i,j \in S} a_{ij}/|S|$ is a computationally intensive problem if A is not non-negative matrix. Therefore, a hill-climbing approach can be utilized. According to this approach, a neighborhood $O(S)$ is pre-defined for every cluster S so that, given an S, only clusters in $O(S)$ are looked at to improve S. Such a computation starts at some S, and then finds the best S' in $O(S)$. If $A(S') > A(S)$, then S is substituted by S', and the process continues. If $A(S') \leq A(S)$, the process stops at S.

The following algorithm uses the neighborhood $O(S)$ defined as the set of all clusters that are obtained by removal from S or addition to S of just one entity. It is a version of algorithm ADDI-S [133].

This procedure involves the average similarity between entity $i \in I$ and cluster $S \subset I$,

$$a(i, S) = \sum_{j \in S} a_{ij}/|S| \tag{7.76}$$

and the average within-cluster similarity $a(S) = \sum_{i \in S} a(i, S)/|S|$.

ADDREMADD(J)

Input: matrix $A = (a_{ij})$ and initial cluster $S = \{j\}$. Output: cluster S containing j, its intensity, the average similarity, λ and contribution g^2 to the A matrix scatter.

1. *Initialization.* Set N-dimensional z to have all its entries equal to -1 except for $z_j = 1$, the number of elements $n = 1$, the average $\lambda = 0$, and contribution $g^2 = 0$. For each entity $i \in I$, take its average similarity to S, $a(i, S) = a_{ij}$.

2. *Selection.* Find i^* maximizing $a(i, S)$.

3. *Test.*

   a. If $z_{i^*} a(i^*, S) < z_{i^*} \lambda/2$,

      i. Change the sign of z_{i^*} in vector z, $z_{i^*} \leftarrow -z_{i^*}$;

      ii. Update:

      $n \leftarrow n + z_{i^*}$ (the number of elements in S),

      $$\lambda \leftarrow \frac{(n - z_{i^*})(n - z_{i^*} - 1)\lambda + 2z_{i^*}a(i^*, S)(n - z_{i^*})}{n(n - 1)}$$ (the average similarity within S), $a(i, S) \leftarrow [(n - z_{i^*})a(i, S) + z_{i^*}a_{ii^*}]/n$ (the average similarity of each entity to S), and $g^2 = \lambda^2 n^2$ (the contribution), and go to 2.

   b. Else: output S, λ and g^2.

   c. End.

The general step is justified by the fact that indeed maximizing $a(i, S)$ over all $i \in I$ does maximize the increment of $A(S)$ among all sets that can be obtained from S by either adding an entity to S or removing an entity from S. Updating formulas can be easily derived from the definitions of the concepts involved.

Obviously, at the resulting cluster, attractions $\beta(i, S) = a(i, S) - a(S)/2$ are positive for within-cluster elements i and negative for out-of-cluster elements i.

The method AddRemAdd(j) can be run in a loop over all $j \in I$ so that of the resulting clusters $S = S(j)$ that one can be chosen that maximizes the overall value $A(S)$—this would not depend on the initial j. Moreover, the resulting clusters $S(j)$ in fact can be considered a fair representation of the cluster structure hidden in the matrix A—they can be combined according to the values of $A(S(j))$ and the mutual overlap.

This algorithm utilizes no ad hoc parameters, except for the similarity shift value, so the cluster sizes are determined by the process of clustering itself. This method, AddRemAdd, works well in application to real world data.

7.4.3 Additive Clustering

In the additive clustering model, each underlying cluster S_k is assumed to be accompanied with a positive real, its intensity weight λ_k. Given a similarity matrix $A = (a_{ij})$ for $i, j \in I$, it is assumed that each of the similarity index values a_{ij} is determined by the summary intensity of clusters containing both i and j. Therefore, the problem of additive clustering: given a similarity matrix, find clusters and their intensities so that the derived summary matrix is as close as possible to the given one.

Let $A = (a_{ij})$, $i, j \in I$ be a given similarity matrix and $\lambda \mathbf{s} = (\lambda s_i s_j)$ a weighted binary matrix defined by a binary membership vector $s = (s_i)$ of a subset $S = \{i \in I : s_i = 1\}$ along with its intensity weight λ. The notion that A can be considered as a noisy representation of a set of weighted "additive clusters" $\lambda_k s_k s_k^T$, $k = 1, \ldots, K$, can be formalized with the following equation:

$$a_{ij} = \lambda_0 + \lambda_1 z_{i1} z_{j1} + \cdots + \lambda_m z_{iK} z_{jK} + e_{ij} \qquad (7.77)$$

where e_{ij} are residuals to be minimized.

The intercept λ_0 can be considered in various perspectives [143]. One of them is that it is the intensity of the universal cluster consisting of all the elements of I, $S_0 = I$, so that $z_{i0} z_{j0} = 1$ for all $i, j \in I$. Another perspective puts λ_0 as the similarity shift value, so that it can be moved onto the left side of the equation, with the minus sign. With this, Equations 7.77 can be considered as an extension of the equations of the spectral decomposition of A, after the shift, to the case at which the scoring vectors z_k may have only $1/0$ values.

This model was introduced, in the English language literature, by Shepard and Arabie in [185], and independently in a more general form embracing other cluster structures as well, by the author in mid-seventies in Russian (see references in [133]). According to the model, similarities a_{ij} are defined by the intensity weights of clusters containing both i and j. Equations 7.77 must hold, generally, for each pair (i, j).

In matrix terms, the model is $A = Z \Lambda Z^T + E$ where Z is an $N \times (K + 1)$ matrix of cluster membership vectors, of which the $(K + 1)$th vector corresponds to λ_0 and has unity in all of its components, Λ is a $K \times K$ diagonal matrix with $\lambda_1, \ldots, \lambda_K$ on its main diagonal, all other entries being zero, and $E = (e_{ij})$ the residual matrix. That is, this equation is a straightforward analogue to the spectral decomposition of a symmetric matrix A in Equation 7.10. The only difference is that here matrix Z is a binary matrix of cluster membership values, whereas in the spectral decomposition Z is matrix of eigenvectors.

With no loss of generality, we accept in this section that a_{ij} are symmetric and not defined at $i = j$ so that only $i < j$ are considered. Indeed, if matrix $A = (a_{ij})$ is not symmetric, it can be converted into a symmetric matrix $(A + A^T)/2$ with elements $(a_{ij} + a_{ji})/2$ that admits the same optimal additive clustering solution [135].

Unfortunately, fitting the model in a situation when clusters, hidden in the data, do overlap is a tricky business. Indeed, an alternating minimization strategy should be invoked. A good news with this strategy is that given a set of clusters $S_1, \ldots, S_K$, finding the least-squares optimal intensity weights is rather simple. Just form an $(N \times N) \times (K+1)$ matrix Γ whose K columns correspond to clusters, and $(K+1)$th column corresponds to the intercept λ_0. The k-th column is a reshaped matrix $s_k s_k^T$ with its (i,j) entry equal to $s_{ik} s_{jk}$ $(i,j \in I, k = 1, 2, \ldots, K+1)$. Take matrix A as an $N \times N$ column as well, and orthogonally project it onto the linear subspace spanning the columns of matrix Γ – the coefficients at the columns of Γ are the least-squares optimal intensity weights. To do the operation of orthogonal projection, form the corresponding operator $P = \Gamma(\Gamma^T\Gamma)^{-1}\Gamma^T$, apply it to the "column" A to obtain PA, and take the $(K+1)$-dimensional vector of coefficients, $\lambda = (\Gamma^T\Gamma)^{-1}\Gamma^T A$. Since all λ_k are supposed to be non-negative, make a further projection by zeroing all the negative components in the vector λ.

Given a set of intensity weights, a further improvement of a set of clusters admits different locally optimal strategies; see, for instance, [32]. One of them would be to check, for every entity, which of the clusters it should belong to, and then do the best change.

In a situation when the clusters are not expected to overlap or, if expected, their contributions ought to much differ, the one-by-one PCA strategy may be applied. The strategy suggests that, provided that A has been preliminarily centered, minimizing $L(E) = \sum_{i,j} e_{ij}^2$ over unknown S_k and λ_k, $k = 1, \ldots, K$, can be done by finding one cluster S at a time. This one cluster S with intensity weight λ is to minimize $L = \sum_{i,j \in I}(a_{ij} - \lambda s_i s_j)^2$ with respect to all possible λ and binary membership vectors $s = (s_i)$ where a_{ij} depends on the previously found clusters. Obviously, given $S \subseteq I$, λ minimizing criterion L is the average similarity within cluster S, that is, $\lambda = a(S)$ where $a(S) = \sum_{i,j \in S; i < j} a_{ij} / [|S|(|S| - 1)/2]$. In matrix terms, $a(S) = s^T A s / s^T s$, a Rayleigh quotient. That is, the criterion in fact should maximize $a(S)|S|$, which is the criterion considered in Section 7.4.2, so that the algorithms described in that section, such as AddRemAdd, apply.

After a cluster S and corresponding $\lambda = a(S)$ are found, the similarities a_{ij} are to be changed for the residual similarities $a_{ij} - \lambda s_i s_j$, and the process of finding a cluster S and its intensity $a(S)$ is reiterated using the updated similarity matrix. In the case of non-overlapping clusters, there is no need to calculate the residual similarity: the process needs only be applied to only those entities that have remained unclustered as yet.

When the clusters are assumed to be mutually non-overlapping (that is, the membership vectors s_k are mutually orthogonal) or when fitting of the model is done using the one-by-one PCA strategy, the data scatter decomposition holds as follows [133]:

$$T(A) = \sum_{k=1}^{K} [s_k^T A_k s_k / s_k^T s_k]^2 + L(E) \qquad (7.78)$$

in which the least-squares optimal λ_k's are taken as the within cluster averages of the (residual) similarities. The matrix A_k in Equation 7.78 is either A unchanged, if clusters are not overlapping, or a residual similarity matrix at iteration k, $A_k = A - \sum_{t=1}^{k-1} \lambda_t s_t s_t^T$.

7.4.4 Agglomeration and Aggregation of Contingency Data

Specifics of contingency and redistribution data have been pointed out in Sections 2.2.3, 2.5.2, and 4.5.2. Basically, they amount to the highest uniformity of all data entries so that they can be meaningfully summed and naturally pre-processed into Quetelet association indexes. These specifics are dwelt on in this section.

Let $P(T, U) = (p_{tu})$ be a contingency matrix with rows and columns representing categories $t \in T$ and $u \in U$, respectively, and $F = \{F_1, \ldots, F_m\}$ a partition of row set T in non-overlapping nonempty classes $F_1, \ldots, F_m$. Since entries p_{tu} can be meaningfully summed all over the table, a partition F can be used not only for clustering but also for aggregating matrix $P(T, U)$ into $P(F, U) = (p_{fu})$ $(f = 1, \ldots, m)$ where $p_{fu} = \sum_{t \in F_f} p_{tu}$ is the proportion of entities co-occurring in class F_f and category u for any $f = 1, \ldots, m$ and $u \in U$.

The row-to-column associations in $P(T, U)$ and $P(F, U)$ can be represented with relative Quetelet coefficients $q_{tu} = p_{tu}/(p_{t+}p_{+u}) - 1$ and $q_{fu} = p_{fu}/(p_{f+}p_{+u}) - 1$ defined in Section 2.2.3. The total difference $e_{tu} = q_{fu} - q_{tu}$ between the aggregate indexes q_{fu} and the individual entry indexes q_{tu}, $t \in F_f$, can be scored with an aggregate weighted square error index, $L(F, U) = \sum_{t \in T} \sum_{u \in U} p_{t+}p_{+u}e_{tu}^2$: the smaller the L, the better $P(F, U)$ aggregates $P(T, U)$.

To support the choice of $L(F, U)$ as an aggregation criterion, the following equation can be derived with elementary algebraic manipulations: $L(F, U) = X^2(T, U) - X^2(F, U)$ where X^2 is the chi-squared Pearson contingency coefficient in the format of equation (2.13) applied to $P(T, U)$ and $P(F, U)$, respectively. This equation can be put as

$$X^2(T, U) = X^2(F, U) + L(F, U) \tag{7.79}$$

and interpreted as a Pythagorean decomposition of the data scatter: $X^2(T, U)$ would be the original data scatter, $X^2(F, U)$ its part taken into account by partition F, and $L(F, U)$ would be the unexplained part. Obviously, minimizing $L(F, U)$ over partition F is equivalent to maximizing $X^2(F, U)$ because of Equation 7.79.

Let $F(k, l)$ be the partition obtained from F by merging its classes F_k and F_l into united class $F_{k \cup l} = F_k \cup F_l$. To score the quality of the merging, one needs to analyze the difference $D = X^2(F, U) - X^2(F(k, l), U)$, according to

Equation 7.79. Obviously, D depends only on items related to k and l:

$$D = \sum_{u \in U} [p_{ku}q_{ku} + p_{lu}q_{lu} - (p_{ku} + p_{lu})q_{k \cup l, u}]$$

$$= \sum_{u \in U} (1/p_{+u})[p_{ku}^2/p_{k+} + p_{lu}^2/p_{l+} - (p_{ku} + p_{lu})^2/(p_{k+} + p_{l+})]$$

By using equation $(x - y)^2 = x^2 + y^2 - 2xy$, one can transform the expression on the right to obtain:

$$D = \sum_{u \in U} (1/p_{+u}) \frac{p_{ku}^2 p_{l+}^2 + p_{lu}^2 p_{k+}^2 - 2p_{ku}p_{lu}p_{k+}p_{l+}}{p_{k+}p_{l+}(p_{k+} + p_{l+})}$$

which leads to

$$D = \frac{p_{k+}p_{l+}}{p_{k+} + p_{l+}} \sum_{u \in U} (1/p_{+u})(p_{ku}/p_{k+} - p_{lu}/p_{l+})^2$$

This is exactly the chi-square distance $\chi(F_k, F_l)$ considered in Sections 2.5.2 and 4.5.2.

Therefore, it is proven that the chi-square distance represents the increase in criterion $L(F, U)$ when two clusters of F are merged. In this aspect, it parallels the conventional Ward distance manifested in the criterion λ_w (7.50).

In fact, all the clustering theory in Sections 7.2 and 7.3 can be extended to the contingency data case via Equation 7.79. Aggregation of columns can be included into the analysis as well, as described in [135], pp. 323–327.

Similar considerations can be applied to the case of an interaction table, that is, a redistribution data table in which the row and column sets coincide, $T = U$ as, for instance, in the Digit confusion data. The agglomeration procedure remains the same as in the case when only rows are aggregated. However, since aggregation of rows here will also involve columns, the change of the Pearson chi-square coefficient cannot be expressed with chi-square distances between rows or columns, which makes computations a bit more complicated.

7.5 Consensus and Ensemble Clustering

7.5.1 Ensemble Clustering

Consensus, or ensemble, clustering is an activity of summarizing a set of clusterings into a single clustering. This subject has become quite popular because of the "popular demand": after applying different clustering algorithms, or even the same algorithm at different parameter or initialization settings, to the same data set, one gets a number of different solutions, a clustering ensemble. The question that arises then: is there a cluster structure behind the wealth of solutions found? And if, as one would hope, yes, what is that structure? There have been a number of approaches developed, especially in bioinformatics (see, e.g., [216]), as well as in general data analysis literature (see, e.g., [201,206]).

Here the problem is considered in the data recovery perspective, with a focus on the case at which all the clusterings in the ensemble are partitions of the entity set. Moreover, the resulting clustering is assumed a partition as well.

Our presentation also covers a somewhat less ordinary case at which the input partitions reflect some features of entities rather than clustering results. For example, in the analysis of family life style patterns, categorical features can be involved to reflect the family structure, the typical leisure activities, father and mother occupations, and so on. Such a combined feature can be represented by a corresponding partition of the entity set. Therefore, one can think that a reasonable partition of families can be found as a consensus partition for all the pattern-of-life-style feature partitions. The consensus partition problem was analyzed by the author using both axiomatic analysis and data recovery approach (see a description and references to earlier work in [135]). We present here the consensus partition problem as a data recovery problem in two perspectives, one related to recovery of a partition behind multiple clustering results and the other to finding a partition representing different types as described by several nominal features. The latter perspective has not received much attention in the literature as yet. This presentation loosely follows that by Mirkin and Muchnik [148].

Consider a partition $S = \{S_1, \ldots, S_K\}$ on I and corresponding binary membership $N \times K$ matrix $Z = (z_{ik})$ where $z_{ik} = 1$ if $i \in S_k$ and $z_{ik} = 0$, otherwise $(i = 1, \ldots, N, k = 1, \ldots, K)$. Obviously, $Z^T Z$ is a diagonal $K \times K$ matrix in which (k, k)th entry is equal to the cardinality of S_k, $N_k = |S_k|$. On the other hand, $ZZ^T = (s_{ij})$ is a binary $N \times N$ matrix in which $s_{ij} = 1$ if i and j belong to the same class of S, and $s_{ij} = 0$, otherwise. Therefore, $(Z^T Z)^{-1}$ is a diagonal matrix of the reciprocals $1/N_k$ and $P_Z = Z(Z^T Z)^{-1} Z^T = (p_{ij})$ is an $N \times N$ matrix in which $p_{ij} = 1/N_k$ if both i and j belong to the same class S_k, and $p_{ij} = 0$, otherwise. The latter matrix is well known in matrix algebra: P_Z is matrix of the operation of orthogonal projection of any N-dimensional vector x onto the linear subspace $L(Z)$ spanning the columns of matrix Z. As described in any text on matrix theory, $L(Z)$ consists of vectors Za for all possible $a = (a_1, \ldots, a_K)$, and, for any x, vector $P_Z x$ belongs to $L(Z)$ and is the nearest to x among all the $L(Z)$. Vector a in this case is determined by equation $a = (Z^T Z)^{-1} Z^T x$ so that $P_Z x = Za$.

A set of partitions R^u, $u = 1, 2, \ldots, U$, along with the corresponding binary membership $N \times L_u$ matrices X^u, found with various clustering procedures, can be thought of as proxies for a hidden partition S, along with its binary membership matrix Z. Each of the partitions can be considered as related to the hidden partition S by equations

$$x_{il}^u = \sum_{k=1}^{K} c_{kl}^u z_{ik} + e_{ik}^u \tag{7.80}$$

where coefficients c_{kl}^u and matrix z_{ik} are to be chosen to minimize the residuals e_{ik}^u.

By accepting the sum of squared errors $E^2 = \sum_{i,k,u}(e_{ik}^u)^2$ as the criterion to minimize, one immediately arrives at the optimal coefficients being orthogonal projections of the columns of matrices X^u onto the linear subspace spanning the hidden matrix Z. More precisely, at a given Z, the optimal $K \times L_u$ matrices $C^u = (c_{kl}^u)$ are determined by equations $C^u = Z(Z^T Z)^{-1} X^u$, as explained above. By substituting these in Equation 7.80, the square error criterion can be reformulated as

$$E^2 = \sum_{u=1}^{U} ||X^u - P_Z X^u||^2 \tag{7.81}$$

where $||.||^2$ denotes the sum of squares of the matrix elements.

Let us first analyze criterion (7.81) in terms of association between given classes R_l^u of partitions R^u, $u = 1, 2, \ldots, U$ and classes S_k to be found.

To do that, let us first take a look at the meaning of the squared difference $\Delta(Z, R) = ||X - P_Z X||^2$ at the binary $N \times L$ matrix X of an arbitrary partition $R = \{R_1, R_2, \ldots, R_L\}$. Class R_l contributes $C_l = \sum_{i \in l}(x_{il} - \sum_j p_{ij} x_{jl})^2$ to the criterion; value p_{ij} is (i, j)th entry in matrix P_Z. Let us keep in mind that $p_{ij} x_{jl}^t$ is either 0 or not, so that it is not zero only in the case at which both of its constituents are not zero, that is, if and only if (i) both i and j are elements of S_k to be found and (ii) $j \in R_l$, that is, $j \in S_k \cap R_l$. In this case, $p_{ij} x_{il}^t = 1/N_k$ where $N_k = |S_k|$ is cardinality of the cluster S_k to be found. Therefore, to compute the full contribution, one needs to take into account clusters S_k of partition S:

$$C_l = \sum_{k=1}^{K} \sum_{i \in S_k}(x_{il} - N_{kl}/N_k)^2$$

where N_{kl} is cardinality of the intersection $S_k \cap R_l$. Indeed, for $i \in S_k$, $\sum_{j \in l} p_{ij} x_{jl} = \sum_{j \in S_k} x_{il}/N_k$ because $p_{ij} = 1/N_k$ for $j \in S_k$ and $p_{ij} = 0$, otherwise. Since $x_{il} = 1$ only at $i \in R_l$, $\sum_{i \in S_k} x_{il} = N_{kl}$. Therefore,

$$C_l = \sum_{i \in l} x_{il} - 2\sum_{k=1}^{K} \sum_{i \in S_k} x_{il} N_{kl}/N_k + \sum_{k=1}^{K} \sum_{i \in S_k}(N_{kl}/N_k)^2$$

The first item is put as x_{il} because it is equal to its square $x_{il} = x_{il}^2$ as a binary $1/0$ value. This can be further reformulated as

$$C_l = N_l - 2 \sum_{k=1}^{K} N_{kl}^2 / N_k + \sum_{k=1}^{K} N_k (N_{kl}/N_k)^2$$

Indeed, $\sum_{i \in S_k} x_{il} = N_{kl}$ in the second item and $\sum_{i \in S_k} 1 = N_k$ in the third item. This expression can be transformed into

$$C_l = N_l - \sum_{k=1}^{K} N_{kl}^2 / N_k$$

By using relative frequencies $p_l = N_l/N$, $p_{kl} = N_{kl}/N$ and conditional frequency of l in cluster S_k, $p(l/k) = N_{kl}/N_k$, one can further express this as

$$C_l = N \left[p_l - \sum_{k=1}^{K} p_{kl} p(l/k) \right]$$

Therefore, the squared difference between X and its projection to Z is equal to

$$\Delta(Z, R) = N \left(1 - \sum_{l} \sum_{k=1}^{K} p_{kl}^2 / p_k \right) = N \left(1 - \sum_{k,l} p_{kl} p(l/k) \right) \qquad (7.82)$$

This has a natural interpretation in terms of the so-called proportional prediction. The proportional prediction mechanism is defined over a stream of entities of which nothing is known beforehand except for the distribution of categories of a feature. According to this mechanism, category l is assigned to incoming entities randomly with its probability p_l. This, obviously, is erroneous in the $1 - p_l$ proportion of the cases. The average error of this mechanism is then $G = \sum_l p_l (1 - p_l) = 1 - \sum_l p_l^2$, the so-called Gini index (see, e.g., [140]). In a more complex situation of two nominal features, the mechanism randomly predicts a category l under the condition that the category k of another feature is known. Then, at a pair (k, l) randomly arriving with probability p_{kl}, the value l is predicted with the probability $p(l/k)$, thus generating the average prediction error $\delta = \sum_{k,l} p_{kl}(1 - p(l/k)) = 1 - \sum_{k,l} p_{kl} p(l/k)$. Therefore, the squared difference Δ in Equation 7.82 has the meaning of the average error of the proportional prediction mechanism.

Expressions (7.82) and (7.81) lead to the following statement.

Statement 7.10. Partition S is a least squares ensemble consensus clustering for the set of partitions $R^1, R^2, \ldots, R^U$ if and only if it minimizes the summary error of proportional prediction

$$E^2 = \Delta(S) = \sum_{u=1}^{U} \Delta(S, R^u) \tag{7.83}$$

or, equivalently, maximizes the weighted sum of conditional probabilities of classes R_l^u:

$$F(S) = \sum_{u=1}^{T} \sum_{l=1}^{L_u} \sum_{k=1}^{K} p_{kl} p(l/k) \tag{7.84}$$

This criterion is akin to the criteria for conceptual clustering derived in Section 7.3.4.3 though somewhat differs from them.

The square error criterion can be reformulated in terms of similarities between entities as well. To this end, let us form $N \times L$ matrix $X = (X^1 X^2 \cdots X^U)$ where $L = \sum_{u=1}^{U} L_u$. The columns of this matrix correspond to clusters R_l that are present in partitions $R^1, \ldots, R^U$. Then the least squares criterion can be expressed as $E^2 = ||X - P_Z X||^2$, or equivalently, as $E^2 = Tr((X - P_Z X)(X - P_Z X)^T)$ where Tr denotes the trace of $N \times N$ matrix, that is, the sum of its diagonal elements, and ^{T}, the transpose. By opening the parentheses in the latter expression, one can see that $E^2 = Tr(XX^T - P_Z XX^T - XX^T P_Z + P_Z XX^T P_Z) = Tr(XX^T - P_Z XX^T)$. Indeed, the operation Tr is commutative, which provides for equations $Tr(P_Z XX^T) = Tr(XX^T P_Z)$ and $Tr(P_Z XX^T P_Z) = Tr(P_Z P_Z XX^T) = Tr(P_Z XX^T)$; the last equation follows from the fact that the projection operator P_Z satisfies what is called idempotence, that is, equation $P_Z P_Z = P_Z$.

Let us denote $A = XX^T$ and take a look at (i, j)th element of this matrix $a_{ij} = \sum_l x_{il} x_{jl}$ where summation goes over all clusters R_l of all partitions $R^1, R^2, \ldots, R^U$. Obviously, $x_{il} x_{jl} = 1$ if both i and j belong to R_l, and $x_{il} x_{jl} = 0$, otherwise. This means that a_{ij} equals the number of those partitions $R^1, R^2, \ldots, R^U$ at which i and j are in the same class. This matrix is referred to in the literature as the (ensemble) consensus matrix [132]. Obviously, $a_{ii} = U$ for all $i \in I$ so that $Tr(XX^T) = NU$. The (i, i)th diagonal element of matrix $P_Z A$ equals $\sum_j p_{ij} a_{ji} = \sum_{j \in S_k} a_{ij}/N_k$ where S_k is that class of partition S to be found that contains entity i. Indeed, $p_{ij} = 1/N_k$ for $j \in S_k$ and $p_{ij} = 0$, otherwise. The following formula for the least-squares criterion (7.81) has been proven:

$$E^2 = NU - \sum_{k=1}^{K} \sum_{i,j \in S_k} a_{ij}/N_k$$

This leads us to the following statement.

Statement 7.11. A partition $S = \{S_1, \ldots, S_K\}$ is an ensemble consensus clustering if and only if it maximizes criterion

$$g(S) = \sum_{k=1}^{K} \sum_{i,j \in S_k} a_{ij}/N_k \tag{7.85}$$

where $A = (a_{ij})$ is the ensemble consensus matrix between entities for the given set of partitions.

This criterion much resembles, and in fact coincides with, that for K-Means if applied to the binary matrix X whose columns correspond to individual clusters of the partitions under consideration, introduced above (see formula (7.69) in Section 7.4.2).

7.5.2 Combined Consensus Clustering

Consider now a symmetric consensus problem: given a set of partitions R^u, $u = 1, 2, \ldots, U$, along with the corresponding binary membership $N \times L_u$ matrices X^u, expressing various features related to different aspects of a combined concept, like the life style, find a combined partition $S = \{S_1, S_2, \ldots, S_K\}$, along with its binary membership matrix Z, to explicitly express the combined concept using equations:

$$z_{ik} = \sum_{l=1}^{L_u} c_{kl}^u x_{il}^u + e_{ik}^u \tag{7.86}$$

where coefficients c_{kl}^u and matrix z_{ik} are to be chosen to minimize the residuals e_{ik}^u. In matrix terms, this equation is $Z = X^u C^{uT} + E^u$.

Using the sum of squared errors $E^2 = \sum_{i,k,u}(e_{ik}^u)^2$ as the criterion to minimize, one immediately arrives at the optimal coefficients being orthogonal projections of the columns of matrix Z in the linear subspaces spanning the matrices X^u of the observed partitions R^u ($u = 1, \ldots, U$). More precisely, at a given Z, the optimal $K \times L_u$ matrices $C^u = (c_{kl}^u)$ are determined by equations $C^{uT} = (X^{uT}X^u)^{-1}X^uZ$ as defined by the projection operators $P_u = X^u(X^{uT}X^u)^{-1}X^{uT}$. (Note, ^{T} here refers to transpose.) By substituting these in Equation 7.86, the square error criterion can be reformulated as

$$E^2 = \sum_{t=u}^{u} ||Z - P_u Z||^2 \tag{7.87}$$

where $||.||^2$ denotes the sum of squares of the matrix elements.

Let us first analyze criterion (7.87) in terms of association between given classes R_l of partitions R^u, $u = 1, 2, \ldots, U$ and classes S_k to be found. Analogously to derivations in the previous section, the criterion (7.87) can be proven to be equal to the summary average error of proportional prediction (7.82)

$$E^2 = \sum_{u=1}^{U} \Delta(R^u, S) = N \sum_{u=1}^{U} \left(1 - \sum_{k=1}^{K} \sum_{l=1}^{L_u} p_{kl} p(k/l) \right) \qquad (7.88)$$

This differs from criterion (7.83) by the direction of prediction: the sought classes S_k are predicted here from classes of R^t rather than the other way around—rather expectably, because of the very definition of the combined consensus.

To derive an equivalent expression for criterion (7.87) by using entity-to-entity similarities, let us consider an item $||Z - P_u Z||^2$ and reformulate it in terms of the matrix trace: $||Z - P_u Z||^2 = Tr((Z - P_u Z)(Z - P_u Z)^T)$ where Tr denotes the trace of an $N \times N$ matrix, that is, the sum of its diagonal elements, and ^{T}, the transpose. By opening the parentheses, one can see that

$$||Z - P_u Z||^2 = Tr(ZZ^T - P_u ZZ^T - ZZ^T P_u + P_u ZZ^T P_u) = Tr(ZZ^T - P_u ZZ^T)$$

This is proven in the same way as a similar equation in the previous section. Therefore, $||Z - P_u Z||^2 = Tr(ZZ^T) - Tr(P_u ZZ^T)$. Then the total criterion value is $E^2 = TN - \sum_{u=1}^{U} Tr(P_u ZZ^T) = UN - Tr(\sum_{u=1}^{U} P_u ZZ^T)$. Let us refer to the summary matrix $P = \sum_{u=1}^{T} P_u$ as the projection consensus matrix. Its (i, j)th element is the sum of reciprocals $1/N_l$ of the cardinalities of those clusters R_l^u that contain both of the entities, i and j. This differs of the ensemble consensus matrix because the latter assigns the same weight to all such clusters, whereas the weight of R_l^u is the inverse of its frequency: the larger the cluster, the lesser the weight.

By using P, the criterion can be reformulated as $E^2 = UN - Tr(PZZ^T)$. The (i, i)th diagonal element of matrix PZZ^T equals $\sum_j p_{ij} z_{ji} = \sum_{j \in S_k} p_{ij}$ where S_k is that class of partition S to be found that contains entity i. Therefore, the following formula for the least-squares criterion (7.87) has been proven: $E^2 = NU - \sum_{k=1}^{K} \sum_{i,j \in S_k} p_{ij}$. That means that the best combined partition is to maximize the summary within cluster similarity criterion

$$f(S) = \sum_{k=1}^{K} \sum_{i,j \in S_k} p_{ij} \qquad (7.89)$$

Since all the elements of matrix $P = (p_{ij})$ are not negative, the optimal partition is quite obvious: it is the universal cluster I consisting of all the entities that maximizes criterion (7.89). What a disappointing result! Superficially, there is a simple explanation that the "combined" universal cluster I is the easiest to predict from any set of clusters R_l.

Yet if one takes a closer look at the mathematical structure of the criterion (7.87), a structural cause of the disappointing solution can be seen. The cause is that however partitions R^u may differ, all the linear subspaces $L(X^u)$ spanning columns of matrices X^u have a common part, on overlap caused by the format of the matrices rather than the substance. Indeed, the columns of each X^u sum to the same vector 1_N, all of the components of which are unities. Therefore, all the subspaces $L(X^u)$ overlap at the bisector one-dimensional space of all the vectors whose all components are equal to each other, just because matrices X^u bear somewhat excessive information.

To properly represent the partitions R^u, the "parasitic" bisector subspace should be removed from $L(X^u)$, which is rather easy to do. Consider the space L_u^- which is complementary to the bisector in $L(X^u)$, that is, the set of all the vectors $y \in L(X^u)$ that are orthogonal to 1_N. These are exactly those that satisfy the additional requirement that the sum of the components is zero. In other words, only centered versions of vectors in $L(X^u)$ are taken to form L_u^-. As is well known from matrix algebra, the projection operator P_u^- to the complementary subspace is equal to the difference between the projection operator P_u to the original subspace and the projection operator to the bisector one-dimensional space. The latter is the $N \times N$ matrix whose all components are equal to $1/N$, which means that the elements of the "right" projection matrix P_u^- are just the original P^u matrix elements with $1/N$ subtracted from them. Curiously, $1/N$ is the average value of all the elements of the projection matrix corresponding to any partition. The elements of the summary matrix $P^- = \sum_{u=1}^{U} P_u^-$, therefore, will be equal to $p_{ij} - U/N$.

After fixing thus the concept of combined consensus, we arrive at the problem of maximizing the summary within-cluster similarities with the average value, $\bar{p} = U/N$, subtracted:

$$f(S, \bar{p}) = \sum_{k=1}^{K} \sum_{i,j \in S_k} (p_{ij} - \bar{p}) \qquad (7.90)$$

This criterion belongs to the class of uniform summary within-cluster criteria, see Section 5.1. The credit for the introduction of the summary reciprocal frequencies to measure the similarity between entities characterized by categorical features should go to a Russian entomologist, E.S. Smirnov (1898-1972), one of the early enthusiasts of the "numerical taxonomy" [191] of which he managed, after all the Stalinist purges of quantitative biology, to publish a monograph [189]. Of course, this measure was introduced in [189] in a purely heuristic way.

7.5.3 Concordant Partition

One more approximation framework, in which the partition consensus problem can be specified, involves the binary equivalence relations on set I

corresponding to partitions R^u, $u = 1, \ldots, U$: they are to be summarized in a consensus partition S. Indeed, any partition $R = \{R_1, R_2, \ldots, R_m\}$ can be represented by an equivalence relation r in the format of $N \times N$ binary matrix $r = (r_{ij})$ in which $r_{ij} = 1$ if i and j belong to the same class in R, and $r_{ij} = 0$, otherwise. This matrix, in fact, can be defined from the binary incidence matrix $X = (x_{il})$ in which $x_{il} = 1$ if $i \in R_l$ ($l = 1, \ldots, m$) and $x_{il} = 0$, otherwise. Indeed, $r = XX^\mathsf{T}$. The distance $d(R, S)$ between partitions R and S is defined as the squared Euclidean distance between the corresponding $N \times N$ matrices r and s: $d(R, S) = \sum_{i,j \in I} (r_{ij} - s_{ij})^2 = \sum_{i,j \in I} |r_{ij} - s_{ij}|$. The latter equation holds because the difference $r_{ij} - s_{ij}$ can have only 1, -1 and 0 values—for these the square and absolute values coincide.

Given a set of partitions R^u, $u = 1, \ldots, U$, let us define the concordant partition S as that minimizing the summary distance $d(S) = \sum_{u=1}^{U} d(S, R^u)$. This concept can be converted to a problem of partitioning over the (ensemble) consensus matrix A introduced in Section 7.5.1 [132]. Indeed, $d(S) = \sum_{i,j \in I} \sum_u (r_{ij}^u - s_{ij})^2 = \sum_{i,j \in I} \sum_u (r_{ij}^u + s_{ij} - 2r_{ij}^u s_{ij})$. Let us denote $a_{ij} = \sum_u r_{ij}^u$. This is the consensus matrix; its (i, j)th element is equal to the number of those of the partitions in which i and j are in the same class. Then $d(S) = \sum_{i,j} a_{ij} - 2 \sum_{i,j} (a_{ij} - U/2) s_{ij}$. Since the first term is constant, this proves that the concordant partition S is to maximize the uniform summary criterion

$$f(S, U/2) = \sum_{k=1}^{K} \sum_{i,j \in S_k} (a_{ij} - U/2) \tag{7.91}$$

Criterion (7.91) looks much similar to criterion of the combined consensus (7.90), except for both the similarity score and the threshold value subtracted. The difference can be crucial as the following benchmark test shows.

7.5.4 Muchnik's Consensus Partition Test

Muchnik [157] proposed the following benchmark test for consensus partitioning models. The test follows the procedure for enveloping any nominal feature into a set of binary attributes, each corresponding to a category. Consider a partition, $R = \{R_k\}$ with K classes R_k, $k = 1, 2, \ldots, K$, to have been pre-specified on I. Now take $U = K$ and define R^k as a two-cluster partition to comprise two clusters, R_k and its complement, $I - R_k$ ($k = 1, \ldots, K$). The pre-specified R should be considered consensus in both of the meanings considered above. On the one hand, the R^ks can be considered ensemble clusterings found in the conditions at which only rather coarse granularity of two-cluster solutions is feasible. On the other hand, R^k can be thought of as different aspects making up the concept represented by partition R. Relating to a criterion of consensus partition, the question is whether that leads to R as the only consensus for the set R^k, $k = 1, \ldots, K$. If Yes, then the test is passed. If the criterion fails this test, there remains little to support it. The Muchnik's test should be considered as a first hurdle to be overcome by a consensus criterion.

Let us apply the test to the three concepts of consensus introduced above: ensemble consensus, combined consensus and concordant partition.

First of all, let us take a look at the consensus matrix $A = (a_{ij})$ for the test. Obviously, if i and j belong to the same class R_k of R then they belong to the same class in every two-class partition R^k so that $a_{ij} = K$ in this case. If, in contrast, i and j belong to different classes, say R_k and R_l, respectively, they will belong to different classes in the corresponding two-class partitions R^k and R^l, whereas they would belong to the same class in all the other two-class partitions. This would make $a_{ij} = K - 2$ in the case when i and j belong to different classes in R.

This implies that the concept of concordant partition fails Muchnik's test at $K \geq 4$. Indeed, in this case all the elements of the consensus matrix A are not negative: $a_{ij} - K/2 \geq (K - 2) - K/2 = (K - 4)/2 \geq 0$. Therefore, the maximum of criterion (7.91) is reached at the universal partition consisting of the only universal cluster I, not at partition R.

In contrast, both of the concepts of consensus partition pass Muchnik's test. Consider first the ensemble consensus criterion $g(S)$ (7.85). At $S = R$, its value is $g(S) = KN$. Indeed, the within cluster sum of similarities $a_{ij} = K$ is KN_k^2, where $N_k = |R_k|$ is the number of entities in R_k. Then the contribution of R_k to $g(R)$ is equal to $KN_k^2/N_k = KN_k$ so that $g(R) = \sum_{k=1}^K KN_k = KN$. Let us now change R by moving an entity, i from its class, say R_1, to another class, say R_2, to arrive at partition denoted as R'. Obviously, $g(R')$ differs from $g(R)$ over only the two classes, $R_1' = R_1 - i$ and $R_2' = R_2 + i$. The sum of within R_1' similarities is equal to $K(N_1 - 1)^2$ so that its contribution to $g(R')$ is $K(N_1 - 1)^2/(N_1 - 1) = K(N_1 - 1)$. The within cluster R_2' similarities differ from those within R_2 by additional set of a_{ij} and a_{ji} similarities for all $j \in R_2$ plus $a_{ii} = K$. Since each of the a_{ij} is $K - 2$ at $j \in R_2$, the total sum of the within R_2' similarities is equal to $KN_2^2 + 2N_2(K - 2) + K = K(N_2 + 1)^2 - 4N_2$. The contribution of R_2' to $g(R')$ is equal to $K(N_2 + 1) - 4N_2/(N_2 + 1)$. Therefore, the total contribution of R_1' and R_2' to $g(R')$ is equal to $K(N_1 + N_2) - 4N_2/(N_2 + 1)$ which is less than the summary contribution of R_1 and R_2 to $g(R)$, $K(N_1 + N_2)$, by the subtracted value $4N_2/(N_2 + 1)$. Obviously, further changes to R will only lessen the value of $g(S)$, which proves that the ensemble consensus clustering criterion $g(S)$ in (7.85) reaches its only maximum at $S = R$ and, thus, does pass the Muchnik's test.

Let us turn now to the combined consensus clustering criterion $f(S, \bar{p})$ in Equation 7.90 and focus, for the sake of simplicity, on the case at which all classes R_k of R have the same number of elements, N/K. In this case, the reciprocal, K/N, is the same for all the classes R_k. This allows us to express the combined within R_k similarities as the product $K(K/N) = K^2/N$. The average $\bar{p} = K/N$ in this case. Then the difference $p_{ij} - \bar{p} = K^2/N - K/N = K(K - 1)/N > 0$ for $K > 1$ at i, j belonging to the same class $R_k, k = 1, \ldots, K$. Let us consider now the case at which i, j are from different R classes. The combined between-class similarities will be all equal to the product $(K - 2)K/(N(K - 1))$ because each of the "not R_k" classes has $N - N/K = N(K - 1)/K$ elements. Then the difference $p_{ij} - \bar{p} = (K - 2)K/(K - 1)N - K/N = (K/N)[(K - 2)/(K - 1) - 1] = -K/(N(K - 1)) < 0$ for $K > 1$. This warrants that R is the only maximizer of $f(S, \bar{p})$ and proves that the combined consensus clustering criterion passes the Muchnik's test too.

7.5.5 Algorithms for Consensus Partition

To find an ensemble consensus partition, one should use algorithms maximizing criterion (7.85) or (7.84). As criterion (7.85) is the same as criterion (7.69) in Section 7.4.2, with the consensus similarity matrix, all the approaches described in that section are applicable. Among them are

1. All clusters at once, including the agglomerative and exchange approaches.
2. Divisive clustering, including the spectral approach, Bisecting K-Means and splitting by Separation.
3. One-by-one clustering with AddRemAdd.

To maximize criterion (7.84), one should apply a conceptual clustering algorithm from Section 7.3.4.3, with an obvious change of the criterion, so that consensus clusters are built by consecutive splitting of the set over most informative clusters R_l^t. The criterion (7.89) of combined consensus clustering is maximized similarly.

The combined consensus uniform criterion (7.90) can be maximized by the algorithms described in Section 7.4.2 such as the agglomerative method or spectral approaches.

7.6 Overall Assessment

An outline of a least-squares theory for data visualization and clustering techniques is presented. The theory embraces such seemingly unrelated methods as Principal component analysis, K-Means clustering, hierarchical clustering, conceptual clustering, consensus clustering and network clustering. All these appear to be methods for fitting the same bilinear model of data at different specifications and local search strategies.

The choice of the least-squares criterion is somewhat substantiated by the properties of solutions, which include averaging quantitative features and taking proportions of categorical features in centroids, using the squared Euclidean distance as the dissimilarity and inner product as the similarity, producing contributions of features to partitions equal to the correlation ratio or Pearson's chi-square coefficient, close relations to spectral decompositions, good tightness properties of the clusters, and so on.

The theory supports

1. A unified framework for K-Means and Ward clustering that not only justifies conventional partitioning and agglomerative methods but extends them to mixed scale and similarity data.

2. Modified versions of the algorithms such as scalable divisive clustering as well as intelligent versions of K-Means mitigating the need in user-driven ad hoc parameters.

3. Compatible measures and criteria for similarity data including uniform and modularity similarity clustering, spectral clustering and additive clustering.

4. Effective measures and clustering criteria for analyzing binary data tables.

5. One-by-one clustering procedures that allow for more flexible clustering structures including single clusters, overlapping clusters and incomplete clustering.

6. Interpretation aids based on decompositions of the data scatter over elements of cluster structures to judge elements' relative contributions.

7. Conventional and modified measures of statistical association, such as the correlation ratio and Pearson's chi-square contingency coefficient, as summary contributions of cluster structures to the data scatter.

8. A related framework for cluster analysis of contingency and redistribution data in which the data scatter and contributions are measured in terms of the Pearson's chi-square contingency coefficient.

9. A reasonable framework for ensemble and combined consensus clustering leading to sound criteria and algorithms.

Some of these relate to currently popular approaches, some are yet to become popular.

References

1. The ACM Computing Classification System. 1998. url = http://www.acm.org/class/1998/ccs98.html.
2. M.A. Aizerman, E.M. Braverman, and L.I. Rozonoer. 1970. *Method of Potential Functions in Machine Learning*, Nauka Phyz-Mat, Moscow (in Russian).
3. K. Ali and M. Pazzani. 1995. Hydra-mm: Learning multiple descriptions to improve classification accuracy, *International Journal on Artificial Intelligence Tools*, 4, 115–133.
4. R.C. Amorim and B. Mirkin. 2012. Minkowski metric, feature weighting and anomalous cluster initializing in K-Means clustering, *Pattern Recognition*, 45, 1061–1074.
5. P. Arabie, L. Hubert, and G. De Soete (Eds.) 1996. *Classification and Clustering*, World Scientific, Singapore.
6. S. Baase. 1991. *Computer Algorithms* (2nd edition), Addison-Wesley, Reading, MA.
7. K.D. Bailey. 1994. *Typologies and Taxonomies: An Introduction to Classification Techniques*, Sage Publications, London.
8. A. Banerjee, I. Dhillon, J. Ghosh, and S. Sra. 2009. Text clustering with mixture of von Mises–Fisher distributions, In: A. Srivastava and M. Sahami (Eds.) *Text Mining: Classification, Clustering, and Applications*, Chapman & Hall/CRC, London, New York, pp. 121–154.
9. V. Barnett and T. Lewis. 1994. *Outliers in Statistical Data*, Wiley, New York.
10. A. Ben-Dor and Z. Yakhini. 1999. Clustering for gene expression data, In: S. Istratil, K. Pevzner, and M. Waterman (Eds.) *RECOMB99: Proceedings of the Third International Conference on Computational Molecular Biology*, Lyon, France, pp. 33–42.
11. D. Beneventano, N. Dahlem, S. El Haoum, A. Hahn, D. Montanari, and M. Reinelt. 2008. Ontology-driven semantic mapping, *Enterprise Interoperability III*, Part IV, Springer, London, pp. 329–341.
12. J.P. Benzecri. 1992. *Correspondence Analysis Handbook*, Marcel Dekker, New York.
13. C. Biernacki, G. Celeux, and G. Govaert. 2000. Assessing a mixture model for clustering with the integrated completed likelihood, *IEEE Transactions on Pattern Analysis and Machine Intelligence*, 22(7), 719–725.
14. M. Berthold and D. Hand. 1999. *Intelligent Data Analysis: An Introduction*, Springer Verlag, Berlin, Heidelberg, New York.
15. J.C. Bezdek, J.M. Keller, R. Krishnapuram, L.I. Kuncheva, and N.R. Pal. 1999. Will the *real* Iris data please stand up?, *IEEE Transactions on Fuzzy Systems*, 7(3), 368–369.
16. L. Billard and E. Diday. 2003. From the statistics of data to the statistic of knowledge: Symbolic Data Analysis, *Journal of the American Statistical Association*, 98, 470-487.

17. H.H. Bock. 1996. Probability models and hypothesis testing in partitioning cluster analysis, In: P. Arabie, C. D. Carroll, and G. De Soete (Eds.) *Clustering and Classification*, World Scientific Publishing, River Edge, NJ, pp. 377–453.

18. H.H. Bock. 1999. Clustering and neural network approaches, In: W. Gaul and H. Locarek-Junge (Eds.) *Classification in the Information Age*, Springer, Berlin-Heidelberg, pp. 42–57.

19. D.L. Boley. 1998. Principal direction divisive partitioning, *Data Mining and Knowledge Discovery*, 2(4), 325–344.

20. L. Breiman, J.H. Friedman, R.A. Olshen, and C.J. Stone. 1984. *Classification and Regression Trees*, Wadswarth International Group, Belmont, CA.

21. T. Calinski and J.A. Harabasz. 1974. Dendrite method for cluster analysis, *Communications in Statistics*, 3, 1–27.

22. Z.-H. Cao and L.-H. Feng. 2003. A note on variational representation for singular values of matrix, *Applied Mathematics and Computation*, 143, 559–563.

23. L.L. Cavalli-Sforza. 2001. *Genes, Peoples, and Languages*, Penguin Books, London.

24. G. Celeux and G. Govaert. 1992. A classification EM algorithm and two stochastic versions, *Computational Statistics and Data Analysis*, 14, 315–332.

25. M. Chiang and B. Mirkin. 2010. Intelligent choice of the number of clusters in K-Means clustering: An experimental study with different cluster spreads, *Journal of Classification*, 27(1), 3–41.

26. S.S. Choi, S.H. Cha, and Ch.C. Tappert. 2010. A survey of binary similarity and distance measures, *Journal of Systemics, Cybernetics and Informatics*, 8(1), 43–48.

27. *Clementine 7.0 User's Guide Package*. 2003. SPSS Inc, Chicago.

28. W.W. Cohen. 1995. Fast effective rule induction, In: A. Prieditis and S. Russell (Eds.) *Proceedings of the 12th International Conference on Machine Learning*, Morgan Kaufmann, Tahoe City, CA, pp. 115–123.

29. T.M. Cover and J.A. Thomas. 1991. *Elements of Information Theory*, Wiley, Hoboken, New Jersey.

30. R.K. Brouwer. 2009. A method of relational fuzzy clustering based on producing feature vectors using FastMap, *Information Sciences*, 179, 3561–3582.

31. S. Deerwester, S.T. Dumais, G.W. Furnas, T.K. Landauer, and R. Harshman. 1990. Indexing by latent semantic analysis, *Journal of the American Society for Information Science*, 41, 391–407.

32. D. Depril, I. Van Mechelen, and B. Mirkin. 2008. Algorithms for additive clustering of rectangular data tables, *Computational Statistics and Data Analysis*, 52, 4923–4938.

33. G. Der and B.S. Everitt. 2001. *Handbook of Statistical Analyses Using SAS, Second Edition*, CRC Press, New York.

34. G. De Soete. 1986. Optimal variable weighting for ultrametric and additive tree clustering, *Quality and Quantity*, 20, 169–180.

35. M. Devaney and A. Ram. 1997. Efficient feature selection in conceptual clustering, In: *Proceedings of the 14th International Conference on Machine Learning*, Morgan Kaufmann, San Francisco, CA, pp. 92–97.

36. S. Dolnicar and F. Leisch. 2000. Getting more out of binary data: Segmenting markets by bagged clustering, Working paper no. 71, Vienna University of Economics and Business Administration, 22 p.

37. D. Dotan-Cohen, S. Kasif, and A.A. Melkman. 2009. Seeing the forest for the trees: Using the Gene Ontology to restructure hierarchical clustering, *Bioinformatics*, 25(14), 1789–1795.

38. R.O. Duda and P.E. Hart. 1973. *Pattern Classification and Scene Analysis*, J.Wiley & Sons, New York.

39. S. Dudoit and J. Fridlyand. 2002. A prediction-based resampling method for estimating the number of clusters in a dataset, *Genome Biology*, 3(7), 1–21.

40. M.H. Dunham. 2003. *Data Mining: Introductory and Advanced Topics*, Pearson Education Inc., Upper Saddle River, NJ.

41. A.W.F. Edwards and L.L. Cavalli-Sforza. 1965. A method for cluster analysis, *Biometrics*, 21, 362–375.

42. B. Efron and R.J. Tibshirani. 1993. *An Introduction to the Bootstrap*, Chapman and Hall, New York, NY.

43. M. Ester, A. Frommelt, H.-P. Kriegel, and J. Sander. 2000. Spatial data mining: Database primitives, algorithms and efficient dbms support, *Data Mining and Knowledge Discovery*, 4, 193–216.

44. B.S. Everitt and G. Dunn. 2001. *Applied Multivariate Data Analysis*, Arnold, London.

45. B.S. Everitt, S. Landau, and M. Leese. 2001. *Cluster Analysis* (4th edition), Arnold, London.

46. F. Farnstrom, J. Lewis, and C. Elkan. 2000. Scalability of clustering algorithms revisited, *2000 ACM SIGKDD*, 2(1), 51–57.

47. U.M. Fayyad, G. Piatetsky-Shapiro, P. Smyth, and R. Uthurusamy (Eds.). 1996. *Advances in Knowledge Discovery and Data Mining*, AAAI Press/The MIT Press, Menlo Park, CA.

48. J. Felsenstein. 1985. Confidence limits on phylogenies: An approach using the bootstrap, *Evolution*, 39, 783–791.

49. D.W. Fisher. 1987. Knowledge acquisition via incremental conceptual clustering, *Machine Learning*, 2, 139–172.

50. K. Florek, J. Lukaszewicz, H. Perkal, H. Steinhaus, and S. Zubrzycki. 1951. Sur la liason et la division des points d'un ensemble fini, *Colloquium Mathematicum*, 2, 282–285.

51. E.B. Fowlkes and C.L. Mallows. 1983. A method for comparing two hierarchical clusterings, *Journal of American Statistical Association*, 78, 553–584.

52. A. Frank and A. Asuncion. 2010. UCI Machine Learning Repository [http://archive.ics.uci.edu/ml]. University of California, School of Information and Computer Science, Irvine, CA.

53. J.M. Freudenberg, V.K. Joshi, Z. Hu, and M. Medvedovic. 2009. CLEAN: CLustering Enrichment ANalysis, *BMC Bioinformatics*, 10, 234.

54. S.L. France, C.D. Carroll, and H. Xiong. 2012. Distance metrics for high dimensional nearest neighborhood recovery: Compression and normalization, *Information Sciences*, to appear.

55. T. Gamblin, B.R. de Supinski, M. Schulz, R. Fowlery, and D.A. Reed. 2010. Clustering performance data efficiently at massive scales, In: T. Boku, H. Nakashima, A. Mendelson (Eds.) *Proceedings of the 24th International Conference on Supercomputing*, Japan, ACM, pp. 243–252.

56. K.R. Gabriel and S. Zamir. 1979. Lower rank approximation of matrices by least squares with any choices of weights, *Technometrics*, 21, 289–298.

57. A.P. Gasch and M.B. Eisen. 2002. Exploring the conditional coregulation of yeast gene expression through fuzzy k-means clustering, *Genome Biology*, 3, 11.

58. J. Ghosh and A. Acharya. 2011. Cluster ensembles, *WIRE Data Mining and Knowledge Discovery*, 1(4), 305–315.

59. M. Girolami. 2002. Mercer kernel based clustering in feature space, *IEEE Transactions on Neural Networks*, 13, 780–784.
60. R. Gnanadesikan, J.R. Kettenring, and S.L.Tsao. 1995. Weighting and selection of variables, *Journal of Classification*, 12, 113–136.
61. G.H. Golub and C.F. Van Loan. 1989. *Matrix Computations*, J. Hopkins University Press, Baltimore.
62. A.D. Gordon. 1999. *Classification* (2nd edition), Chapman and Hall/CRC, Boca Raton.
63. J.C. Gower. 1967. A comparison of some methods of cluster analysis, *Biometrics*, 23, 623–637.
64. J.C. Gower. 1971. A general coefficient of similarity and some of its properties, *Biometrics*, 27, 857–872.
65. J.C. Gower and G.J.S. Ross. 1969. Minimum spanning trees and single linkage cluster analysis, *Applied Statistics*, 18, 54–64.
66. S.B. Green and N.J. Salkind. 2003. *Using SPSS for the Windows and Macintosh: Analyzing and Understanding Data* (3rd Edition), Prentice Hall, Upper Saddle River, NJ.
67. A. Guenoche. 2011. Consensus of partitions: A constructive approach, *Advances in Data Analysis and Classification*, 5(3), 215–229.
68. S. Guha, R. Rastogi, and K. Shim. 2000. ROCK: A robust clustering algorithm for categorical attributes, *Information Systems*, 25(2), 345–366.
69. J. Han and M. Kamber. 2001. *Data Mining: Concepts and Techniques* (3d edition), Morgan Kaufmann Publishers, San Francisco, CA.
70. J.A. Hartigan. 1967. Representation of similarity matrices by trees, *Journal of American Statistical Association*, 62, 1140–1158.
71. J.A. Hartigan. 1972. Direct clustering of a data matrix, *Journal of American Statistical Association*, 67, 123–129.
72. J.A. Hartigan. 1975. *Clustering Algorithms*, J. Wiley & Sons, New York.
73. J.A. Hartigan. 1977. Asymptotic distributions for clustering criteria, *Annals of Statistics*, 6, 117–131.
74. P. Hansen and N. Mladenovic. 2001. J-means: A new local search heuristic for minimum sum-of-squares clustering, *Pattern Recognition*, 34, 405–413.
75. T. Hastie, R. Tibshirani, M.B. Eisen, A. Alizadeh, R. Levy, L. Staudt, W.C. Chan, D. Botstein, and P. Brown. 2000. 'Gene shaving' as a method for identifying distinct sets of genes with similar expression patterns. *Genome Biology*, 1(2):research0003.1–0003.21.
76. T. Hastie, R. Tibshirani, and J.R. Fridman. 2001. *The Elements of Statistical Learning*, Springer, New York.
77. S. Haykin. 1999. *Neural Networks* (2nd edition), Prentice Hall, New Jersey.
78. L.J. Heier, S. Kruglyak, and S. Yooseph. 1999. Exploring expression data: Identification and analysis of coexpressed genes, *Genome Research*, 9, 1106–1115.
79. K.J. Holzinger and H.H. Harman. 1941. *Factor Analysis*, University of Chicago Press, Chicago.
80. J.Z. Huang, J Xu, M. Ng, and Y. Ye. 2008. Weighting method for feature selection in K-Means, In: H. Liu and H. Motoda (Eds.) *Computational Methods of Feature Selection*, Chapman & Hall/CRC, Boca Raton, London, New York, pp. 193–209.
81. L.J. Hubert and P. Arabie. 1985. Comparing partitions, *Journal of Classification*, 2, 193–218.

82. P. Jaccard. 1908. Nouvelles recherches sur la distribution florale, *Bulletine de la Société Vaudoise de Sciences Naturelles*, 44, 223–370.

83. A.K. Jain and R.C. Dubes. 1988. *Algorithms for Clustering Data*, Prentice Hall, Englewood Cliffs, NJ.

84. A.K. Jain, M.N. Murty, and P.J. Flynn. 1999. Data clustering: A review, *ACM Computing Surveys*, 31(3), 264–323.

85. C.V. Jawahar, P.K. Biswas, and A.K. Ray. 1995. Detection of clusters of distinct geometry: A step toward generalized fuzzy clustering, *Pattern Recognition Letters*, 16, 1119–1123.

86. R. Jenssen, D. Erdogmus, J.C. Principe, and T. Eltoft. 2005. The Laplacian PDF distance: A cost function for clustering in a kernel feature space, In: *Advances in Neural Information Processing Systems*, 17, MIT Press, Cambridge, pp. 625–632.

87. W.S. Jevons. 1958. *The Principles of Science*, Dover Publications, New York.

88. I.T. Jolliffe. 1986. *Principal Component Analysis*. Springer-Verlag, New York.

89. G. John, R. Kohavi, and K. Pfleger. 1994. Irrelevant features and the subset selection problem, In: *Proceedings of the Eleventh International Machine Learning Conference*, Morgan Kaufmann, New Brunswick, NJ, pp. 121–129.

90. S.C. Johnson. 1967. Hierarchical clustering schemes, *Psychometrika*, 32, 241–245.

91. L. Kaufman and P. Rousseeuw. 1990. *Finding Groups in Data: An Introduction to Cluster Analysis*, Wiley, New York.

92. M.G. Kendall and A. Stuart. 1979. *The Advanced Theory of Statistics*, 2 (4th edition), Hafner, New York.

93. M.K. Kerr and G.A. Churchill. 2001. Bootstrapping cluster analysis: Assessing the reliability of conclusions from microarray experiments, *Proceedings of the National Academy of Science USA*, 98(16), 8961–8965.

94. H.A.L Kiers. 1997. Weighted least squares fitting using ordinary least squares algorithms, *Psychometrika*, 62, 251–266.

95. D. Klein and M. Randic. 1993. Resistance distance, *Journal of Mathematical Chemistry*, 12, 81–95.

96. W. Klösgen. 1996. Explora—A multipattern and multistrategy discovery assistant, In: U.M. Fayyad, G. Piatetsky-Shapiro, P. Smyth, and R. Uthurusamy (Eds.) *Advances in Knowledge Discovery and Data Mining*, AAAI Press/The MIT Press, Menlo Park, CA, pp. 249–271.

97. T. Kohonen. 1995. *Self-Organizing Maps*, Springer-Verlag, Berlin.

98. B. Korte, L. Lovász, and R. Schrader. 1991. *Greedoids*, Springer-Verlag, New York.

99. B. Kovalerchuk and E. Vityaev. 2000. *Data Mining in Finance: Advances in Relational and Hybrid Methods*, Kluwer Academic Publishers, Boston/Dordrecht/London, 308 p.

100. E.V. Kovaleva and B. Mirkin. 2012. Bisecting K-Means, principal directions and random directions in divisive clustering (submitted).

101. D.E Krane and M.L. Raymer. 2003. *Fundamental Concepts of Bioinformatics*, Pearson Education, San Francisco, CA.

102. W. Krzanowski and Y. Lai. 1985. A criterion for determining the number of groups in a dataset using sum of squares clustering, *Biometrics*, 44, 23–34.

103. W.J. Krzanowski and F.H.C. Marriott. 1994. *Multivariate Analysis*, Edward Arnold, London.

104. L.I. Kuncheva and D.P. Vetrov. 2005. Evaluation of stability of k-means cluster ensembles with respect to random initialization, *IEEE Transactions on Pattern Analysis and Machine Intelligence*, 28, 1798–1808.

105. L.I. Kuncheva, S.T. Hadjitodorov, and L.P. Todorova. 2006. Experimental comparison of cluster ensemble methods, *Proceedings FUSION* Florence, Italy, pp. 105–115.
106. S. Laaksonen. 2000. Regression-based nearest neighbour hot decking, *Computational Statistics*, 15, 65–71.
107. G. Lakoff. 1990. *Women, Fire, and Dangerous Things: What Categories Reveal About the Mind*, University of Chicago Press, Chicago.
108. W. Lam, C.-K. Keung, and D. Liu. 2002. Discovering useful concept prototypes for classification based on filtering and abstraction, *IEEE Transactions on Pattern Analysis and Machine Intelligence*, 24(8), 1075–1090.
109. G.N. Lance and W.T. Williams. 1967. A general theory of classificatory sorting strategies: 1. Hierarchical Systems, *The Computer Journal*, 9, 373–380.
110. M.H. Law, M.A.T. Figueiredo, and A.K. Jain. 2004. Simultaneous feature selection and clustering using mixture models, *IEEE Transactions on Pattern Analysis and Machine Intelligence*, 26(9), 1154–1166.
111. L. Lebart, A. Morineau, and M. Piron. 1995. *Statistique Exploratoire Multidimensionnelle*, Dunod, Paris.
112. E. Levine and E. Domany. 2001. Resampling method for unsupervised estimation of cluster validity, *Neural Computation*, 13, 2573–2593.
113. T.S. Lim, W.Y. Loh, and Y.S. Shih. 2000. A comparison of prediction accuracy, complexity, and training time of thirty three old and new classification algorithms, *Machine Learning*, 40, 203–228.
114. R.J.A Little and D.B Rubin. 1987. *Statistical Analysis with Missing Data*, John Wiley and Sons, New York.
115. H. Liu and H. Motoda (Eds.) 2008. *Computational Methods of Feature Selection*, Chapman & Hall/CRC, Boca Raton, London, New York.
116. W.Y. Loh and Y.S. Shih. 1997. Split selection methods for classification trees, *Statistica Sinica*, 7, 815–840.
117. U. von Luxburg. 2007. A tutorial on spectral clustering, *Statistics and Computing*, 17(4), 395–416.
118. V. Makarenkov and P. Legendre. 2001. Optimal variable weighting for ultrametric and additive trees and K-Means partitioning, *Journal of Classification*, 18, 245–271.
119. T. Margush and F.R. McMorris. 1981. Consensus n-trees, *Bulletin of Mathematical Biology*, 43, 239–244.
120. R. Mazza. 2009. *Introduction to Information Visualization*, Springer, London.
121. R.M. McIntyre and R.K. Blashfied. 1980. A nearest-centroid technique for evaluating the minimum variance clustering procedure, *Multivariate Behavioral Research*, 22, 225–238.
122. G. McLachlan and K. Basford. 1988. *Mixture Models: Inference and Applications to Clustering*, Marcel Dekker, New York.
123. G.J. McLachlan and N. Khan. 2004. On a resampling approach for tests on the number of clusters with mixture model-based clustering of tissue samples, *Journal of Multivariate Analysis*, 90, 990–1005.
124. J.B. MacQueen. 1967. Some methods for classification and analysis of multivariate observations, In: L. Lecam and J. Neymen (Eds.) *Proceedings of 5th Berkeley Symposium*, 2, University of California Press, Berkeley, pp. 281–297.

125. L.M. McShane, M.D. Radmacher, B. Freidlin, R. Yu, M.-C. Li, and R. Simon. 2002. Methods for assessing reproducibility of clustering patterns observed in analyses of microarray data, *Bioinformatics*, 18, 1462–1469.

126. M. Meila. 2007. Comparing clusterings—An information based distance, *Journal of Multivariate Analysis*, 98, 873–895.

127. R.S. Michalski. 1980. Knowledge acquisition through conceptual clustering: A theoretical framework and an algorithm for partitioning data into conjunctive concepts, *International Journal of Policy Analysis and Information Systems*, 4, 219–244.

128. G.W. Milligan. 1981. A Monte-Carlo study of thirty internal criterion measures for cluster analysis, *Psychometrika*, 46, 187–199.

129. G.W. Milligan. 1989. A validation study of a variable weighting algorithm for cluster analysis, *Journal of Classification*, 6, 53–71.

130. G.W. Milligan and M.C. Cooper. 1985. An examination of procedures for determining the number of clusters in a data set, *Psychometrika*, 50, 159–179.

131. G.W. Milligan and M.C. Cooper. 1988. A study of standardization of the variables in cluster analysis, *Journal of Classification*, 5, 181–204.

132. B.G. Mirkin. 1974. Approximation problems in relation space and the analysis of nonnumeric variables, *Automation and Remote Control*, 35, 1424–1431.

133. B. Mirkin. 1987. Additive clustering and qualitative factor analysis methods for similarity matrices, *Journal of Classification*, 4, 7–31; Erratum (1989), 6, 271–272.

134. B. Mirkin. 1990. Sequential fitting procedures for linear data aggregation model, *Journal of Classification*, 7, 167–195.

135. B. Mirkin. 1996. *Mathematical Classification and Clustering*, Kluwer Academic Press, Dordrecht.

136. B. Mirkin. 1997. L_1 and L_2 approximation clustering for mixed data: scatter decompositions and algorithms, In: Y. Dodge (Ed.) L_1-*Statistical Procedures and Related Topics*, Institute of Mathematical Statistics (Lecture Notes-Monograph Series), Hayward, CA, pp. 473–486.

137. B. Mirkin. 1999. Concept learning and feature selection based on square-error clustering, *Machine Learning*, 35, 25–40.

138. B. Mirkin. 2001. Eleven ways to look at the chi-squared coefficient for contingency tables, *The American Statistician*, 55(2), 111–120.

139. B. Mirkin. 2001. Reinterpreting the category utility function, *Machine Learning*, 45, 219–228. .

140. B. Mirkin. 2011. *Core Concepts in Data Analysis: Summarization, Correlation, Visualization*, Springer, London, 390 p.

141. B. Mirkin. 2011. Choosing the number of clusters, *WIRE Data Mining and Knowledge Discovery*, 3, 252–260.

142. B. Mirkin and R. Brooks. 2003. A tool for comprehensively describing class-based data sets, In: J.M. Rossiter and T. Martin (Eds.) *Proceedings of 2003 UK Workshop on Computational Intelligence*, University of Bristol, UK, pp. 149–156.

143. B.G. Mirkin, R. Camargo, T. Fenner, G. Loizou, and P. Kellam. 2010. Similarity clustering of proteins using substantive knowledge and reconstruction of evolutionary gene histories in herpesvirus, *Theoretical Chemistry Accounts: Theory, Computation, and Modeling*, 125(3–6), 569–581.

144. B. Mirkin and E. Koonin. 2003. A top-down method for building genome classification trees with linear binary hierarchies, In: M. Janowitz, J.-F. Lapointe, F. McMorris, B. Mirkin, and F. Roberts (Eds.) *Bioconsensus*, DIMACS Series, V. 61, Providence: AMS, pp. 97–112.

145. B. Mirkin, M. Levin, and E. Bakaleinik. 2002. Intelligent K-Means clustering in analysis of newspaper articles on bribing, *Intelligent Systems*, Donetsk, Ukraine, 2, 224–230.

146. B. Mirkin and L. Cherny. 1970. Axioms for measuring distances between partitions of a finite set, *Automation and Remote Control*, 5, 120–127.

147. B. Mirkin, T. Fenner, S. Nascimento, and R. Felizardo. 2011. How to visualize a crisp or fuzzy topic set over a taxonomy, In: Sergei, K., Mandal, D.P., Kundu, M.K., and Pal, S.K. (Eds.) *Proceedings of the 4th International Conference on Pattern Recognition and Machine Intelligence (PReMI)*, 3–12. Lecture Notes in Computer Sciences, Springer-Verlag, Berlin Heidelberg, Vol. 6744.

148. B. Mirkin and I. Muchnik. 1981. Geometric interpretation of clustering criteria, In: B. Mirkin (Ed.) *Methods for Analysis of Multidimensional Economics Data*, Nauka Publishers (Siberian Branch), Novosibirsk, pp. 3–11 (in Russian).

149. B. Mirkin and I. Muchnik. 2002. Layered clusters of tightness set functions, *Applied Mathematics Letters*, 15, 147–151.

150. B. Mirkin and I. Muchnik. 2008. Some topics of current interest in clustering: Russian approaches, *Electronic Journal for History of Probability and Statistics*, 4(2), 1960–1985.

151. B. Mirkin and S. Nascimento. 2012. Additive spectral method for fuzzy cluster analysis of similarity data including community structure and affinity matrices, *Information Sciences*, 183, 16–34.

152. B.G. Mirkin, S. Nascimento, and Pereira L.M. 2010. Cluster-lift method for mapping research activities over a concept tree, In: J. Koronacki, Z. Ras, and S. Wierzchon (Eds.) *Advances in Machine Learning II*, Springer, Berlin-Heidelberg, pp. 245–257.

153. B. Mirkin and A. Shestakov. 2012. Least squares consensus clustering: Criteria, methods, experiments (in progress).

154. M. Mitchell. 1998. *An Introduction to Genetic Algorithms (Complex Adaptive Systems)*, MIT Press, Cambridge, MA.

155. D.S. Modha and W.S. Spangler. 2003. Feature weighting in k-means clustering, *Machine Learning*, 52, 217–237.

156. S. Monti, P. Tamayo, J. Mesirov, and T. Golub. 2003. Consensus clustering: A resampling-based method for class discovery and visualization of gene expression microarray data, *Machine Learning*, 52, 91–118.

157. I.B. Muchnik. 1970. A test for consensus partition criteria (Personal communication).

158. J. Mullat. 1976. Extremal subsystems of monotone systems, *Automation and Remote Control*, 37, 758-766 (I), 1286–1294 (II).

159. F. Murtagh. 1985. *Multidimensional Clustering Algorithms*, Physica-Verlag, Heidelberg.

160. F. Murtagh. 1996. Neural networks for clustering, In: P. Arabie, L. Hubert, and G. De Soete (Eds.) *Combinatorial Data Analysis*, World Scientific, River Edge, NJ.

161. S. Nascimento and P. Franco. 2009. Unsupervised fuzzy clustering for the segmentation and annotation of upwelling regions in sea surface temperature images, In: J. Gama (Ed.) *Discovery Science*, LNCS 5808, Springer-Verlag, Heidelberg, pp. 212–226.

162. S. Nascimento, B. Mirkin, and F. Moura-Pires. 2003. Modeling proportional membership in fuzzy clustering, *IEEE Transactions on Fuzzy Systems*, 11(2), 173–186.

163. M. Nei and S. Kumar. 2000. *Molecular Evolution and Phylogenetics*, Oxford University Press, Oxford, UK.

164. M.E.J. Newman. 2006. Modularity and community structure in networks, *Proceedings of the National Academy of USA*, 103(23), 8577–8582.

165. M. Newman and M. Girvan. 2004. Finding and evaluating community structure in networks, *Physical Review E*, 69, 026113.

166. A. Ng, M. Jordan, and Y. Weiss. 2002. On spectral clustering: analysis and an algorithm, In: Ditterich, T.G., Becker, S., and Ghahramani, Z. (Eds.) *Advances in Neural Information Processing Systems*, 14, MIT Press, Cambridge, MA, pp. 849–856.

167. K. Pearson. 1901. On lines and planes of closest to systems of points in space, *The London, Edinburgh and Dublin Philosophical Magazine and Journal of Science*, Sixth Series, 2, 559–572.

168. M. Perkowitz and O. Etzioni. 2000. Towards adaptive Web sites: Conceptual framework and case study, *Artificial Intelligence*, 118, 245–275.

169. S.M. Perlmutter, P.C. Cosman, C.-W. Tseng, R.A. Olshen, R.M. Gray, K.C.P. Li, and C.J. Bergin. 1998. Medical image compression and vector quantization, *Statistical Science*, 13, 30–53.

170. K.S. Pollard and M.J. van der Laan. 2002. A method to identify significant clusters in gene expression data, *U.C. Berkeley Division of Biostatistics Working Paper Series*, 107.

171. J.R. Quinlan. 1993. *C4.5: Programs for Machine Learning*, Morgan Kaufmann, San Mateo.

172. W.M. Rand. 1971. Objective criteria for the evaluation of clustering methods, *Journal of American Statistical Association*, 66, 846–850.

173. P.N. Robinson and S. Bauer. 2011. *Introduction to Bio-Ontologies*, Chapman & Hall/CRC, Boca Raton FL.

174. R.J. Roiger and M.W. Geatz. 2003. *Data Mining: A Tutorial-Based Primer*, Addison Wesley, Pearson Education, Inc.

175. C.H. Romesburg. 1984. *Cluster Analysis for Researchers*, Lifetime Learning Applications, Belmont, CA. Reproduced by Lulu Press, NC, 2004.

176. G. Salton. 1989. *Automatic Text Processing: The Transformation, Analysis, and Retrieval of Information by Computer*, Addison-Wesley, Reading, MA.

177. S.L. Salzberg. 1998. Decision trees and Markov chains for gene finding, In: S.L. Salzberg, D.B. Searls, and S. Kasif (Eds.) *Computational Methods in Molecular Biology*, Elsevier Science B.V., Amsterdam, pp. 187–203.

178. *SAS/ETS User's Guide*, Version 8. 2000. Volumes 1 and 2, SAS Publishing, Cary, NC, USA.

179. G.A. Satarov. 1991. Conceptual control: A structural method for knowledge testing, Personal communication.

180. S.M. Savaresi and D. L. Boley. 2004. A comparative analysis on the bisecting K-means and the PDDP clustering algorithms, *Intelligent Data Analysis*, 8(4), 345–362.

181. J.L. Schafer. 1997. *Analysis of Incomplete Multivariate Data*, Chapman and Hall, London.

182. B. Schoelkopf and A.J. Smola. 2001. *Learning with Kernels: Support Vector Machines, Regularization, Optimization, and Beyond*, The MIT Press, Cambridge MA, USA.

183. C. Seidman. 2001. *Data Mining with Microsoft SQL Server 2000 Technical Reference*, Microsoft Corporation, Redmond, WA.

184. X. Sevillano, F. Alias, and J.C. Socoro. 2007. BordaConsensus: A new consensus function for soft cluster ensembles, *SIGIR '07: Proceedings of the 30th Annual International ACM SIGIR Conference on Research and Development in Information Retrieval*, Amsterdam.

185. R.N. Shepard and P. Arabie. 1979. Additive clustering: representation of similarities as combinations of overlapping properties, *Psychological Review*, 86, 87–123.

186. J. Shi and J. Malik. 2000. Normalized cuts and image segmentation, *IEEE Transactions on Pattern Analysis and Machine Intelligence*, 22(8), 888–905.

187. R. Shope. 2001. *Pluto: Ninth Planet or Not!* http://www.jpl.nasa.gov/ice_fire/outreach/PlutoN.htm.

188. B. Silverman. 1986. *Density Estimation for Statistics and Data Analysis*, Chapman and Hall, New York.

189. E.S. Smirnov. 1969. *Taxonomy Analysis*, Moscow State University Press, Moscow (in Russian).

190. A.J. Smola and R. Kondor. 2003. Kernels and regularization on graphs, *Proceedings of the Annual Conference on Computational Learning Theory and Kernel Workshop COLT2003*, Lecture Notes in Computer Science, 2777, Springer, Heidelberg, pp. 144–158.

191. P.H.A. Sneath and R.R. Sokal. 1973. *Numerical Taxonomy*, W.H. Freeman, San Francisco.

192. SNOMED Clinical Terms, http://www.nlm.nih.gov/research/umls/Snomed, US NLM NIH, accessed 8 January 2012.

193. S. Sosnovsky, A. Mitrovic, D. Lee, P. Prusilovsky, M. Yudelson, V. Brusilovsky, and D. Sharma. 2008. Towards integration of adaptive educational systems: Mapping domain models to ontologies, In: Dicheva, D., Harrer, A., and Mizoguchi, R. (Eds.) *Proceedings of the 6th International Workshop on Ontologies and Semantic Web for ELearning (SWEL'2008)*, Montreal, Canada.

194. M. Sonka, V. Hlavac, and R. Boyle. 1999. *Image Processing, Analysis and Machine Vision*, Brooks/Cole Publishing Company, Pacific Grove, CA.

195. J.A. Sonquist, E.L. Baker, and J.N. Morgan. 1973. *Searching for Structure*, Institute for Social Research, University of Michigan, Ann Arbor.

196. H. Spaeth. 1985. *Cluster Dissection and Analysis*, Ellis Horwood, Chichester.

197. R. Spence. 2001. *Information Visualization*, ACM Press/Addison-Wesley, New York.

198. R. Srikant, Q. Vu, and R. Agraval. 1997. Mining association rules with the item constraints, In: D. Heckerman, H. Manilla, D. Pregibon, and R. Uthrusamy (Eds.) *Proceedings of Third International Conference on Knowledge Discovery and Data Mining*, AAAI Press, Menlo Park, CA, pp. 67–73.

199. M. Steinbach, G. Karypis, and V. Kumar. 2000. A comparison of document clustering techniques, KDD Workshop on Text Mining.
200. D. Steinley and M. Brusco. 2007. Initializing K-Means batch clustering: A critical evaluation of several techniques, *Journal of Classification*, 24, 99–121.
201. A. Strehl and J. Ghosh. 2002. Cluster ensembles: A knowledge reuse framework for combining multiple partitions, *Journal of Machine Learning Research*, 3, 583–617.
202. C.A. Sugar and G.M. James. 2002. Finding the number of clusters in a data set: An information theoretic approach, *Journal of American Statistical Association*, 98(463), 750–778.
203. W. Tang, Z. Lu, and I.S. Dhillon. 2009. Clustering with multiple graphs, In: *ICDM '09: Proceedings of the 2009 Ninth IEEE International Conference on Data Mining.* IEEE Computer Society, Washington, DC, USA, pp. 1016–1021.
204. S.K. Tasoulis, D.K. Tasoulis, and V.P. Plagianakos. 2010. Enhancing principal direction divisive clustering, *Pattern Recognition*, 43(10), 3391–3411.
205. R. Tibshirani, G. Walther, and T. Hastie. 2001. Estimating the number of clusters in a dataset via the Gap statistics, *Journal of the Royal Statistical Society B*, 63, 411–423.
206. A. Topchy, A.K. Jain, and W. Punch. 2005. Clustering ensembles: Models of consensus and weak partitions, *IEEE Transaction on Pattern Analysis and Machine Intelligence*, 27(12), 1866–1881.
207. O. Troyanskaya, M. Cantor, G. Sherlock, P. Brown, T. Hastie, R. Hastie, R. Tibshirani, D. Botsein, and R.B. Altman. 2001. Missing value estimation methods for DNA microarrays, *Bioinformatics*, 17, pp. 520–525.
208. R.C. Tryon. 1939. *Cluster Analysis*, Edwards Bros., Ann Arbor.
209. M. Turk and A. Pentland. 1991. Eigenfaces for recognition, *Journal of Cognitive Neuroscience*, 3(1), 71–86.
210. S. Van Buuren and W.J. Heiser. 1989. Clustering N objects into K groups under optimal scaling of variables, *Psychometrika*, 54, 699–706.
211. J.H. Ward, Jr. 1963. Hierarchical grouping to optimize an objective function, *Journal of American Statistical Association*, 58, 236–244.
212. A. Webb. 2002. *Statistical Pattern Recognition*, John Wiley & Sons, Chichester, England.
213. A. Weingessel, E. Dimitriadou, and S. Dolnicar. 1999. An examination of indexes for determining the number of clusters in binary data sets, Working Paper No. 29, Vienna University of Economics, Wien, Austria.
214. A. Weingessel, E. Dimitriadou, and K. Hornik. 2003. An ensemble method for clustering. *Proceedings of International Conference on Data Mining*, pp. 3–10.
215. D. Wishart. 1999. *The ClustanGraphics Primer*, Clustan Limited, Edinburgh.
216. G. Xiao and W. Pan. 2007. Consensus clustering of gene expression data and its application to gene function prediction, *Journal of Computational and Graphical Statistics*, 16(3), 733–751.
217. K.Y. Yeung, C. Fraley, A. Murua, A.E. Raftery, and W.L. Ruzzo. 2001. Model-based clustering and data transformations for gene expression data, *Bioinformatics*, 17(10), 977–987.
218. R. Xue and D.C. Wunsch II. 2009. *Clustering*, IEEE Press and Wiley, Hoboken NJ.
219. H. Zha, X. He, C.H.Q. Ding, M. Gu, and H.D. Simon. 2001. Spectral relaxation for K-means clustering, *Proceedings of NIPS'2001*, pp. 1057–1064.

220. L. Zhang, J.S. Marron, H. Shen, and Z. Zhu. 2007. Singular value decomposition and its visualization, *Journal of Computational and Graphical Statistics*, 16(4), 833–854.

221. T. Zhang, R. Ramakrishnan, and M. Livny. 1996. Birch: An efficient data clustering method for very large data bases. *Proceedings of the ACM International Conference on Management of Data*, 103–114.

222. Y. Zhao and G. Karypis. 2002. Evaluation of hierarchical clustering algorithms for document datasets, *Technical Report 02-022*, Department of Computer Science, University of Minnesota.

223. S. Zhong and J. Ghosh. 2003. A unified framework for model-based clustering, *Journal of Machine Learning Research*, 4, 1001–1037.

Index